The Myth of Continents

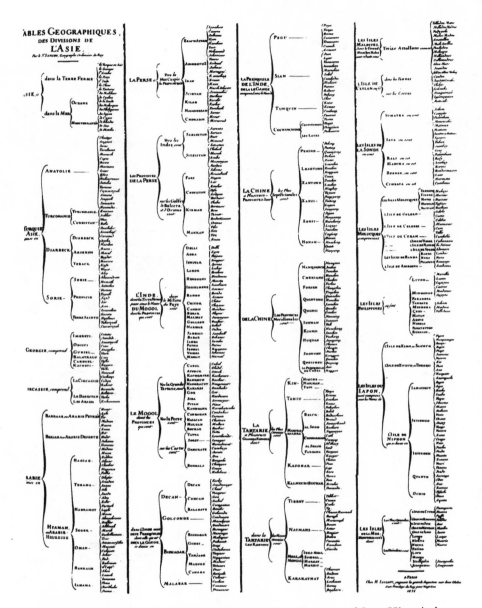

Nicolas Sanson's Geographical Divisions of Asia, 1674. (Courtesy of State Historical Society of Wisconsin.)

The Myth of Continents

A Critique of Metageography

Martin W. Lewis
Kären E. Wigen

UNIVERSITY OF CALIFORNIA PRESS

Berkeley / Los Angeles / London

University of California Press
Berkeley and Los Angeles, California

University of California Press, Ltd.
London, England

© 1997 by
The Regents of the University of California

Library of Congress Cataloging-in-Publication Data

Lewis, Martin W.
    The myth of continents : a critique of metageography / Martin
W. Lewis, Kären E. Wigen.
        p.   cm.
    Includes bibliographic references and index.
    ISBN 978-0-520-20743-1 (pbk. : alk. paper)
    1. Geographical perception. I. Wigen, Kären, 1958– . II. Title.
G71.5.L48   1997
304.2—dc20   96-30294
                                                            CIP

Printed in the United States of America

19   18   17   16   15   14
16   15   14   13   12

Contents

Maps

Preface

Foreign Minister Gareth Evans was determined to put Australia on the map. Literally. He arrived in Brunei in late July for talks centered around the annual ASEAN [Association of Southeast Asian Nations] foreign ministers' meeting armed with a new map depicting the land Down Under as smack in the heart of the East Asian hemisphere.

The persuasive diplomat failed to convert everyone to his world view, however. "If I look at a map, I will immediately say that Australia is not part of Asia," Malaysia's Foreign Minister Datuk Abdullah Ahmad Badawi said under questioning from Australian journalists. "You don't know your geography."

—Murray Hiebert[1]

Every global consideration of human affairs deploys a metageography, whether acknowledged or not. By *metageography* we mean the set of spatial structures through which people order their knowledge of the world: the often unconscious frameworks that organize studies of history, sociology, anthropology, economics, political science, or even natural history.[2] This book has been written in the belief that a thorough examination of these frameworks is long overdue.

With the end of the Cold War, popular conceptions of global geography in the English-speaking world have grown ever more tenuous and uncertain, and standard methods of organizing and classifying spatial phenomena no longer seem adequate for the task. During the period of Cold War rivalry, Americans relied heavily on a tripartite divisional scheme to give order to the map;[3] now, however, the communist Second World has all but collapsed, and comfortable distinctions between First and Third Worlds are being shaken by differential rates of economic growth. In the case of continents, which educated Europeans have taken for centuries as forming the earth's fundamental geographical structure, the erosion of confidence is more subtle; today we still employ continental divisions but with increasing uneasiness about where they lie and what they signify. The notion of world regions or "areas" is also under siege. Scholars grew accustomed in the 1960s and 1970s to organizing their investigations of the globe around the area studies concept; now, they find the salience of world areas called into question, as funding agencies threaten to withdraw support from area-focused research. Even the nation-state, which had acquired the status of a foundational and seemingly permanent territorial entity, suddenly appears vulnerable; cartographers have had to revise the basic political map of the world several times since 1989. But perhaps most problematic of all is our simplest, highest-order geographical concept, distinguishing something called the West from the rest of the world. While this terminology continues to be used even by its critics, the notion of dividing the globe in this way has been loudly—and incontestably— denounced as bigoted and distorting.

While the resulting crisis in conceptualization has yet to be fully acknowledged, various groups are already proposing new ways of cutting up the world, competing to come up with more appropriate geographical categories for the twenty-first century. As yet, however, there is no clear consensus about how to proceed. Many writers replace the paradigm of First, Second, and Third Worlds with a bipolar scheme, opposing a wealthy North to an impoverished South. But the use of these terms is neither precise nor consistent; as recently as 1991, a gathering of scholars concerned with geopolitics rather than social development deployed the category North—without comment—to refer to the former Soviet Union and its onetime allies.[4] Another approach would divide the world into core, semiperiphery, and periphery. This terminology is used most conspicuously in the works of world-systems theorists and historians, who argue that the basic structural building blocks of the modern world are these interlocking political-economic zones.[5] By contrast, a group of political theorists led by Samuel Huntington claims that the relevant units in the emerging world order are a handful of age-old civilizations, each of which is purportedly reasserting its identity and perhaps even preparing to engage its neighbors in combat.[6]

As this last example suggests, even as new proposals are being floated, older geographical schemes are being given a new spin. The widespread paradigm of Afrocentrism is an interesting case here, predicated as it is on the idea that social life is fundamentally organized along the lines of continents. In the more extreme versions of this thesis, Africa forms not only a contiguous landmass but a coherent human community as well: one utterly distinct from all others—and utterly central to the human experience. Stalwart defenders of so-called Western civilization, led by Lynne Cheney, often adopt a similar continental framework, although they naturally identify centrality with a different section of the earth's terrain.[7] Some optimists in this camp, claiming that the downfall of communism spells the "end of history," argue that the twenty-first century will see the ascendance of specifically Western liberalism and democracy throughout the world.[8] Others take a more pessimistic view, seeing both liberalism and democracy as imperiled if not doomed. Benjamin Barber envisages a future "McWorld" united by the networks of an antidemocratic and morally bankrupt global capitalism but simultaneously rent by proliferating "jihads" of ethnic nationalism.[9] In an influential article in the *Atlantic Monthly,* Robert Kaplan similarly prophesies a "coming anarchy": a world dominated, not by stable capitalist centers or ancient civilizations, but rather by the "shadowy tentacles" of "drug cartels, mafias, and private security agencies." Kaplan proceeds to call for a new kind of cartography that can serve as "an ever-mutating representation of chaos."[10] Still other thinkers, however, believe that chaos is so pervasive as to negate the possibility of meaningful mapping altogether. Riding the postmodernist whirlwind, their growing body of scholarship proclaims the rise of "nomad cultures" and boundary-dissolving transgressions, suggesting that metageographical distinctions have already dissolved into an indeterminate flux.[11]

Whatever their differences, all of these approaches share one attribute: a profound skepticism toward received metageographical constructs.[12] Such skepticism is merited for two reasons. First, our guiding vision of the basic spatial patterning of human societies is clearly flawed, with problematic consequences for study after study, in every field of human inquiry. Second, beyond considerations of sheer accuracy in spatial representation, metageographies also constitute ideological structures. It is no coincidence that sea changes in ideology are generally accompanied by a questioning of metageographical categories—or that those attempting consciously to formulate new visions of the globe often do so as part of a campaign to promote new patterns of belief. Precisely because of their ideological power, however, hoary geographical ideas about the earth's division have proved remarkably tenacious, even among those who are trying to shake them off. Moreover, while it may be increasingly recognized that particular concepts are inadequate, the problem has only been addressed in an ad hoc and piecemeal fashion; metageography *as a system* has yet to emerge as a topic of sustained intellectual discussion and debate. In the absence of a systematic and forceful effort to expose their inadequacies and to replace them with something better, the old geographical concepts continue to hold our imagination in thrall.

If metageography has not been prominent on the national agenda, one important cause lies in the institutional weakness of geography as a discipline. The neglect of geography in this country is so pervasive that the crumbling of our global geographical concepts is obscured by sheer geographical illiteracy.[13] Since the reigning consensus in postwar American education has held geography of minor account, it is hardly surprising that so many know so little about the world and stumble so quickly when attempting to understand its basic structures. Sensational headlines in the American press decrying our students' fundamental ignorance of the world map are not exaggerated. Even at prestigious universities one can find seniors who, when provided with an outline map of the world, will unwittingly locate Asia in the Iberian Peninsula.[14]

Students are not entirely to blame. Most college-aged Americans have never explicitly studied geography, since most primary and secondary schools discontinued teaching the subject in the 1960s on the theory that memorization would stultify young minds. The sorry results of this misguided experiment are now fully evident, and broad-based efforts at reform are under way. Thanks in part to the National Geographical Society and the Association of American Geographers, basic geographical education is gradually returning to American elementary and high schools.[15] Parents also seem to be responding, and cartographic toys and games now form a minor growth industry. At the university level, however, the situation remains grim. Geography is a marginalized discipline, absent from many of this country's top-ranked universities and threatened at others. In consequence, the need to reconceive our basic vision of world geography comes at a time when geography as an academic discipline lacks the institutional support to respond effectively.

When one turns to world regional or "global" geography, the subdiscipline that ought to be directly concerned with how we conceive the globe, the problem is particularly acute. Except at the lowest level of pedagogy, world geography is simply not viewed as an intellectually defensible academic subject. It does not merit inclusion in the Association of American Geographers' lengthy roster of specialty groups—a roster that serves as a guide to the active frontiers of geographical research. While such subfields as "the geography of aging and the aged," "Bible geography," and "the geography of recreation, tourism, and sport" are institutionalized as legitimate research areas, global geography is not. In most of the country's top-ranked geography departments, world regional courses are viewed as suitable only for remedial instruction to beginning students; in a few schools' bulletins, they do not exist at all. And because global geography is ignored as a research field, even when the subject *is* offered to students it tends to be taught in an outmoded fashion. World regional geography textbooks are, at their worst, repositories of the discipline's past mistakes, constructing 1950s-style catalogs of regional traits over unacknowledged substrata of 1920s-style environmental determinism. It is little wonder that most American college graduates have such a fuzzy conception of the world.

Institutional failures of this order are not easily addressed. What is possible, however, is to expose the fault lines in Americans' guiding notions of the world: to trace how conventional metageographies emerged and developed, and to explore how they continue to lead us astray. That is what has been attempted here. Our starting point is the premise that laypersons and scholars alike have uncritically accepted a series of convenient but stultifying geographical myths, based on unwarranted simplifications of global spatial patterns. In particular, we identify four related errors that lie at the root of metageographical confusion in the English-speaking world: the myth of continents, the myth of the nation-state, the myth of East and West, and the myth of geographical concordance (i.e., the idea that disparate phenomena exhibit the same variation in space). We further argue that such notions survive not merely as naive "mistakes," but often as instruments of ideological power. Diplomats, politicians, and military strategists employ a metageographical framework no less than do scholars and journalists. Such political actors have also had a far larger role in formulating global constructs for the public imagination than scholars have cared to recognize; as we shall see, some of the most basic and taken-for-granted "regions" of the world were first framed by military thinkers.

The bulk of the present work is accordingly devoted to examining the intellectual underpinnings of the standard metageographies employed in the United States today. Our aim is to make the case that concepts of global geography matter, not merely for how they influence discourse about the world, but for how they guide policy as well. To bring this point home, we scrutinize large-scale geographical concepts, trace their intellectual histories, and consider their present uses in both scholarly and popular discourse. We would be the first to acknowledge the incompleteness of this initial effort, however. For one thing, our coverage is limited primarily to the dominant Euro-American modes of geographical thought, and even within that tradition we concentrate primarily on representations of the Eurasian world. Additionally, the ways in which metageographical ideas play out in the realm of international power politics, while touched upon at several points, deserve much fuller treatment. Finally, the development of an alternative metageographical framework still awaits further work. While our conclusion sketches out a set of guidelines for reforming macroregional frameworks, as well as a number of particularly promising areas for future research, the task of generating a truly critical metageography will require a broad collective effort. In many ways, then, the present study is incomplete. Its goal is to draw attention to long-neglected questions, more than to proffer definitive answers. If this book succeeds in provoking debate and prodding further inquiry, it will have served its purpose.

During the years when this book was written, we were involved in the intellectual life of three different universities. The book would not have taken its present form were it not for contributions from all three communities in which we found ourselves.

At George Washington University, Joel Kuipers made valuable early comments on the project's conceptual direction, and Don Vermeer expressed warm interest in its completion. Marie Price was especially helpful, suggesting new sources and conceptual refinements for our discussions of Africa and Latin America, and sharing insights from her own work in progress. Our gratitude goes to each of these GW colleagues.

Our brief stay at the University of Wisconsin in Madison also proved significant for this project. Thanks are due to the staff of the Wisconsin state historical library, who facilitated access to their tremendous collection of world atlases. In the Wisconsin geography department, we benefited from discussions with Dan Doeppers, Robert Sack, Yi-Fu Tuan, Gerry Kearns (now at Cambridge), and especially Mark Bassin (now at University College, London), who enthusiastically shared his analysis of Russia's metageographical position. We are also appreciative of the graduate students who provided intellectual stimulation, particularly Christian Brannstrom and Valentin Bogorov. Other Wisconsin colleagues we would like to single out for thanks include André Wink, for his encyclopedic knowledge of the Indian Ocean world; Ed Friedman, for his incisive analysis of Chinese and global political economy; and Colleen Dunlavy and Ronald Radano, for their thoughtful questions and general succor.

Our greatest debts have been incurred at Duke University. The Workshop in Comparative Area Studies provided a primary forum for airing the basic metageographical critique forwarded here. While not all of its members would endorse the positions we have taken, the CAS workshop and the informal reading group that preceded it helped shape our agenda by seriously posing the question of what comparative area studies should mean. Thanks to Connie Blackmore for her good will in facilitating those meetings and to Bruce Lawrence for giving institutional support to the forum. We also want to thank Bruce for his supportive and substantive response to an early draft, offering insights on the geography of Islamic civilization as well as on the intellectual project of Marshall Hodgson. John Richards shared his knowledge and astute interpretations of both area studies in general and South Asia in particular. James Rollston provided valuable comments on our metageographical interpretation of Germany, and Walter Mignolo taught us much about colonial conceptions of Latin America. Among many other colleagues and friends at Duke who have given parts of our argument a critical hearing, we would especially like to thank Anne Allison, Vincent Cornell, Arif Dirlik, Vasant Kaiwar, Rebecca Karl, Sucheta Mazumdar, Charles Piot, Alex Roland, Susan Thorne, Kristina Troost, and Peter Wood.

Numerous others around the country and the world have supported this project as well. Norm Levitt and Meera Nanda have initiated ongoing discussions about science, reason, and "Western culture," and Victor Lieberman has shared his remarkable erudition on the geography of early modernity. Paul Starrs has been a great long-distance friend and colleague, as has James Hynes, who has helped us maintain perspective on the academic endeavor. Les Rowntree has offered crucial support for the project of reconceiving global geography at the pedagogical level, as well as warm friendship. Members of the Bulgarian Academy of Geographers (especially Marianna Nikolova) generously enlightened us on the (meta)geographical situation of their country.

The final shape of the book owes much to the care and guidance of the University of California Press. Alexander Murphy and two anonymous reviewers provided probing questions and thoughtful suggestions for the manuscript's improvement, and Sheila Levine, Laura Driussi, Sue Heinemann, and Rachel Berchten guided the book expertly to production. Thanks also to Joe School and the cartography staff at UMBC.

Finally, deepest thanks to our families, as always.

Introduction

In a country where high school graduates strain to locate Australia on a globe, conveying basic information about the world has become the overriding pedagogical imperative for university-level courses in global geography. But fulfilling that imperative is harder than it might appear. For it is precisely the most basic information—the highest level of our geographical taxonomy—that is the most problematic. Whether we parcel the earth into half a dozen continents, or whether we make even simpler distinctions between East and West, North and South, or First, Second, and Third Worlds, the result is the same: like areas are inevitably divided from like, while disparate places are jumbled together.

Such niceties are beside the point when geography is being introduced to ten-year-olds. Constructs of utmost simplicity are essential starting points for learning the map of the world. But to continue teaching these categories at the university level as if they were nonproblematic is to deny our students the tools they need to think clearly about the complicated patterns that actually mark the earth's surface. Even less excusable is the continued recourse to simplistic geographical frameworks at the highest levels of scholarly discourse. Otherwise sophisticated and self-critical works habitually essentialize continents, adopting their boundaries as frameworks for analyzing and classifying phenomena to which they simply do not apply. Dividing the world into a handful of fundamental units in this way may be convenient, but it does injustice to the complexities of global geography, and it leads to faulty comparisons. When used by those who wield political power, its consequences can be truly tragic.

The Myth of Continents

The myth of continents is the most elementary of our many geographical concepts. Continents, we are taught in elementary school, form the basic building blocks of world geography. These large, discrete landmasses can be easily discerned by a child on a map of the earth. One has simply to spin the globe and watch them pass by: the massive triangles of North and South America, tenuously linked by the Panamanian isthmus; the great arch of Africa, neatly sundered from Europe and Asia by the Mediterranean and Red Seas; the squat bulk of Australia, unambiguously disjoined from other lands; the icy wastes of Antarctica, set alone at the bottom of the world.

But continents are much more than the gross elements of global cartography. The continental structure also guides our basic conceptions of the natural world. We talk of African wildlife as if it constituted a distinct assemblage of animals, and we commonly compare it with the fauna of Asia or South America. The continents are also held to reveal fundamental geological processes, the "fit" between Africa and South America being the prime visual evidence for geology's unifying theory of plate tectonics. Even more important is our tendency to let a continental framework structure our perceptions of the human community. Thus Africans become a distinct people, who can be usefully contrasted with Asians or Europeans, and we imagine Africa's problems to be unique to its landmass, as though tied to it by some geographical necessity. Similarly, the cultural distinction between Europe and Asia has long guided our historical imagination. Each continent is accorded its own history, and we locate its essential nature in opposition to that of the other continents.

Perhaps because continents are such obvious visual units, their utility is seldom questioned. The continental scheme is reproduced and reinforced ubiquitously in atlases, encyclopedias, and bibliographic reference tools, virtually all of which routinely employ these divisions as their organizing geographical framework. The signal role of continents in American education is nowhere more evident than in Rand McNally's *Educational Publishing Catalog*, a primary source of maps, globes, and other geographical paraphernalia for the classroom.[1] Here one finds not only that a "beginner's political" map of the world prominently marks each standard continent with a bright color, but that all of the more specialized political and physical maps are designed within the continental scheme as well. Europe and Asia are thus each accorded one identically sized map (50 by 68 inches) in the Level III series—ensuring that Europe will be mapped in far greater detail.

Despite its ubiquity and commonsensical status, there are many reasons to believe that the standard seven-part continental scheme employed in the United States obscures more than it reveals. An obsolete formulation, this framework is now wholly inadequate for the load it is routinely asked to carry. Equally in the realms of natural history and human geography, the most important distributional patterns and structuring processes are not based on continental divisions. The Isthmus of Panama, separating North from South America, is of little importance for either social history or the animal and plant kingdoms; most of what is unique about Africa begins south of the Sahara Desert, not south of the Mediterranean Sea; and the division between Europe and Asia is entirely arbitrary. Only by discarding the commonplace notion that continents denote significant biological or cultural groupings can a sophisticated understanding of global geography be reached.

The World's Worlds:
Global Economic and Political Divisions

Twofold or threefold economic partitions of the world are less objectionable than the continental system in the important sense that they are overtly limited in scope. The "Third World" is essentially a political-economic category, and no one would presume the existence of Third World vegetation or Third World geology (although there has been vociferous debate about the category of Third World literature).[2] But even as a politico-economic category, the Third World is unduly monolithic, and its boundaries too simply drawn to be of much utility.[3] Most attempts at global economic mapping place the relatively poor nations of Portugal and Greece within the First World, while labeling the dynamic states of Singapore, Taiwan, and South Korea as Third World, despite their higher levels of economic activity.[4] Such erroneous distinctions in part reflect inertia; it would be a taxing endeavor to revise a basic classification system quickly enough to reflect the latest data on changing national fortunes. But residual continental thinking is also partly to blame.[5] Greece and Portugal lie within Europe, and since Europe is a wealthy "continent," its constituent states are often considered "developed" by definition, regardless of their actual circumstances. Owing to such difficulties, Carl Pletsch not unfairly argues that "the scheme of three worlds is perhaps the most primitive system of classification in our social science discourse."[6]

Problems of categorization within the three-worlds model have given rise to heated debates concerning how to categorize any number of borderline cases. In a recent exchange in the *New York Review of Books*,[7] for instance, economist Robert Solow excoriated author Edward Luttwak for implying that Chile and Argentina are not members of the developed world; "surely," he informs the reader, these are not "third world countries." In many ways, Solow is right. Chile has experienced dramatic economic growth in recent years, and Argentina has long boasted relatively high levels of social welfare. Yet neither is routinely counted among the members of the First World, and for good reason. If Solow is able to argue that Argentina does not belong in the Third World, it is mainly because the latter is ultimately a highly fungible category, amenable to any author's desire for enlargement or reduction.[8] Such imprecision may prove polemically convenient, but it does not necessarily advance geographic understanding.

Paring the globe into First, Second, and Third Worlds is now anachronistic in any case, since the political criterion used to define the Second World (namely, government by a communist regime) has all but disappeared.[9] Yet the categorization scheme has survived far longer than was ever warranted, in large part because it served the ideological needs of both Cold War American partisans and, on the opposite side of the political spectrum, the most vigorous opponents of American neo-imperialism.[10] Moreover, the emerging preferred alternative (see map 1), distinguishing a developed North from a less developed (or actively underdeveloped) South, is equally problematic.[11] To begin with, the labels *North* and *South* are fantastically imprecise. China, reaching 58 degrees north latitude, is

---

Map 1. *Cold War Metageography: North, South, and the Three Worlds.* On this 1993 base map, the heaviest line separates the "developed" First World and Second World (the North) from the "less-developed" or "underdeveloped" Third World (the South). A secondary boundary distinguishes the First World from the Second World.

Three problematic countries are indicated with small dots. Turkey is usually considered Third World but is sometimes annexed to the First World in geopolitical discussions. South Africa has occasionally been appended to the First World, while the former Yugoslavia occupied an unstable position between the Second World and the First World.

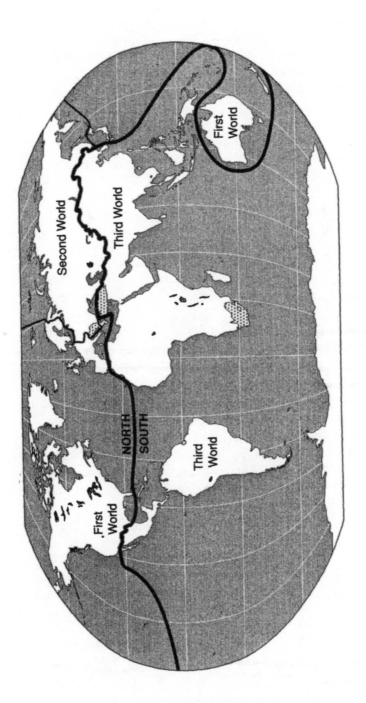

routinely placed within the South, while Australia and New Zealand are commonly grouped with the North.[12] The more fundamental problem, however, is again that of joining unlike entities into massively agglomerated categories.[13] The category North was originally based on the notion that industrial economies of the Soviet Union, Europe, Japan, and North America were fundamentally similar. With the collapse of the Soviet state—and economy—such a system of categorization no longer makes any sense.[14] It requires some stretch of the imagination to regard the economies of Russia and Japan as essentially similar enough to warrant the inclusion of both into a meaningful North, but it requires concerted ignorance to place Tajikistan and Albania in the same category with them—all the while relegating Taiwan, Hong Kong, and the United Arab Emirates to the South. Nor is it instructive (as Robert Solow would insist) to regard the southern nation of Argentina as having more in common with Lesotho and Mozambique than with Portugal or southern Italy. Nonetheless, many scholars are so taken by the fantastic simplicity of the North-South scheme that they have come to believe that the Southern *Hemisphere* actually constitutes the uniform zone of global poverty.[15]

The North-South distinction, like that between the First, Second, and Third Worlds, is essentially defined in economic terms. Cultural and political matters are more often framed as East versus West: a far older and more important division. Indeed, the notion of a First World is itself deeply rooted in the idea of a distinctive Western realm; in many works, the Third World is thus contrasted, not to the First or Second Worlds, but rather simply to the West.[16] When such a scheme is carried to its logical extreme, the world is again divided into two sections: the West and the non-West. As Bernard Cohn remarks, this maneuver entails "a neat ethnocentrism which defines nine-tenths of the people of the world in a single negative term."[17]

While the "West" is often contrasted simply with the "rest," its historical counterpart is of course the "East." The myth of continents is also implicated in this binary longitudinal division, for the West is conventionally defined as Europe (plus its direct colonial offshoots), while the East in many instances is simply a proxy for Asia—with Africa, in this view, threatening to fall off the map altogether. The East-West opposition maps a huge array of human attributes onto a stupendously simplified set of geographical coordinates, but its staple feature has historically been the linking of the West with reason and progress and the East with spirituality and stagnation. Baseless though it may be, this purported correspondence ultimately forms a central structure of our metageographical mythology.

Like other metageographical concepts, the East-West split is remarkably protean, and on certain occasions a completely different referent system is implicated by these terms: one differentiating eastern from western Europe. Inevitably, however, the two referents of *East* tend to be conflated, implying that eastern Europe is somehow Asian in its essence. While one can argue that Russia shares certain characteristics with Central Asia, no criterion of Asianness can reasonably be extended to Slovenia, Bohemia, or Thuringia. Yet the geopolitical category East does precisely this, riding roughshod over previous cultural divisions by giving undue weight to a political grouping that existed only between 1946 and 1989.

The metageographical distinction between the West and the rest of the world is particularly debilitating when married to a key metahistorical concept: the notion that the West is coincident with modernity and that the non-West can enter the modern world only to the extent that it emulates the norms established in Europe and northern North America. In a powerful exposé, J. M. Blaut labels these linked constructs "the colonizer's model of the world" and shows that they rest on a rarely acknowledged substrate of "geographical diffusionism" (where progress is seen as flowing endlessly out of the center [Europe] toward the otherwise sterile periphery).[18] But while Blaut convincingly argues that this is a central geographical myth of the modern age, it is hardly the only one. Ultimately, all received metageographical constructs need to be subjected to similarly sustained geographical and historical scrutiny.[19] Likewise, while Blaut is to be commended for showing that the "colonizer's model" has often been embraced by Marxists no less than by liberals and conservatives, we would add that similar geohistorical visions of the world are not uncommonly encountered even in post-Marxist social theory.[20] Like the classical Left, the cultural Left of poststructuralists, postmodernists, and radical environmentalists often perpetuates the West-rest binarism—only in the form of rhetoric that disparages the West and celebrates the rest.

The Myth of the Nation-State and the Eurocentric Perspective

Similarly simplistic assumptions are also encountered at other levels of geographical analysis. After the myth of continents and the fable of an enduring East-West division, the most debilitating geographical misconception is probably the myth of the nation-state.[21] The nation-state idea—i.e., the assumption that cultural identities (nations) coincide with politically sovereign entities (states) to create a series of internally unified and essentially equal units—replicates at a smaller scale many of the errors found in continental thinking. To be sure, countries, unlike continents, are real entities, with armies to prove the point. For this reason alone, the global framework of sovereign states is essential for analyzing political affairs. But very few countries are nation-states in the strict sense of the term; seldom is an independent political territory coterminous with the territory of a self-consciously united people. That we elect to call such internally divided countries as India, Nigeria, or even Switzerland nation-states shows a determined desire to will uniformity out of diversity. In the process, states—like continents—become reified as natural and fundamental building blocks of global geography, rather than being recognized as the constructed, contingent, and often imposed political-geographical units that they are.[22]

The inevitable result is that state boundaries are evoked in innumerable arenas where their usefulness is truly circumscribed. Features of the natural world seldom conform to political terrains, and even patterns of human culture more often than not crosscut country boundaries. In few parts of the world, for example, does a map of language distribution bear much resemblance to the political map, and only in exceptional areas like Japan or Iceland do the two correlate more or less precisely. (Moreover, in the case of Japan, it has required centuries of concerted state effort to make them correspond.) Yet most of our encyclopedias, textbooks, atlases, and almanacs portray states as holistic entities, unified and distinct. While this may seem an innocuous device for classifying knowledge, it can lead to real mischief when its limitations are ignored. A country like Sudan is simply not a basic unit of the human community, despite its ubiquitous portrayal as such; northern Sudan has far more in common with Egypt than it does with southern Sudan, which has greater affinity with Uganda. To assume that northern and southern Sudan form a "Sudanese nation," as the myth of the nation-state leads us to do, is to reduce the struggle currently being waged there to the status of a civil war—an internal affair that need not concern the global community to any great extent. In fact, the Sudanese war is a rather clear case of one people (those in the north) brutally attempting to impose its will on another. Clearly, countries do not cohere on all the levels we commonly imagine. At its best, the myth of the nation-state obscures internal difference. At its worst, it can become a tool of genocide.

In economic terms as much as cultural ones, countries are not necessarily the essential units that we often imagine them to be. It is some-times observed, for instance, that while northern Italy belongs to the First World, southern Italy might be better classified with the Third. Similar divisions may be seen in many other states. Disparities of economic activity can be particularly extreme within poor countries; India's relatively well-off Punjab is a world away from Bihar, for instance, while the booming Guangdong region of China contrasts dramatically with poor Gansu. Meanwhile, contemporary economic evolution is making state boundaries highly porous. Capital can be shifted from one world-city to another with alacrity, while trading blocks and currency unions increasingly challenge national economic sovereignty. To make meaningful maps of economic activity and socioeconomic development requires violating the principle of state indivisibility. While this might seem an obvious point, it is routinely ignored in both journalism and scholarship. After the breakup of Yugoslavia, for example, the American public was informed that the new country of Slovenia faced "special problems of economic viability" because of its limited territorial extent.[23] Yet there is absolutely no evidence of any correlation between a country's size and its economic success; what matters for a small state's survival is whether it occupies a viable niche in the global division of labor. Luxembourg, Singapore, and even Liechtenstein have hardly been economically paralyzed by their exiguous areal extent.

For these and related reasons, the nation-state is a dubious concept for economic as much as for cultural analysis. But the mischief of this myth is compounded when nation-states and continents are misunderstood as constituting elements of the same, naturalized geographical taxonomy. Political maps that also highlight continents encourage us to imagine that nation-states are the building blocks of these higher-order entities, and therefore must nest neatly within them. States, in a word, are seen as constituent units of continents; South America can thus be defined simply as the collection of South American states. Such thinking leads to two errors. The first is a tendency to see "transcontinental" countries (i.e., states like Turkey, Egypt, Russia, and Kazakhstan, all of which straddle the conventional continental divides) as unduly anomalous. The second and more serious error, however, is what might be called the fallacy of unit comparability. Because France and Germany are to Europe as India and China are to Asia (i.e., large and important countries within their respective continents), a fundamental parity is assumed between these four countries. But the analogy is deadly false. In physical, cultural, and historical diversity, both China and India are comparable to the entire European landmass, not to a single European country. A better (if still imperfect) analogy would compare France, not to India as a whole, but to a single Indian state, such as Uttar Pradesh.

This comparison demonstrates a third pernicious by-product of the continental and nation-state myths: together, they work to consistently and unduly exaggerate the importance of Europe. By elevating Europe's position on the world map, metageographic categories make more plausible the notion of Europeans' priority in the history of human affairs, effectively serving as visual propaganda for Eurocentrism. Throughout the pages that follow, we shall see how these taken-for-granted and seemingly innocuous notions of world geography covertly function to magnify the "Western" portions of the globe and correspondingly to reduce all others. Geographical myopia is the inevitable result; Europe and the United States appear in swollen importance, while the rest of the world is shrunk into a distorting miniature.

In one sense there is nothing unusual about Eurocentrism; all geographical traditions are rooted in local concerns and ethnocentric conceits,[24] and had China emerged as the hegemon of the modern world system, our metageographical concepts would surely reflect Sinocentrism. (Actually, our common ideas about Central and East Asia *do* powerfully reflect Sinocentrism.) We are accordingly suspicious of any historical reading that would regard Eurocentric global geographies as stemming from some sort of grand intellectual conspiracy. But what we do strenuously argue is that it is now necessary to relinquish all such notions and to reach instead for a cosmopolitan and ecumenical perspective. By continuing to employ Eurocentric concepts uncritically, even at the highest levels of academic discourse, scholars only perpetuate a conception of the earth that is both arrogant and faulty. An increasingly integrated world demands a more modest, honest, and accurate geographical depiction.

The Roots of Geographical Confusion

To this point we have looked at three specific kinds of metageographical concepts: continents, nation-states, and supracontinental blocks, such as East and West. Despite their differences, we would argue that all three of these conceptual formations share two fundamental characteristics. The first is a jigsaw-puzzle view of the world; the second is an assumption that geographic phenomena are necessarily and neatly hierarchically ordered. Together, these intellectual errors constitute the source of much metageographical mischief.

By a jigsaw-puzzle view of the world, we mean the expectation that a proper map will always show a set of sharply bounded units that fit together with no overlap and no unclaimed territory. The paradigm for such a picture may well be the map of American states, which is taught to every American child. Having never been exposed to a critique of the simple geography taught in elementary school, we continue to seek correspondingly simple maps for much more problematic phenomena, expecting to see their spatial patterns conveniently arrayed in large, contiguous, colorful blocks. Implicated in this jigsaw-puzzle image are two further expectations: (1) that the discrete pieces thus delineated are fully comparable and can be abstracted from their contexts for analytical purposes, and (2) that the world order thus described is essentially stable. For underlying the myth of continents, the myth of the nation-state, the overgeneralization of world regions, and other related errors is a conception of global geography as fundamentally static. Mapping is accordingly seen as a purely technical exercise, a matter of simple description; while the boundaries may need to be located with more precision, the world (in this view) can still be mapped by drawing simple lines around preexisting units. Although such a perspective seemed supportable to many in the postwar decades, the years since 1989 have revealed it as bankrupt.

A second and related error of conventional geographical thinking is to treat the earth's surface as if it were amenable to taxonomic classification in neat hierarchies of territorial units. Such a vision may be traced back to the Enlightenment cartographic tradition of the late seventeenth and eighteenth centuries, and it has probably never been expressed so completely as in Nicolas Sanson's world atlas of 1674. To use the language of contemporary social theory, what we have inherited from the Enlightenment is a totalizing spatial framework, a "metastructure" that strives to impose its own rigid order. To use an older formulation, our map of the earth is a Procrustean bed, in which the complexities that make real places interesting have been violently deformed to fit a set of standardized shapes that do not accommodate them well.

As with the jigsaw-puzzle view of the world, the taxonomic principle comes with a host of corollary expectations. For one, viewing the world as divided into a hierarchy of fundamentally equal units suggests comparability among what are often grossly unequal terrains. Consider, for instance, the odd balance of coverage in popular U.S. atlases, which often devote one page to each state—with the result that Connecticut is shown at more than ten times the scale of Texas.[25] The same sort of problem takes a more insidious form when it distorts our perceptions of spatial relations among countries and continents. It is in world almanacs that such unit-egalitarianism reaches its extreme, with microstates like Monaco and Liechtenstein routinely receiving almost the same print space as China or India.[26] In all such cases, the expectation of geographical comparability paradoxically yields extreme bias.

The source of this bias ultimately lies in the unwarranted application of natural science models to the study of spatial phenomena. Plant and animal species may be assimilable into a neat sequence of species, genus, family, order, class, phylum, and kingdom (although even this taxonomy, as experts know, is far from perfect).[27] But places, unlike birds or trees, simply do not relate to each other in straightforward hierarchical terms. To be sure, there are exceptions. The territorial units of political structures (states, provinces, districts, and the like) do characteristically form nesting hierarchies, and the same is also true, although to a lesser extent, for economically defined regions. But even economic linkages have horizontal components that often crosscut the vertical connections of place-based hierarchies, and many other equally important facets of global geography, such as the distribution of cultural patterns, are not so neatly structured at all. Linguistic geography, long thought of as a straightforward matter of mapping nested groups of language "families" or "subfamilies," individual languages, and dialects, is now understood to be much more complex; the simple spatial projection of evolutionary relationships is confounded by long-term borrowing and recombinations.[28]

The ultimate problem for the taxonomic model, however, stems from the fact that global geography can never be derived from a single phenomenon, for the simple reason that there are no geographical phenomena per se. Since everything that varies over space has a geographical dimension (just as everything that varies over time has a historical dimension), global geography encompasses an enormous array of disparate elements, all of which have their own distributional patterns. Understanding the patterning of the earth's surface thus requires synthesizing all manner of disjunct distributions and developing a sense of the interaction between spatial patterns in a whole range of fields. For this reason above all, our expectation of a neat taxonomy is foiled at every turn. Like the jigsaw-puzzle view of discrete, bounded territories, the assumption that spatial relationships will naturally assume a hierarchical form must also be overturned if we are to think clearly about the globe.

The main problem with abandoning a set structure of nonproblematic geographical entities, in exchange for an open-ended mélange of over-lapping and incommensurable distributional patterns, is the danger of losing our ability to talk about the world effectively. For all its artificiality, a geographical shorthand of large-scale and comprehensive territorial blocks must be delineated simply to facilitate communication. Moreover, a quasi-taxonomic organization, one based on smaller regions nesting within larger ones, is also often appropriate. Where possible, regional-ization schemes should be context-specific; when analyzing economic is-sues, for instance, it is best to employ a system of global division based on genuinely economic criteria. Still, there are occasions when it is use-ful to refer to a general-purpose scheme as well: one that can be invoked for heuristic purposes, when discussing the interaction of disparate spa-tial phenomena, or when specificity is not required.[29] We believe that the best available alternative for such general-purpose frameworks is a refined version of the so-called world regional scheme.

World Regions: A Way Out?

World regions are multicountry agglomerations, defined not by their supposed physical separation from one another (as are con-tinents), but rather (in theory) on the basis of important historical and cultural bonds. Common world regional categories include such entities as Latin America, sub-Saharan Africa, and South Asia, all of which are al-ready widely used in the popular press. Area studies centers at universi-ties across the country are usually organized around such groupings, as are many college geography courses. Such schemes have two features to their credit: (1) they ignore the dictates of landmass shape, and (2) they demote Europe to its proper place (Europe being usually represented as one or two world regions, while Asia is partitioned into as many as six).

While some form of world regional categorization is appropriate for basic geographical education and for academic organizational purposes, the schemes currently in use remain weakened by a number of faulty pre-suppositions. For one thing, world regions are usually distorted by the reification of the nation-state. Since states are still considered primary units, world regional configurations are almost never allowed to violate their borders. Thus all of China is routinely classified as part of East Asia, even though its Islamic, Turkish-speaking northwest quadrant more properly belongs with the south-central portion of the former USSR in the distinctive region of Central Asia. Second, world regions are often overgeneralized, taken as all-purpose frameworks for analysis. If they are to help us understand our planet, these entities must be treated, not as natural or suprahistorical entities, but as approximate intellectual constructs, imperfectly reflecting cultural and historical relationships. It is also crucial not to regard separate world regions as fully comparable with each other. In fact, different regions are inevitably defined on the basis of different and incommensurable criteria. South Asia, as we shall see, is in fundamental respects a different *kind* of region than is Southeast Asia. While it defies our desire for logical consistency, this incommensurability between world regions is one of the most important lessons that global geography should be teaching. Finally, it is essential to maintain a cross-regional comparative perspective. The various area studies complexes at American universities sometimes encourage a certain insularity in scholarship, making it unnecessarily difficult for scholars to investigate processes that transcend conventional world regional boundaries.

Despite these caveats, however, we would insist that world regions—more or less boundable areas united by broad social and cultural features—do exist and that their recognition and delineation are essential for geographical understanding. Such macrogeographical units may not be *fundaments,* in the sense that they can simply be taken as given. But they are nonetheless *fundamental* to understanding how the world is put together. Abstracting a world region called South Asia may be a tricky intellectual exercise, but it is necessary if we are to come to terms with the spatial unfolding of world history. If recent history tells us anything, it is that globalization does not obviate such regional culture blocks;[30] in some ways, it has even heightened their distinctiveness as a reaction to "Western" cultural hegemony. In a word, while macroregions are imperfect intellectual constructs, they are not entirely made up; real patterns in the world precede the attempt to understand them. And it is only by reference to those real patterns that we can judge the resulting attempts.

World Geography and the Postmodern Mood

It is here that we part ways with the academic movement known by the name of postmodernism. Scholarship in a postmodernist (or, more precisely, poststructuralist) vein has thoroughly reconfigured the humanities over the past decade and has had a major impact on geography in the process. More a sensibility than a coherent theoretical stance, postmodernism in geography emphasizes fluidity, contingency, movement, and multiplicity, questioning the rigid spatial frameworks that have limited and constrained our geographical imagination. Geographers who adopt this approach have made many of the general points that are argued here, particularly regarding the need to break down conventional, static, and objectified regional schemes. Here is one important segment of the discipline calling for liberation from the myth of continents, the myth of the nation-state, and the imprisoning thesis of European priority—all heretofore taken-for-granted frameworks for geographical inquiry.

The postmodernist approach, however, presents pitfalls as well as promises. In the more extreme versions of poststructuralism, any attempt to come to a systematic understanding of the world is attacked, and any notion of unity, similarity, or locality-transcending process becomes suspect. Not surprisingly, many geographers, beholden to the insights of postmodernist scholarship, "appear to have lost the desire (and the confidence?) to say anything about this empirical social world."[31] The alternative is to pursue the deconstruction not only of categories but of their referents in the physical world. As one recent text would have it, the natural environment itself "is first and foremost an artifact of language."[32] Another article claims that "modern human geography is the pulped corpse of a paranoid and schizophrenic masochist" that "must be deconstructed in order to live."[33] This might be dismissed as a morbid joke, though it is hardly one to inspire much mirth. More worrisome, and ironic, is the tendency of postmodern scholars to mirror the very theories that they arraign themselves against.[34] Not uncommonly, contemporary critics accept the Eurocentric conceit that science, technology, and "rationality" are defining and characteristic features of the West. Only the ethical signposts are reversed, with the supposedly nonrational ways of the non-West being upheld as morally and intellectually superior.

We would contend that such Eurocentric concepts are best avoided, in their denigrating as well as their celebratory guises. While following the postmodern impulse in our concern to expose the socially and culturally constructed nature of taken-for-granted geographical categories, we resist the notion that the attempt at geographical classification should be abandoned altogether as merely another instance of Western objectification. To borrow the words of historian William Cronon, "Social constructionism tells us something important about the world, but it does not tell us everything, for the redoubt of realism is not quite so ruined as [the postmodernists] would have us believe. The realist-idealist, objectivist-relativist dualism has been around since at least the fifth century B.C., and despite the best efforts of postmodern critics it is not likely to be vanquished simply by amputating its polarities. One transcends such paradoxes by passing through them, not by wishing they would go away."[35]

The Ironies of Geographical Ecumenicalism

It is our aim here to attempt to transcend the *sic et non* paradox of global divisions primarily by historicizing the categories through which we think about the world. To do this it is necessary to go back at least to the fifth century B.C.E., for just as in the case of epistemology, the foundational ideas of world geography are rooted in the debates of the ancient Greeks. This is not merely true, it is essential to note, for contemporary European and American societies; Greek geography has been to a certain extent globalized, and what were originally Western categories—such notions as Europe, Asia, and Africa—are now employed throughout the world. This was, of course, not always the case, and investigating alternative systems of global division is an essential project for a critical metageography. But to transcend Eurocentrism initially requires very close engagement with the history of European thought.

Nor is this the only irony of our project. Equally problematic is the fact that we cannot but continue to use terms that we make great efforts to debunk. While arguing, for example, that there is no logically constituted geographical category called Asia, we find this term an indispensable element of many world regional labels (South Asia, Southeast Asia, Central Asia, and the like), and occasionally we write about "Asia" as if it were a whole—sometimes even forgoing the use of the quotation marks. In some cases this is simply a matter of referring to a set of geographical ideas that has a very real existence, even if the qualities and phenomena that they purport to locate do not; Asia has an important position in the map of intellectual history, if not on the map of the world. At other times, however, we continue to employ problematic terms because it is our intention to debunk only certain of their connotations. "Europe," for example, may not be a continent, but it does effectively label an area that can be defined as a cultural region or a civilization. Similarly, we believe it is defensible to isolate an intellectual tradition that may be called Western, so long as we realize, first, that there is no corresponding Eastern tradition, and second, that the roots of this tradition are far more entangled and geographically complex than is commonly imagined.

We therefore attempt here both to deconstruct and to conserve; to highlight fluidity and indeterminacy, but also to map out real geographical structures that we consider independent of anyone's attempt to understand them; to point out the conventional and constructed nature of the fundamental ideas of global geography, while yet denying that they are *nothing but* social constructs; and finally, to uncover the political motivation behind metageographical conceptualizations, without implying that they are all reducible to strategic interests. Problems of language are inevitable in such a project; in order to continue talking about the world, we must have the cake of metageography while deconstructing it, too. Moreover, such tensions are heightened by dual authorship. While this book is wholly a joint product, its authors are not in complete agreement on all the issues it raises, and some formulations represent the exigencies of compromise. Whether such compromises fundamentally vitiate our analysis is for the reader to judge.

The empirical exposition of our argument begins in the following chapter, where we trace the history of the idea of continents, the most basic of all global divisions. This is followed by a two-chapter look at the notion of binary opposition, focusing on the division of the world into a West and an East. Chapter 4 examines how two particular continents, Europe and Africa, have been construed by contending intellectual camps as the primary reference points for human history, while chapter 5 explores the framework of "civilizations" that is traditionally invoked by world historians. The final chapter examines the system of world regions employed in contemporary American area studies. Before proceeding with this analysis, however, a number of caveats are in order.

First, we want to acknowledge the pioneering work of several scholars who precede us: William McNeill, John Steadman, Andrew March, Jack Goody, and especially the late Islamicist Marshall Hodgson. After having completed roughly half of the initial manuscript for this book, we were thrilled to discover Hodgson's early essays on global history, which propounded a vision in close concordance with our own. Excitement was accompanied by bemusement, however, as we could not help wondering whether there was any point in making these arguments again. If several key passages now seem (to those familiar with Hodgson's work) merely to reiterate and extend his arguments, our defense is that these ideas deserve a much broader public debate than they have so far been accorded. Although his novel approach to world history is at long last receiving some recognition, Hodgson's equally bold ideas about world geography remain almost unnoticed. The evening news offers a poignant reminder of his failure to reach a broad audience, for proudly displayed behind the newscasters on all three major American networks is a distorted version of the Mercator projection: a map Hodgson denounced some forty years ago as a "Jim Crow" depiction of the world.[36]

A second caveat concerns a bias in our coverage. No effort is made in this book to grant equal consideration to all important metageographical constructs—or to all portions of the globe. As the reader will soon discover, we dwell at most length on the concept of European, or Western, society. In chapter 1, extended consideration is given to the notion of Europe as a continent; both chapters 2 and 3 are concerned with the distinction between the West and the rest of the world; the ideology of Eurocentrism occupies a large portion of chapter 4; and repeated reference is made in chapter 5 to the notion of European civilization. While all major portions of the earth are examined to a certain extent, Europe clearly occupies a prominent position here. The reason for this gambit is straightforward: more than anything else, we believe it is the reification of the Occidental realm, as geographically no less than culturally and historically distinct from all others, that warps the prevailing geographical imagination.

We would also warn the reader that the following survey of metageographical thought is suggestive rather than exhaustive. Our concern here is to identify the foundational ideas of global human geography, to trace the main outlines of their development, and to examine their implications for present-day conceptualizations of the earth. We do not study the precise pathways along which such ideas were passed down, nor do we delve into their many branching byways or comb through their many nuanced expressions. We even largely bypass in the substantive chapters some of the key metageographical constructs of the modern age, notably the political-economic distinction between a realm of power and a realm of poverty (North versus South, First World versus Third World). While these divisions are essential components of contemporary global discourse, they are also of very recent coinage, are growing rapidly obsolete, and are ultimately rooted in the East-West divide, which is explored here in detail. Nor do we dwell at length on the myth of the nation-state, another key metageographical idea of our time. This construct is in itself complex enough to warrant an entire book-length exploration. All that we can do here is identify those instances in which nation-state ideology conspires with other metageographical formulations to further confuse the received ideas of global geography.

In sum, ours is neither the first word on this subject, nor the last. If we have nonetheless been emboldened to venture into this difficult terrain, it is because of a growing sense that its importance is inadequately appreciated. While we cannot present a definitive history of metageographical ideas, we can at least try to stir broader interest in the subject, in the hope that more complete historical expositions—and a more accurate geographical schema—will eventually follow.

CHAPTER I

The Architecture of Continents

The Development of the Continental Scheme

In contemporary usage, continents are understood to be large, continuous, discrete masses of land, ideally separated by expanses of water. Although of ancient origin, this convention is both historically unstable and surprisingly unexamined; the required size and the requisite degree of physical separation have never been defined. As we shall see, the sevenfold continental system of American elementary school geography did not emerge in final form until the middle decades of the present century.

CLASSICAL PRECEDENTS

According to Arnold Toynbee, the original continental distinction was devised by ancient Greek mariners, who gave the names *Europe* and *Asia* to the lands on either side of the complex interior waterway running from the Aegean Sea through the Dardanelles, the Sea of Marmara, the Bosporus, the Black Sea, and the Kerch Strait before reaching the Sea of Azov.[1] This water passage became the core of a continental system when the earliest Greek philosophers, the Ionians of Miletus, designated it as the boundary between the two great landmasses of their world. Somewhat later, Libya (or Africa) was added to form a three-continent scheme.[2] Not surprisingly, the Aegean Sea lay at the heart of the Greek conception of the globe; Asia essentially denoted those lands to its east,[3] Europe those lands to its west and north, and Libya those lands to the south.

A seeming anomaly of this scheme was the intermediate position of the Greeks themselves, whose civilization spanned both the western and the eastern shores of the Aegean. Toynbee argued that the inhabitants of central Greece used the Asia-Europe boundary to disparage their Ionian kin, whose succumbing to "Asian" (Persian) dominion contrasted flatteringly with their own "European" freedom.[4] Yet not all Greek thinkers identified themselves as Europeans. Some evidently employed the term *Europe* as a synonym for the northern (non-Greek) realm of Thracia.[5] In another formulation, Europe was held to include the mainland of Greece, but not the islands or the Peloponnesus.[6] Still others — notably Aristotle — excluded the Hellenic "race" from the continental schema altogether, arguing that the Greek character, like the Greek lands themselves, occupied a "middle position" between that of Europe and Asia.[7] In any case, these disputes were somewhat technical, since the Greeks tended to view continents as physical entities, with minimal cultural or political content.[8] When they did make generalizations about the inhabitants of different continents, they usually limited their discussion to the contrast between Asians and Europeans; Libya was evidently considered too small and arid to merit more than passing consideration.

Twofold or threefold, the continental system of the Greeks clearly had some utility for those whose geographical horizons did not extend much beyond the Aegean, eastern Mediterranean, and Black Seas. But its arbitrary nature was fully apparent by the fifth century B.C.E. Herodotus, in particular, consistently questioned the conventional three-part system, even while employing it. Criticizing the overly theoretical orientation of Greek geographers, who attempted to apprehend the world through elegant geometrical models, he argued instead for an "empirical cartography founded on exploration and travel."[9] One problematic feature of the geography that Herodotus criticized was its division of Asia and Africa along the Nile, a boundary that sundered the obvious unity of Egypt.[10] After all, as he noted, Asia and Africa were actually contiguous, both with each other and with Europe: "Another thing that puzzles me is why three distinct women's names should have been given to what is really a single landmass; and why, too, the Nile and the Phasis — or, according to some, the Maeotic Tanais and the Cimmerian Strait — should have been fixed upon for the boundaries. Nor have I been able to learn who it was that first marked the boundaries, or where they got their names from."[11]

Similar comments, suggesting a continued awareness that these were constructed categories, echoed throughout the classical period. Strabo, writing in the first century B.C.E., noted that there was "much argument respecting the continents," with some writers viewing them as islands, others as mere peninsulas. Furthermore, he argued, "in giving names to the three continents, the Greeks did not take into consideration the whole habitable earth, but merely their own country, and the land exactly opposite. . . . "[12]

Under the Romans, the continental scheme continued to be employed in scholarly discourse, and the labels *Europe* and *Asia* were sometimes used in an informal sense to designate western and eastern portions of the empire.[13] In regard to military matters, the term *europeenses* was deployed rather more precisely for the western zone.[14] *Asia* was also used in a more locally specific sense to refer to a political subdivision of the Roman Empire in western Anatolia.

MEDIEVAL AND RENAISSANCE CONSTRUCTIONS

For almost two millennia after Herodotus, the threefold division of the earth continued to guide the European scholarly imagination. The continental scheme was reinforced in late antiquity when early Christian writers mapped onto it the story of Noah's successors. According to St. Jerome (who died circa A.D. 420), translator of the Vulgate Bible, "Noah gave each of his three sons, Shem, Ham, and Japheth, one of the three parts of the world for their inheritance, and these were Asia, Africa, and Europe, respectively."[15] This new theological conception had the merit of explaining the larger size of the Asian landmass by reference to Shem's primogeniture.[16] It also infused the Greeks' tripartite division of the world with religious significance. This sacralized continental model would persist with little alteration until the early modern period.

Medieval Europe thus inherited the geographical ideas of the classical world, but in a calcified and increasingly mythologized form. Whereas the best Greek geographers had recognized the conventional nature of the continents—and insisted that the Red Sea made a more appropriate boundary between Asia and Africa than the Nile River[17]—such niceties were often lost on their counterparts in late antiquity and the early Middle Ages. Martianus Capella, whose compilation of knowledge became a standard medieval text,[18] took it as gospel that the world was divided into Europe, Asia, and Africa, with the Nile separating the latter two landmasses.[19] Other influential encyclopedists of the period, including Orosius and Isidore of Seville, held similar views.[20]

During the Carolingian period, by contrast, the inherited framework of Greek geography began to recede from view. The term *Europe* (in one form or another) was sometimes used to refer to the emerging civilization in the largely Frankish lands of Latin Christendom, which were occasionally contrasted with an increasingly fabulous Asia to the east.[21] In fact, proponents of both Carolingian and Ottonian (German) imperialism, as well as the papacy, employed the concept of Europe as "a topos of panegyric, [and] a cultural emblem."[22] But until the late Middle Ages, reference to the larger formal continental scheme was largely limited to recondite geographical studies, finding little place in general scholarly discourse.[23] Africa in particular did not figure prominently in the travel lore and fables of medieval Europeans. The southern continent at the time was dismissed as inferior, on the mistaken grounds that it was small in extent and dominated by deserts.[24]

Scholarly geographical studies, of course, were another matter. Here the tripartite worldview of the Greeks was retained, but transposed into an abstract cosmographical model, abandoning all pretense to spatial accuracy. The famous "T-O" maps of the medieval period, representing the earth in the form of a cross, reflect the age's profoundly theological view of space. The cross symbol (represented as a T within the circle of the world) designated the bodies of water that supposedly divided Europe, Asia, and Africa; these landmasses in a sense served as the background on which the sacred symbol was inscribed. The Nile remained, in most cases, the dividing line between Africa and Asia. Classical precedence joined here with theological necessity, converting an empirical distortion into an expression of profound cosmographical order.[25]

With the revival of Greek and Roman learning in the Renaissance, the older continental scheme was revived as well, becoming endowed with an unprecedented scientific authority.[26] The noted sixteenth-century German geographer Sebastian Münster, for example, invoked "the ancient division of the Old World into three regions separated by the Don, the Mediterranean, and the Nile."[27] Despite the considerable accumulation of knowledge in the centuries since Herodotus, few Renaissance scholars questioned the boundaries that had been set in antiquity. On the contrary, it was in this period that the continental scheme became the authoritative frame of reference for sorting out the differences among various human societies.[28]

The elevation of the continental scheme to the level of received truth was conditioned in part by an important historical juncture. In the fourteenth and fifteenth centuries, just as classical writings were being revalued, the geography of Christianity was in flux on several fronts at once. Turkish conquests at its southeastern edge were causing the remaining Christian communities in Asia Minor to retreat, while Christian conquests and conversions in the northeast were vanquishing the last holdouts of paganism in the Baltic region. Meanwhile, the rise of humanism was challenging the cultural unity of the Catholic world from within. These historical circumstances combined to give the Greek continental scheme new salience. On the one hand, as Christianity receded in the southeast and advanced in the northeast, the boundaries of Christendom increasingly (although never perfectly) coincided with those of the Greeks' Europe. On the other hand, humanist scholars began to search for a secular self-designation. As a result, these centuries saw Europe begin to displace Christendom as the primary referent for Western society.[29]

As Western Christians began to call themselves Europeans in the fifteenth century, the continental schema as a whole came into widespread use. But it was not long before the new (partial) geographical fit between Europe and Christendom was once again offset. Continuing Turkish conquests, combined with the final separation of the Eastern and Western Christian traditions, pulled southeastern Europe almost completely out of the orbit of the increasingly self-identified European civilization.[30]

OLD WORLDS, NEW CONTINENTS

Once Europeans crossed the Atlantic, they gradually discovered that their threefold continental system did not form an adequate world model. Evidence of what appeared to be a single "new world" landmass somehow had to be taken into account. The transition from a threefold to a fourfold continental scheme did not occur immediately after Columbus, however. First, America had to be intellectually "invented" as a distinct parcel of land—one that could be viewed geographically, if not culturally, as equivalent to the other continents.[31] According to Eviatar Zerubavel, this reconceptualization took nearly a century to evolve, in part because it activated serious "cosmographic shock."[32] For a long time, many Europeans simply chose to ignore the evidence; as late as 1555, a popular French geography text entitled *La Division du monde* pronounced that the earth consisted of Asia, Europe, and Africa, making absolutely no mention of the Americas.[33] The Spanish imperial imagination persisted in denying continental status to its transatlantic colonies for even longer. According to Walter Mignolo, "The Castilian notion of 'the Indies' [remained] in place up to the end of the colonial empire; 'America' [began] to be employed by independentist intellectuals only toward the end of the eighteenth century."[34] Yet by the early sixteenth century, the Portuguese cosmographer Duarte Pacheco and his German counterpart Martin Waldseemüller had mapped the Americas as a continent.[35] While cartographic conventions of the period rendered the new landmass, like Africa, as distinctly inferior to Asia and Europe,[36] virtually all global geographies by the seventeenth century at least acknowledged the Americas as one of the "four quarters of the world."

As this brief account suggests, accepting the existence of a transatlantic landmass required more than simply adding a new piece to the existing continental model. As Edmundo O'Gorman has brilliantly demonstrated, reckoning with the existence of previously unknown lands required a fundamental restructuring of European cosmography.[37] For in the old conception, Europe, Africa, and Asia had usually been envisioned as forming a single, interconnected "world island," the *Orbis Terrarum*. The existence of another such "island" in the antipodes of the Southern Hemisphere — an *Orbis Alterius* — had often been hypothesized, but it was assumed that it would constitute a world apart, inhabited, if at all, by sapient creatures of an entirely different species. Americans, by contrast, appeared to be of the same order as other humans,[38] suggesting that their homeland must be a fourth part of the human world rather than a true alter-world. Thus it was essentially anthropological data that undermined the established cosmographic order.

In the long run, the discovery of a distant but recognizably human population in the Americas would irrevocably dash the world island to pieces. Over the next several centuries the fundamental relationship between the world's major landmasses was increasingly seen as one of separation, not contiguity. In 1570 Ortelius divided the world into four constituent parts, yet his global maps did not emphasize divisional lines, and his regional maps sometimes spanned "continental" divisions.[39] By the late seventeenth century, however, most global atlases unambiguously distinguished the world's main landmasses and classified all regional maps accordingly.[40] The Greek notion of a unitary human terrain, in other words, was disassembled into its constituent continents, whose relative *isolation* was now ironically converted into their defining feature. Although the possibility of an *Orbis Alterius* was never again taken seriously, the boundaries dividing the known lands would henceforth be conceived in much more absolute terms than they had been in the past. Even as the accuracy of mapping improved dramatically in this period, the conceptualization of global divisions was so hardened as to bring about a certain conceptual deterioration.

NEW DIVISIONS

As geographical knowledge increased, and as the authority of the Greeks diminished, the architecture of global geography underwent more subtle transformations as well. If continents were to be meaningful geographical divisions of human geography, rather than mere reflections of an ordained cosmic plan, the Nile and the Don obviously formed inappropriate boundaries. Scholars thus gradually came to select the Red Sea and the Gulf and Isthmus of Suez as the African-Asian divide. Similarly, by the sixteenth century, geographers began to realize that Europe and Asia were not separated by a narrow isthmus, that the Don River did not originate anywhere near the Arctic Sea, and that the Sea of Azov was smaller than had previously been imagined. While the old view was remarkably persistent, a new boundary for these two continents was eventually required as well.[41]

The difficulty was that no convenient barrier like the Red Sea presented itself between Europe and Asia. The initial response was to specify precise linkages between south- and north-flowing rivers across the Russian plains; by the late seventeenth century, one strategy was to divide Europe from Asia along stretches of the Don, Volga, Kama, and Ob Rivers.[42] This was considered an unsolved geographical issue, however, and geographers vied with each other to locate the most fitting divisional line. Only in the eighteenth century did a Swedish military officer, Philipp-Johann von Strahlenberg, argue that the Ural Mountains formed the most significant barrier. Von Strahlenberg's proposal was enthusiastically seconded by Russian intellectuals associated with Peter the Great's Westernization program, particularly Vasilii Nikitich Tatishchev, in large part because of its ideological convenience.[43] In highlighting the Ural divide, Russian Westernizers could at once emphasize the European nature of the historical Russian core while consigning Siberia to the position of an alien Asian realm suitable for colonial rule and exploitation.[44] (Indeed, many Russian texts at this time dropped the name *Siberia* in favor of the more Asiatic-sounding *Great Tartary*.)[45] Controversy continued in Russian and German geographical circles, however, with some scholars attempting to push the boundary further east to the Ob or even the Yenisey River, while others argued for holding the line at the Don.[46]

Tatishchev's and von Strahlenberg's position was eventually to triumph not only in Russia but throughout Europe. After the noted French geographer M. Malte-Brun gave it his seal of approval in the nineteenth century, the Ural boundary gained near-universal acceptance.[47] Yet this move necessitated a series of further adjustments, since the Ural Mountains do not extend far enough south—or west—to form a complete border. In atlases of the eighteenth and early nineteenth centuries, the old and new divisions were often combined, with Europe shown as separated from Asia by the Don River, a stretch of the Volga River, and the Ural Mountains.[48] From the mid-1800s on the most common, although by no means universal,[49] solution to this problem was to separate Asia from Europe by a complex line running southward through the Urals, jumping in their southern extent to the Ural River, extending through some two-thirds the length of the Caspian Sea, and turning in a sharp angle to run north-westward along the crest of the Caucasus Mountains.[50] Indeed, as recently as 1994, the United States Department of State gave its official imprimatur to this division.[51] The old usage of the Don River, arbitrary though it might have been, at least required a less contorted delineation. Moreover, the new division did even more injustice to cultural geography than did the old, for it included within Europe such obviously "non-European" peoples as the Buddhist, Mongolian-speaking Kalmyks.

While this geographical boundary between Europe and Asia is now seldom questioned and is often assumed to be either wholly natural or too trivial to worry about, the issue still provokes occasional interest. In 1958, for example, a group of Russian geographers argued that the true divide should follow "the eastern slope of the Urals and their prolongation the Mugodzhar hills, the Emba River, the northern shore of the Caspian Sea, the Kumo-manychskaya Vpadina (depression) and the Kerchenski Strait to the Black Sea"[52]—thus placing the Urals firmly within Europe and the Caucasus within Asia. Other writers have elected to ignore formal guidelines altogether, placing the boundary between the two "continents" wherever they see fit. The 1963 edition of the *Encyclopedia Britannica,* for example, defines the Swat district of northern Pakistan as "a region bordering on Europe and Asia"[53]—"Europe" perhaps connoting, in this context, all areas traversed by Alexander the Great. Halford Mackinder, on the other hand, selected a "racial" criterion to divide Europe from Africa (although not from Asia), and thus extended its boundaries well to the south: "In fact, the southern boundary of Europe was and is the Sahara rather than the Mediterranean, for it is the desert land that divides the black man from the white."[54]

THE CONTINUING CAREER
OF THE CONTINENTAL SCHEME

Despite the ancient and ubiquitous division of the earth into Europe, Asia, and Africa (with the Americas as a later addition), such "parts" of the earth were not necessarily defined explicitly as continents prior to the late nineteenth century. While the term *continent*—which emphasizes the contiguous nature of the land in question—was often used in translating Greek and Latin concepts regarding the tripartite global division, it was also employed in a far more casual manner. In fact, in early modern English, any reasonably large body of land or even island group might be deemed a continent. In 1599, for example, Richard Hakluyt referred to the West Indies as a "large and fruitfull continent."[55] Gradually, however, geographers excluded archipelagos and smaller landmasses from this category, adhering as well to a more stringent standard of spatial separation. By 1752 Emanuel Bowen was able to state categorically: "A continent is a large space of dry land comprehending many countries all joined together, without any separation by water. Thus Europe, Asia, and Africa is one great continent, as America is another."[56]

The division of the world into two continents certainly forces one to recognize, as Herodotus did many centuries earlier, that Europe, Asia, and Africa are not separated in any real sense. Indeed, perspicacious geographers have always been troubled by this division. As early as 1680, the author of *The English Atlas* opined: "The division seems not so rational; for Asia is much bigger than both of the others; nor is Europe an equal balance for Africa."[57] Several prominent nineteenth-century German geographers, Alexander von Humboldt and Oskar Peschel among them, insisted that Europe was but an extension of Asia; many Russian Slavophiles, perennial opponents of the more influential Westernizers, concurred.[58] Such clear-headed reasoning was not to prevail, however. By the late nineteenth century the old "parts of the earth" had been definitively named "continents," with the separation between Europe and Asia remaining central to the scheme. The *Oxford English Dictionary* (compiled in the decades bracketing the turn of the twentieth century) recounts the transition as follows: "Formerly two continents were reckoned, the Old and the New; the former comprising Europe, Asia, and Africa, which form one continuous mass of land; the latter, North and South America, forming another. These two continents are strictly islands, distinguished only by their extent. Now it is usual to reckon four or five continents, Europe, Asia, Africa, and America, North and South; the great island of Australia is sometimes reckoned as another."[59]

Regardless of the term used to denote them, the standard categories of antiquity, with the addition of the "new world(s)," continued to comprise the fundamental framework within which global geography and history were conceived.[60] Yet minor disagreements persisted as to the exact number of units one should count. In eighteenth- and nineteenth-century world atlases, which generally printed the world's major units in different colored inks, one can find fourfold, fivefold, and sixfold divisional schemes. North and South America might be counted as one unit or two, while Australia ("New Holland") was sometimes colored as a portion of Asia, sometimes as a separate landmass, and sometimes as a mere island.[61] All things considered, however, the fourfold scheme prevailed well into the 1800s.

Whatever the exact form it took on maps, the division of the world into great continents became an increasingly important metageographical concept in the eighteenth and nineteenth centuries. Montesquieu, the foremost geographical thinker of the French Enlightenment, based his social theories on the absolute geographical separation of Europe from Asia, the core of his fourfold continental scheme.[62] The most influential human geographer of the mid–nineteenth century, Carl Ritter, similarly argued (in his signature teleological style): "Each continent is like itself alone . . . each one was so planned and formed as to have its own special function in the progress of human culture."[63] Ritter also attempted to ground the entire scheme in physical anthropology. Conflating continents with races, he viewed Europe as the land of white people, Africa that of black people, Asia of yellow people, and America of red people[64]—a pernicious notion that still lingers in the public imagination.

It was with Arnold Guyot, the Swiss scholar who introduced Ritter's version of geography to the United States in the mid-1800s, that continent-based thinking reached its apogee. Guyot saw the hand of Providence in the assemblage of the continents as well as in their individual outlines and physiographic structures. The continents accordingly formed the core of Guyot's geographical exposition—one aimed at revealing "the existence of a general law, and disclos[ing] an arrangement which cannot be without a purpose."[65] Not surprisingly, the purpose Guyot discerned in the arrangement of the world's landmasses entailed the progressive revelation of a foreordained superiority for Europe and the Europeans. From his position on the faculty of Princeton University, Guyot propagated his views on the subject for many years, influencing several generations of American teachers and writers.

As the continental system was thus formalized in the nineteenth century, its categories were increasingly naturalized, coming to be regarded, not as products of a fallible human imagination, but as real geographical entities that had been "discovered" through empirical inquiry.[66] E. H. Bunbury, the leading Victorian student of the history of geographic thought, went so far as to label Homer a "primitive geographer" for his failure to recognize "the division of the world into three continents."[67]

Bunbury also took Herodotus to task for his "erroneous notion" that Europe was of greater east-west extent than Asia and Libya [Africa] combined. Herodotus came to this conclusion, however, not because his spatial conceptions were any less accurate than those of his peers, but because he eschewed using the north-south trending Tanais (Don) as the continental border, preferring instead east-west running rivers such as the Phais and Araxes (in the Caucasus region). To the Victorian Bunbury, this was not an issue on which educated people could disagree.[68] What nineteenth-century geographers had lost was Herodotus's sense that the only reason for dividing Europe and Asia along a north-south rather than an east-west axis was convention. In fact, by scientific criteria, Herodotus probably had the better argument. Certainly in physical terms, Siberia has much more in common with the far north of Europe—where Herodotus's boundary would have placed it—than with Oman or Cambodia.

INTO THE TWENTIETH CENTURY

Since the early eighteenth century, one of the most problematic issues for global geographers was how to categorize Southeast Asia, Australia, and the islands of the Pacific. Gradually, a new division began to appear in this portion of the world. According to one popular Victorian work of world history, "It was usual until the present century to speak of the great divisions of the earth as the Four Quarters of the World, VIZ; Europe, Asia, Africa, and America," while insisting that a "scientific distribution" of the world's "terrestrial surfaces" would have to include Australia and Polynesia as separate divisions.[69] By the middle of the nineteenth century, Australia was usually portrayed as a distinct part of the world, albeit often linked with the islands of the Pacific.[70] The notion of Oceania as a fifth (or sixth, if the Americas were divided) section of the world grew even more common in the early twentieth century, when several cartographers marked off insular Southeast Asia from Asia and appended it to the island world.[71]

In the early twentieth century, world geography textbooks published in Britain and the United States almost invariably used the continental system as their organizing framework, typically devoting one chapter to each of these "natural" units. This pattern may be found in works on the natural world as well as in those concerned with human geography. Scanning through these textbooks, one notices only slight deviations from the standard model. *The International Geography,* edited by Hugh Robert Mill,[72] for example, places Central and South America in a single chapter, while devoting another to the polar regions. Leonard Brooks, in *A Regional Geography of the World,* follows the conventional scheme—with successive chapters on Europe, Asia, North America, South America, Africa, and Australia—but devotes an additional chapter to the British Isles alone.[73] Here Eurocentrism yields pride of place to Britanocentrism, suggesting the emergence of a new virtual continent in the north Atlantic.

Yet not all geographical writers in the early twentieth century viewed continents as given and unproblematic divisions of the globe. In the popular *Van Loon's Geography* of 1937, for example, the author describes the continental scheme with a light and almost humorous touch, concluding that one might as well use the standard system so long as one remembers its arbitrary foundations. Van Loon viewed the standard arrangement as including five continents: Asia, America, Africa, Europe, and Australia.[74] While it might seem surprising to find North and South America still joined into a single continent in a book published in the United States in 1937, such a notion remained fairly common until World War II.[75] It cannot be coincidental that this idea served American geopolitical designs at the time, which sought both Western Hemispheric domination and disengagement from the "Old World" continents of Europe, Asia, and Africa.[76]

By the 1950s, however, virtually all American geographers had come to insist that the visually distinct landmasses of North and South America deserved separate designations. This was also the period when Antarctica was added to the list, despite its lack of human inhabitants,[77] and when Oceania as a "great division" was replaced by Australia as a continent along with a series of isolated and continentally attached islands.[78] The resulting seven-continent system quickly gained acceptance throughout the United States. In the 1960s, during the heyday of geography's "quantitative revolution," the scheme received a new form of scientific legitimization from a scholar who set out to calculate, through rigorous mathematical equations, the exact number of the world's continents. Interestingly enough, the answer he came up with conformed almost precisely to the conventional list: North America, South America, Europe, Asia, Oceania (Australia plus New Zealand), Africa, and Antarctica.[79]

Despite the implicit European bias of the continental scheme, its more recent incarnations have been exported to the rest of the world without, so far as we are aware, provoking any major critical response or local modification. In the case of Japan, a European-derived fourfold continental schema came into use in the 1700s and was ubiquitous by the middle 1800s.[80] Subsequent changes in Japanese global conceptualization closely followed those of Europe—with the signal difference that Asia almost always ranked as the first continent.[81] Geographers in the Islamic realm, for their part, had adopted the ancient threefold global division from the Greeks at a much earlier date,[82] although the continents generally played an insignificant role in their conceptions of the terrestrial order before the twentieth century.[83] South Asians and others influenced by Indian religious beliefs employed a very different traditional system of continental divisions, one much more concerned with cosmographical than with physical geographical divisions.[84] With the triumph of European imperialism, however, the contemporary European view of the divisions of the world came to enjoy near-universal acceptance. Scholars from different countries may disagree over the exact number of continents (in much of Europe, for instance, a fivefold rather than a sevenfold scheme is still preferred), but the basic system has essentially gone unchallenged.

Paradoxically, almost as soon as the now-conventional seven-part continental system emerged in its present form, it began to be abandoned by those who had most at stake in its propagation: professional geographers. Whereas almost all American university-level global geography textbooks before World War II reflected continental divisions, by the 1950s most were structured around "world regions" (discussed in chapter 6).[85] Yet the older continental divisions have persisted tenaciously in the popular press, in elementary curricula, in reference works, and even in the terminology of world regions themselves. Anyone curious about the contemporary status of the continental scheme need only glance through the shelves of cartographic games and products designed for children.[86] Nor is such pedagogy aimed strictly at the young. A recently published work designed primarily for adults, entitled *Don't Know Much about Geography*, locates the "nations of the world" according to their "continental" positions. The author further informs us that cartographers only "figured out" that Australia "was a sixth continent" in 1801. And his repetition of the familiar claim that Australia is at once "the world's smallest continent and its largest island"[87] confirms as well the continuing invisibility of the "world island," encompassing Europe, Asia, and Africa.

The Modern Continental Scheme— and Its Exception

THE IRRELEVANCE OF CONTINENTS

When it comes to mapping global patterns, whether of physical or human phenomena, continents are most often simply irrelevant. In regard to the distribution of life-forms, for instance, most contemporary continental boundaries are trivial. The animal communities of North and South America were indeed dissimilar when the two land-masses were actually separate in the Tertiary period, but with the emergence of an isthmus between them several million years ago, they melded together in what paleontologists call "the great faunal interchange." The more important modern-day zoogeographical boundary in the Americas, separating the Nearctic from the Neotropical faunal realms, lies well to the north, in central Mexico. Similarly, the fauna of northern Africa is more closely allied to that of northern Eurasia (the two areas together constituting the zoogeographer's Palearctic region) than it is to the "Ethiopian" fauna of sub-Saharan Africa and southern Arabia. Madagascar, on the other hand, while conventionally classified as merely a large African island, unambiguously forms its own faunal realm. Floral realms, too, fail to conform to the structure of continents.[88]

Even in the field of geology, continental divisions have only minor utility. Immediate visual evidence notwithstanding, tectonic plates—the true physical building blocks of the earth's surface—do not respect the geographer's continental framework. India is tectonically linked, not to its neighbors in Asia, but to distant Australia, which lies on the same "Indo-Australian" piece of lithic crust. Africa, on the other hand, is in the process of splitting in two along the Rift Valley. Geologically speaking, continents are momentary assemblages of land that continually grow, divide, and re-form; the visually obvious major landmasses of the globe thus only partially reveal the underlying processes of tectonic motion.[89] In fact, the term *continental* is used in geology in a technical sense to describe blocks of granitic crust separated by expanses of "oceanic" (basaltic) crust. By this definition, such "islands" as Madagascar and New Zealand[90] should have continental status, while there is a less-clear continental divide between North America and Eurasia (which are connected by an expansive, if submerged, shelf of "continental" rock under the Bering Sea).[91]

Explanations are never offered as to why Madagascar—or certain other large, discrete areas of land composed of continental rock—are routinely considered islands and not continents in their own right. Greenland,[92] Borneo, and New Guinea may reasonably be denied continental status due to their connection, via continental shelves, with much larger landmasses. Such a distinction, however, cannot be applied to Madagascar, New Zealand, or even New Caledonia. Indeed, the noted paleontologist Björn Kurtén has pronounced that "Madagascar is a minor continent rather than an island,"[93] but his remains distinctly the minority view. Evidently, some unspecified minimum size implicitly differentiates an island from a continent. Yet in practice few scientists seem concerned with the issue or even with maintaining consistent usage. Edward O. Wilson, for example, writes that "in biogeographical terms Australia is only an extremely large island," but then goes on to assert in the same text that Madagascar—which is less biogeographically distinct than Australia—should be considered "a small continent."[94]

THE EUROPEAN ANOMALY

If continents are simply irrelevant for physical geography, however, they can be positively pernicious when applied to human geography. Pigeonholing historical and cultural data into a continental framework fundamentally distorts basic spatial patterns, leading to misapprehensions of cultural and social differentiation. Nowhere is such misrepresentation more clearly exemplified than in the supposed continental distinction between Europe and Asia.

In current usage, continents are defined not as absolutely distinct bodies but as *more or less* discrete masses of land. North and South America, for instance, have been accorded continental status (at least by U.S. geographers) on the grounds that they are *almost* distinct landmasses (with the additional pedigree of separate geological histories). Their connection at the Isthmus of Panama—like the narrow linkage tying Africa to Asia—can be easily overlooked or dismissed as insignificant. Indeed, over time, the continental scheme has grown increasingly faithful to this fundamentally visual definition. Shifting the boundary between Africa and Asia from the narrow Nile to the broad Red Sea, acknowledging the distinctiveness of both Australia and Antarctica, and dividing North from South America have all made the continental classification system increasingly congruent with the basic patterns of land and sea that spring to the eye from a world map.

The one glaring exception to this rule is the boundary between Asia and Europe. Since Europe is by no stretch of the imagination a discernible landmass, it can hardly be reckoned a continent according to the dictionary definitions of that term. The Ural and Caucasus ranges, which are said to form its eastern border, are separated by an embarrassing 600-mile gap. Moreover, the Urals themselves are hardly a major barrier. (The Cossacks managed to invade Siberia by carrying their river boats over a brief portage "across the Urals crest.")[95] As a result, conscientious geographers sometimes group Europe and Asia together as the single continent of Eurasia, whittling down the list of major landmasses from seven to six.[96] It was the growing popularity of this view that drove Oscar Halecki—a determined champion of European civilization—to lament in 1950 that it had already become "commonplace to say that Europe is nothing but a peninsula of Asia."[97]

But Halecki's lament was premature. While a few professionals may regard Europe as a mere peninsula of Asia (or Eurasia),[98] most geographers—and almost all nongeographers—continue to treat it, not only as a full-fledged continent, but as the *archetypal* continent. The *Encyclopedia Britannica* is a prime case in point. While admitting that Europe forms an anomalous landmass, the encyclopedia nonetheless explicitly deems its civilization distinctive enough to warrant extended consideration as a continent.[99] Likewise, world atlases, the source of our most enduring continental imagery, virtually never portray Eurasia as a single division of the earth.[100] Although it creates considerable awkwardness in dealing with Russia (a state that contains large portions of both supposed continents), cartographic practice stubbornly persists in keeping Asia and Europe categorically distinct. Nor is it only the staid publishing establishment that participates in policing this boundary. Even the most au courant postmodern geographers sometimes treat Europe as a distinct continent.[101] In short, despite the pragmatic adjustments that have been made elsewhere to an increasingly rationalized continental scheme, Europeans and their descendants continue blithely to exempt their own homeland from its defining criterion.

That Europe's continental status may be denied with a wink but then continually confirmed in practice does not indicate a simple oversight. Nor can it be dismissed as a mere convenience, a simplification necessary for making sense of a complex world. Rather, Europe's continental status is intrinsic to the entire conceptual scheme. Viewing Europe and Asia as parts of a single continent would have been far more geographically accurate, but it would also have failed to grant Europe the priority that Europeans and their descendants overseas believed it deserved. By positing a continental division between Europe and Asia, Western scholars were able to reinforce the notion of a cultural dichotomy between these two areas—a dichotomy that was essential to modern Europe's identity as a civilization. This does not change the fact, however, that the division was, and remains, misleading. Not only do Europe and Asia fail to form two continents, they are not even comparable portions of a greater Eurasian landmass. Europe is in actuality but one of half a dozen Eurasian subcontinents, better contrasted to a region such as South Asia than to the rest of the landmass as a whole. (It would be just as logical to call the Indian peninsula one continent while labeling the entire remainder of Eurasia—from Portugal to Korea—another.)

Granted that Europe is not a separate landmass, however, it can still be argued that it does form a coherent cultural region. It is on these grounds, as noted, that the *Encyclopedia Britannica* tells us Europe is to be regarded as a continent. But to define Europe as a continent in cultural terms is to imply that the other continents can be similarly defined — which would require that Asia, too, be united by a distinctive culture or civilization. Unfortunately, identifying this common culture has not proven an easy task. As Élisée Reclus, an encyclopedic French geographer of the nineteenth century, recognized, Asia is internally divided to an extraordinary degree: "Nor does it, like Europe, present the great advantage of geographical unity. . . . Asia may have given birth to many local civilizations, but Europe alone could have inherited them, by their fusion raising them to a higher culture, in which all of the peoples of the earth may one day take part. . . . Isolated from each other by plateaux, lofty ranges or waterless wastes, the Asiatic populations have naturally remained far more distinct than those of Europe."[102]

As this passage suggests, of all the so-called continents, Asia is not only the largest but also the most fantastically diversified, a vast region whose only commonalities—whether human or physical—are so general as to be trivial.[103] Yet clever geographers have turned this around, seeing such diversity either as a kind of fault (as in the case of Reclus) or as the essence of Asian identity.[104] On the one hand, this is easier than looking for substantive traits that could be said to characterize such diverse places as Saudi Arabia, India, Thailand, Korea, Tibet, Uzbekistan, and Yakutia in northern Siberia. On the other hand, it has allowed Europeans to see the disproportionate diversity of the Asian "continent" as a challenge *for Asian civilization,* rather than as a challenge to their own system of geographical classification. Jean-Jacques Rousseau himself forwarded the truly remarkable argument that one of the things that made Europe "so special" was the fact that its various nations constituted a real society, whereas the other continents were but collectivities with nothing but a "name in common."[105] As Andrew March brilliantly argues, this intellectual maneuver says far more about the psychology of European scholars than it does about the geographical entity known as Asia.[106]

If Asia's internal cohesion has been difficult to ascertain, specifying its geographical limits has proven problematic as well. The conventional southeastern boundary of this so-called continent, while perhaps more obscure than that separating it from Europe, is no less contrived. Extending east-southeast from the Malay Peninsula is a continuous chain of islands, large and small, which eventually attenuates in eastern Melanesia. The western portion of this island group, contemporary Indonesia, is conventionally included as part of Asia (although in former times this was not always the case), while the eastern portion, Melanesia, is excluded.[107] On cultural and historical grounds such a division might be supportable, but in practice the boundary between the two zones is not consistently dictated by cultural criteria. Rather, New Guinea is typically sliced cleanly down the middle, along the political boundary between Indonesia and Papua New Guinea, and the western half of this unambiguously Melanesian island is ceded to Asia.[108] Nor is such a cartographic absurdity limited to political atlases; among other manifestations, it is enshrined on the walls of no less a venerated cultural institution than the Smithsonian's Sackler Museum of Asian Art. This continental bifurcation of New Guinea can be justified neither on physical nor on cultural grounds. Its sole claim to legitimacy is the political incorporation of western New Guinea (Irian Jaya) into the "Asian" state of Indonesia. Given that most of the people of Irian Jaya resent their subjugation (and that many are in open rebellion against it), this easily overlooked cartographic maneuver has troubling political implications.

CONTINENTAL REORIENTATIONS
IN THE POPULAR IMAGINATION

Few Americans, of course, notice such niceties. In cases where the gap between official boundaries and popular conceptions is particularly large, many people increasingly ignore the official continental scheme altogether (see map 2). Despite the fact that our encyclopedias lump

---

Map 2. *The Seven "Continents" and Their Displacement in the Popular Imagination.* Heavy lines mark the official continental boundaries of U.S. geography, while lighter lines denote the areas commonly associated with these labels in the popular imagination. Thus Asia as commonly perceived includes only the southeastern portion of the standard continent (although it is extended in some contexts to encompass Australia and New Zealand, as indicated by the broken line). Likewise, the official Europe-Asia boundary at the Urals is often shifted westward in popular use to exclude Russia, the Ukraine, and Belarus from Europe. Africa, as a category in U.S. journalism, usually excludes the northern tier of African countries, while North America is conceptually truncated far north of the Panama Isthmus (either at Mexico's northern or southern boundary)—excising large areas that are officially within the continent's boundaries.

Problematic "islands" have been shaded with small dots. New Guinea is typically counted as half in Asia and half in Oceania, but by geological criteria it is unambiguously part of Australia (a continent with which it is virtually never associated). New Zealand, on the other hand, is usually grouped with Australia, if it is not overlooked altogether. Madagascar is almost always linked with Africa, though the reason for this maneuver is never spelled out. Greenland is officially classified as part of North America, but it is often forgotten when that "continent" is discussed.

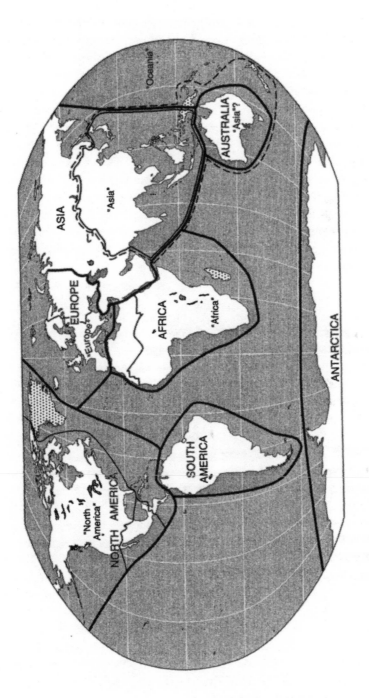

all "Asian" lands and peoples together, for instance, most Americans routinely excise large sections of the conventional continent from their mental maps. Few today would identify the residents of Syria and Saudi Arabia, much less Israel, as Asians. Even the Association for Asian Studies, the principal academic group devoted to the study of the area, excludes Southwest Asia—as well as North Asia (Siberia)—from its scholarly purview. The academic journal *Modern Asian Studies,* according to a map on its advertising flier, similarly excludes the southwestern quadrant of the official continent, although it appears to include eastern Siberia.[109] Journalistic conventions as well often limit Asia to the southeastern half of the official continent. In the *Economist,* for example, Pakistan counts as part of Asia, but stories on Iran are published under the catchall heading "international." In the most extreme reduction, "Asia" becomes essentially limited to East and Southeast Asia—and even then such countries as North Korea and Myanmar may be excluded if convenient. The Asia-Europe summit (or "Asem") of March 2, 1996, for example, included on the Asia side only the seven countries in the Association of Southeast Asian Nations (ASEAN) along with Japan, South Korea, and China.[110]

In regard to North America one can detect a similar shift between official designation and popular conception. Strictly speaking, the North American continent includes Panama and all points north, but in common parlance Central America is usually excluded, while in some circumstances Mexico is deleted as well.[111] Most Spanish-speaking peoples of the Western Hemisphere likewise reserve the term *Norte America* for the United States and Canada. Less noticed is the fact that, when deployed in this way, North America is no longer a continental category.

In many respects, Australia is the only unambiguous inhabited continent; it alone is surrounded by water. Moreover, until the coming of Europeans, Australian peoples were nearly isolated, maintaining only a tenuous connection with New Guinea through the Torres Strait Islands. Aboriginal Australia formed a coherent (if diverse) cultural region, and one can easily argue that modern-day Australia, transformed though it may be, retains such a distinction. In contemporary usage, however, Australia is often accorded less than full continental status. The joining of Australia with various Pacific islands to form the quasi continent of Oceania is an old gambit. More recently, news magazines like the *Economist* have begun to classify Australia as part of Asia, presumably by virtue of its growing economic ties to Tokyo, Hong Kong, and Singapore. Many Australian politicians, seeking to intensify such connections, concur.

These various adjustments to received continental categories have incrementally improved the fit between major geographical divisions and sociocultural features. But distortions stubbornly persist. In particular, the new conception of the world does not eliminate the problem of false comparability between Europe and Asia. While careful writers no longer elevate the European peninsula to a position of equivalence with the massive zone extending from the Bosporus to Kamchatka, Europe does continue to be juxtaposed with a much-reduced "monsoon Asia," anchored by India and China. This may be a step in the right direction, but the comparison still does not wash. The historically constituted cultural region of far western Eurasia simply cannot usefully be compared with the vast and heterogeneous swath of terrain from Afghanistan to Japan. And even the new Asia of popular imagination, pared down though it may be, still lacks the unifying features that are expected to characterize a human-geographical region. In essence, it remains little more than a flattering mirror to Europe, conceptualized more by its supposed lack of Europeanness than by any positive attributes of its own.

Finally, as the boundaries of the continents have come loose from their geophysical moorings, these categories have become increasingly vague in the public imagination, reducing their usefulness even as locating devices. A survey one of us conducted at the beginning of an introductory world geography course at Duke University indicates the pervasiveness of the problem. When asked to identify Europe on an unlabeled political map of the Eastern Hemisphere, only thirty percent of the students circled the standard continent; more than half excluded all of Russia, while a few excluded such areas as southern Europe, the British Isles, Scandinavia, and even (in one case) France. A larger proportion (forty percent) correctly identified the standard continent of Asia, but those who did not offered a much more wildly divergent set of readings. Nor were such disagreements limited to distant places. When asked to draw a line around North America on a map of the Western Hemisphere, a third of the students circled only the United States and Canada, and another third included Mexico but not Central America. While this is hardly a scientific sample, it does raise very real doubts about the uniformity of the categories that underlie our inherited geographical scheme.

The Roots of Geographical Determinism

CONTINENTS AND REGIONS

What ultimately damns the continental system, however, is not its vagueness or its tendency to mislead us into making faulty associations among human cultural groupings. Most insidious in the long run is the way in which this metageographical framework perpetuates a covert form of environmental determinism.

Environmental (or geographical) determinism is the belief that social and cultural differences between human groups can ultimately be traced to differences in their physical environments. As this philosophy took definitive shape in the Anglo-American academy at the turn of this century, it tended to support the self-serving notion that temperate climates alone produced vigorous minds, hardy bodies, and progressive societies, while tropical heat (and its associated botanical abundance) produced races marked by languor and stupefaction. Such overtly racialist claims disappeared several generations ago from respectable works.[112] Yet we would argue that a more subtle and largely unrecognized variant of environmental determinism lurks behind the myth of continents.

The reason for this is simple. In practice, the continental system continues to be applied in such a way as to suggest that continents are at once physically and culturally constituted—i.e., that natural and human features somehow correspond in space. Nineteenth-century geographers regarded this notion as a virtual article of faith; the long-running debate over Russia's true continental position was animated by precisely this assumption.[113] It is hardly surprising that the same idea doggedly persists in the public imagination. Having been taught that continents are the basic building blocks of global geography, our students slide easily into assuming that the configuration of landmasses must correspond to the distribution of cultural traits and social forms. Surely there must be something identifiably African about all people who live in Africa, as distinct from the Asianness of those who inhabit Asia. This slippage of categories suggests that the continent itself, through some unspecified process, imparts an essence to its human inhabitants. The result is that actual cultural connections and distinctions across the complexly variegated human landscape are made to seem pale before the arbitrary divisions of continental terrain.

Once this natural-seeming conception takes hold, it becomes a Herculean task to dislodge it. Arnold Toynbee certainly attempted to do so, devoting strenuous arguments to the task. Concerning the distinction between Europe and Asia, Toynbee rightly pointed out that "the geographers' error here lay in attempting to translate a serviceable piece of navigational nomenclature into political and cultural terms." He attempted to disabuse scholars of this notion by insisting that "the historian cannot lay his finger on any period at all, however brief, in which there was any significant cultural diversity between 'Asiatic' and 'European' occupants of the all but contiguous opposite banks of a tenuous inland waterway."[114] Toynbee's work, however, is seldom read these days, and his geographical arguments have never received much attention. Like other attempts to expose the faulty logic behind the continental scheme, his too evidently met with very little success.

As a result, contemporary geographers, while distancing themselves deliberately from the racialist thinking that once dominated the discipline, sometimes fall back into an environmentalist position simply by remaining faithful to continental categories. This may be seen, for instance, in the persistent idea that a distinctly "Asiatic Mode of Production" formerly prevailed all the way from the Ottoman Empire to China. Even in the 1990s, a prominent scholar can argue, following Karl Marx and Karl Wittfogel, that the need for large-scale irrigation—an imperative ultimately attributed to physical geography—was significantly responsible for the development of Asia's "despotic" forms of rule.[115] As will be demonstrated at greater length in chapter 5, careful scholarship has thoroughly discredited this thesis; just as there is no Asia, neither is there an Asiatic Mode of Production or a characteristically Asian form of despotic power. More subtle examples of modern geodeterminism may be found in the introductory chapters of textbooks on Asian history and culture. A recent work by Rhoads Murphey, for instance—in many ways a fine piece of historical geographic synthesis—begins by specifying its field of reference as South, Southeast, and East Asia, positing that this constitutes a coherent frame for historical analysis based on climatic criteria (namely, monsoonal circulation).[116] While one could quibble with Murphey's climatic regionalization,[117] the more important point is that "monsoon Asia" cannot be regarded as a primary cultural or historical region unless one accepts the basic tenets of environmental determinism.[118]

EUROPE AND METAGEOGRAPHICAL DETERMINISM

As Andrew March shows, however, proponents of geographical determinism have often construed the intensity of environmental influence as varying according to continental location—opening the way for exempting Europeans from the strict rule of nature. Since at least the time of Montesquieu, Europe has been pictured as a land of moderate climate and diverse landforms, allowing unusual scope for human freedom. In other words, Europe has been depicted as the arena of environmental *possibilism*. Asia and Africa, by contrast, have been often viewed as continents of climatic rigor and physiographic uniformity, whose people have been subject to a corresponding set of "iron physical laws."[119] In this view, the bonds of geographical concordance, especially those linking human developments with physiographic features, can be asserted to be much stronger in Asian countries than in those of Europe.[120]

This kind of theorizing reached its peak in the Victorian period, when Henry Thomas Buckle wrote his massively influential *History of Civilization in England*. The cornerstone of Buckle's history was the supposed fact that the "feebleness" of nature in Europe allowed for the development of "thought," whereas on other continents a rougher nature held humanity in its thrall.[121] Buckle also suggested an additional reason why "it was easier for Man to discard the superstitions which nature suggested to his imagination" in Europe: namely, the European continent was "constructed upon a smaller plan" than the other landmasses.[122] (Such a line of thinking would lead us to suppose that Madagascar should enjoy even greater advantages—if only it too could be defined as a continent.) This notion—that diversity in a small place somehow promoted cultural development—has often been repeated by nationalistic geographers on behalf of individual European states as well. No less a scholar than Paul Vidal de la Blache, often considered the founder of French geography, could baldly claim that "because of the extremely varied physical environment of Europe in general, and *France in particular,* higher civilization came to exist in these places."[123]

The idea that Europe alone escaped geographical determination persists to this day, albeit in more subtle forms. Europe's physiographic and climatic diversity are now sometimes viewed merely as having prevented the consolidation of large empires and allowed scope for the development of a market-driven economy. Paul Kennedy, in his widely acclaimed book *The Rise and Fall of the Great Powers,* expresses this view succinctly:

> For [its] political diversity Europe had largely to thank its geography. There were no enormous plains over which an empire of horsemen could impose its swift domination; nor were there broad and fertile river zones like those around the Ganges, Nile, Tigris and Euphrates, Yellow and Yangtze, providing the food for masses of toiling and easily conquered peasants. Europe's landscape was much more fractured, with mountain ranges and large forests separating the scattered population centers in the valleys; and its climate altered considerably from north to south and west to east. . . .
> Europe's differentiated climate led to different products, suitable for exchange; and in time, as market relations were developed, they were transported along the rivers or the pathways which cut through the forest be-

tween one area of settlement and the next. . . . Here again geography
played a crucial role, for water transport of these goods was so much more
economical and Europe possessed so many navigable rivers.[124]

The many misconceptions in this brief passage betray the geographical myopia associated with the myth of continents. From Kennedy's
avowedly Eurocentric perspective,[125] Europe's geographical features are
seen in fine detail, suggesting great diversity across the region. The rest
of the world, by contrast, appears on the edges of his mental map as a
vague blur, looking highly monotonous. The discrepancy becomes evident as soon as one looks carefully at a map of southern and eastern Eurasia, focusing on precisely the features Kennedy emphasizes. To begin with,
both South and East Asia show at least as much topographic diversity as
does Europe. While both subsume large expanses of flat land, neither the
north Indian nor the north Chinese plain dwarfs the great European plain
(which extends, after all, from Aquitaine to the Urals). Climatic variation
is also comparable in all three regions; China's climate, in fact, exhibits
greater differentiation than does Europe's, ranging as it does from truly
tropical to subarctic. Similarly, all three areas feature navigable rivers, those
of China in particular having been more highly developed for transportation than their counterparts in Europe in premodern times.[126] And
as for Kennedy's claim that Europe's forests served as an impediment to
conquest, it is hard to imagine how this could have been true after the
"great age of forest clearance" in the Middle Ages—a period of massive
deforestation such as South Asia, at least, did not experience until modern times.

If the passage quoted above nonetheless remains persuasive, even to
a college-educated American audience, it does so in part because most of
Kennedy's readers have only the sketchiest knowledge of the global environment. At the level of continental units, the kinds of spatial correlations that he and others assert between features of the physical and
the human worlds are simply insupportable. For late-twentieth-century
Americans to sustain belief in a sweeping fit between cultural and natural
features requires turning a blind eye to the most basic findings of geographical research.

The standard sevenfold continental division of the world, commonsensical though it may appear, obscures rather than clarifies the essential
patterns of global geography. It represents a parochial conception of the
world, rooted in the limited ecumene of the classical Mediterranean world
and elaborated by a European culture that was as proud of its conquests

as of its cultural accomplishments. For a global community seeking a truly cosmopolitan conceptual scheme, the continental formula has clearly outlived its usefulness.

If Americans are to think clearly about the world and about our place within it, we must relinquish the final vestiges of environmental determinism, especially in our definition of sociocultural units. Our division of the human community into large-scale regional aggregations must be based on criteria appropriate to humankind, rather than those suggested by the configurations of the physical world. Human history is no more molded by the rigid framework of landmasses and ocean expanses than it is determined by the distribution of "ideal climates." As scholars in many disciplines are now arguing, the imperative of the moment is to "denaturalize" the categories through which we apprehend the human experience.[127] It is time for geographers to join in this multidisciplinary endeavor by dismantling the myth of continents.

CHAPTER 2

The Spatial Constructs
of Orient and Occident,
East and West

The publication of Edward Said's *Orientalism* in 1978 was a landmark in cultural studies. In this seminal work, the author—a Palestinian schooled in European literary history—argued that the Orient was essentially an elaborate construct of the European imagination. Focusing on analysis of British and French texts, Said criticized eighteenth- and nineteenth-century European scholars both for denying the Islamic world its dynamic history and for ascribing to it a bogus cultural unity. Despite vehement criticisms from regional specialists,[1] the impact of Said's vision across the humanities has been enormous. In the nearly twenty years since his book appeared in print, the dissecting of Occidental discourses on the Orient has become a major intellectual project, broadening out from comparative literature to include contributions from anthropology, history, and art history, among other disciplines.

As copious and diverse as the literature spawned by Said's work has become, however, there is one aspect of Orientalist discourse that has yet to be subjected to sustained scrutiny: namely, its geography. In Said's original contribution, the actual spatial referent of the term *Orient*— the crucial question of what it encompasses and what it excludes—is barely touched upon.[2] In some ways this is not surprising. Said is a literary scholar, not a geographer; moreover, as he is the first to concede, his book focuses narrowly on conceptions of the Levant and adjacent areas. But the same geographical lacuna continues to characterize the post-Orientalist literature as a whole. To date, the spatial contours of Orient and Occident—as well as their relationship to the closely related but never identical categories of East and West—have yet to be adequately mapped.

Two objections might be raised against such an exercise. First, from a planetary perspective, all East-West distinctions are clearly arbitrary. On a rotating sphere, east and west are directional indicators only and can be used to divide the entire planetary surface into distinct regions only if there is an agreed-upon point of reference; "since the world is round, what can this word [the East] mean?"[3] (Indeed, in early Japanese world maps, the Americas often constituted the Eastern Hemisphere.)[4] Second, it might be objected that East and West were never meant to be precisely mappable categories. Vagueness is of their essence; both are simply convenient rhetorical labels without any rigorous geographical underpinning.

We believe, however, that such objections miss a crucial point. Like the myth of continents, the East-West myth suggests that the globe is divided into fundamental and ultimately comparable groupings of humanity. And as with the Asia-Europe divide on which it is ultimately based, this false binarism plays to a sense of European exceptionality, reinforcing the untenable distinction between Europe and Asia while doing nothing to solve the imbalance between the two. It is our conviction that these issues are due for extended examination.

The present chapter is the first of two in this book that will explore the vicissitudes of this single most important pair of metageographical categories in use today. Our concern in this initial chapter is to trace the shifting geography of the longitudinal divide between East and West. The discussion opens by tracing the original locus of the terms *Occident* and *Orient*. After a detailed consideration of Russia's liminal position, we then turn our attention to the complex spatial shifts that both East and West have undergone in recent decades. A second major section of the chapter takes up the question of the so-called Middle East, a portion of the world that is particularly problematic in regard to the East-West divide. Finally, we shift our gaze away from the European intellectual tradition to consider the geographical worldviews of non-Occidental peoples. After a brief look at alternative global divides that have been constructed in other portions of the world, the chapter concludes by surveying the continuing career of the categories East and West in modern Indian, Chinese, Japanese, and Turkish thought. The cultural, social, and political characteristics said to distinguish the social worlds on either side of this mutable boundary are the subject of the following chapter.

The Shifting Boundaries of East and West

WHERE IS THE WEST?

As anyone familiar with the "canon wars" of the past decade is aware, academic opinion on Western civilization is deeply divided. Traditionalists celebrate the West as the wellspring of everything progressive in human history; critical revisionists disparage the same region as the primary font of imperialism and repression. Yet few, it would seem, doubt that the West is a legitimate conceptual category.[5] Twenty-five years ago, Stephen Hay confidently asserted that "the idea of 'the East' is all but extinct [and] the idea of 'the West' is fast following it into limbo,"[6] but such a prognostication now seems premature. The notion of a fundamental longitudinal divide severing West from East—and ultimately, the West from every other place—retains currency in the national media, in popular literature, and in academic discourse.[7]

The East-West division is many centuries old,[8] and has had at least three distinct referents. While these referents have followed each other in historical development, all remain in current use. Like other metageographical categories, the spatial designation of the West remains unstable and can be subtly shifted by different authors to fit their particular arguments (see map 3).[9]

The original and persistent core of the West has always been Latin Christendom, derived ultimately from the Western Roman Empire—with (ancient) Greece included whenever the search for origins goes deeper.[10] As the Hungarian scholar Jenö Szücs shows, the most significant historical divide across Europe was that separating the Latin church's *Europa Occidens* from the Orthodox lands of the Byzantine and Russian spheres. Since shortly after its inception in the Middle Ages, the "Western" cultural area associated with Latin Christianity has encompassed central as well as western Europe.[11] But as we will see below, central Europe's status in the West has been unstable. In particular, the far eastern frontier of church lands (i.e., Poland, Hungary, Croatia, and environs) has often been seen as a transitional zone between West and East,[12] and one can trace back to the Enlightenment the notion that all of Europe lying to the east of Germany constitutes a separate buffer zone, intermediary between Asia and the West—and between barbarism and civilization.[13]

Following the European diaspora of the sixteenth through nineteenth centuries, in any case, divisions within European Christendom began to recede in importance.[14] In their stead, the idea of a supra-European West,

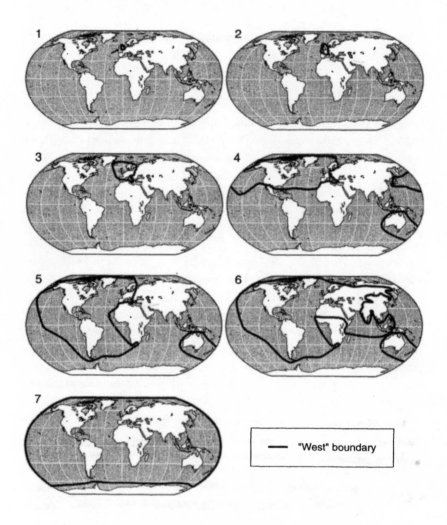

encompassing European settler colonies across the Atlantic, increasingly took hold. This sense of an expanded West was greatly strengthened after World War II. The United States and Canada had long been regarded as an overseas annex of European civilization (just as Australia and New Zealand formed distant outposts), but with Europe now sundered by the Iron Curtain, the Atlantic community began to displace western Europe as the primary geographical referent of the West.[15] Leftist opponents of Euro-American neo-imperialism promoted this vision as much as Cold Warriors, shifting the value signs but retaining the category.[16] At the height of the Cold War, at least one geographer went so far as to encompass all American "allies" into the West—including South Vietnam and most of Africa—while excluding such "neutral" countries as Sweden and Switzerland.[17] At the same time, however, as David Slater notes, the vision of the West being propagated was based essentially on a "selective reading of the history of [only] the United States and Britain."[18]

Finally, a third and still broader notion of the West has arisen since the 1960s to become widespread in journalism and popular use. This version casts aside all geographical moorings to become simply a proxy for the developed world. Newspaper headlines, for example, occasionally refer to the heads of state in the G-7 forum as "Western powers"—overlooking the fact that one member, Japan, is physically and culturally rooted in what used to be considered the extreme East. The implicit contention is that

---

Map 3. *Seven Versions of the "West."* The portion of the earth denoted by the term *West* varies tremendously from author to author and from context to context (the area enclosed by a heavy black line is what has been called the West): 1) One extreme incarnation, where the West includes only England ("The Wogs begin at Calais," as an old racist, xenophobic refrain has it). 2) The standard minimal West, which is essentially Britain, France, the Low Countries, and Switzerland. As interpreted by Thomas Mann, this West is basically centered on France. 3) The historical West of medieval Christendom, circa 1250. 4) The West of the Cold War Atlantic alliance, or Europe and its "settler colonies" (with Japan often included as well). 5) The greater "cultural" West. By the criteria of language, religion, and "high culture," Latin America and the areas of concentrated European settlement in South Africa are added to the West. The Philippines is sometimes included here as well. (Those more concerned with "race," on the other hand, are inclined to add only Argentina, Uruguay, and southern Brazil.) 6) The maximum West of the eco-radical and New Age spiritual imagination. In this formulation, all areas of Christian and Islamic heritage are included. 7) The global (future?) West of modernization. See, for instance, Arnold Toynbee's cartography showing the entire globe as under Western hegemony in one form or another, whether political, "associative" (India, Iran, Ethiopia), or "in the heterodox form of Communism" (Toynbee 1934–61, volume 11 [1959], pages 192–93).

Japan has been Westernized simply by becoming rich and powerful.[19] Similar assumptions lie behind characterizations of modern technological artifacts as items of "Western culture"—as though automobiles and soft drinks were inherently of the West, wherever they might be produced or consumed. A recent newspaper story, for example, claims that "Western culture flourishes in a changed Cambodia" on the grounds that "Mercedes, Peugeots and Toyotas glide alongside trishaws and water buffalo, running almost all the Spartan Soviet-made Ladas out of town."[20] The only way to make sense of this passage is to accept a dubious definition of non-Westernness as backwardness (manifested in everything from animal-drawn carts to cars made in Russia), while glossing all things modern and sophisticated, including Japanese Toyotas, as Western.[21]

The same conceptual maneuver by which Japan is offered (honorary?) membership in the West often requires the exclusion of Latin America from the same metageographical category. When writers use the term West as a substitute for the "developed world"—contrasting it not with the East but rather with the Third World[22]—all American territory south of the Rio Grande is excluded. Yet in terms of their cultural and social background, the inhabitants of countries like Argentina and Uruguay are arguably more closely connected to western Europe than are their counterparts in such diverse societies as the United States and Canada. The position of most other Latin American countries is more ambiguous, but certainly the elite cultural traditions throughout the region are profoundly Western in inspiration. Owing to such discrepancies, George Yúdice cogently argues that the concept of the West must be reformulated to fit Latin American realities:

> It may seem odd that someone who subscribes to multiculturalism should put forth the argument that Latin America is not non-Western. In the first place, mine is not a defense of "Western Civilization" in its imperial forms. It is, however, a challenge to the monolithic straw man that has been constructed as the West, as if there were no variants in it. Not everything that forms part of Western cultures can be reduced to imperialism. I follow the thinking of those Latin Americans who . . . [advocate rather] an alternative occidentality emerging from Latin America's cultural heterogeneity.[23]

That Latin American countries are rarely counted as part of the West in contemporary discourse shows the radical extent to which economic and geopolitical indices have displaced the cultural criteria of the original formulation. The resulting conflation of the West with the modern is a long-standing Euro-American conceit—and one with a number of questionable connotations. It effectively massages the egos of western Europeans and Americans in two ways: first, by insinuating that their culture is somehow single-handedly responsible for the shape of the modern world, and second, by suggesting that the only way for other peoples of the world to attain economic, political, and even personal success is to abandon their indigenous social and cultural patterns and adopt the cultural forms prevalent in western Europe and the United States. Recent history flatly contradicts both notions. As anyone familiar with East Asian development will aver, that region's stunning economic, technological, and scientific successes in the twentieth century cannot seriously be ascribed to cultural Europeanization; nor can East Asia's role in the creation of the modern world be downplayed. As Mimi Hall Yiengpruksawan cogently argues, Japan has had an intrinsic part in the formation of modernity, and indeed, in the creation of the postmodern condition as well.[24]

Even as Westernness has shifted from a purely spatial to a quasi-temporal category, the other half of this global pair has undergone a similar transmutation, giving rise to that essentially aspatial abstraction, the Third World. This term has its own complicated history, and deserves a more extended treatment than is possible here. Our concern is rather to trace what happened to the original geographical complement of the Occident. For despite the rise of categories like the Third World, the older and more definably spatial category has never entirely disappeared. Whenever it is convenient, the term *West* is still contrasted with a supposed *East,* whether that be defined in cultural or geopolitical terms. It is to the tortured history of the latter category that our attention now turns.

EUROPE'S FIRST OTHER: THE ORIENT

Corresponding to the gradual expansion of the European West in recent centuries has been an expansion of its counterpart, the Asiatic Orient or East. As explained in chapter 2, *Asia* originally referred to a small area in what is now northwest Turkey. It was extended eastward and southward by Greek geographers to encompass the Levant, was subsequently expanded all the way to the Bering Strait, and is now in the process of being conceptually pushed out of its original range altogether and into the southeastern quadrant of the Eurasian landmass. A similar movement has occurred in the deployment of the term *Orient,* the historian's counterpart to the geographer's *Asia.* If anything, the dislocation of the Orient has been even more pronounced than that of Asia.

The Orient began its career in the eastern Mediterranean, at a time when India was to Europeans the eastern limit of the known world and China little more than a rumor.[25] Its original referent consisted essentially of Southwest Asia. Prior to the arrival of Islam, the Orient effectively comprised the eastern variant of a common cultural and economic region centered on the Mediterranean Sea.[26] After the Arab conquests of the seventh and eighth centuries, however, the Orient took on new meaning as the alien cultural realm against which Europeanness was defined. Its physical location did not immediately change; from the Middle Ages through the Enlightenment, Orientalists were typically philologists who worked with Arabic, Syriac, Coptic, and Hebrew sources (only the more adventuresome setting their sights as far east as Persia). But as the Orient became synonymous with Islam, its referent began to expand out of the eastern Mediterranean. Only thus could Morocco, most of which lies to the west of England, be subsumed under the rubric of Oriental civilization.

It was with the expansion of European colonial networks into the Indian Ocean and South China Sea that the conceptual Orient began to push eastward. To be sure, what is now called Southwest Asia remained for a long time the primary focus of Orientalist scholarship. As recently as 1924, a book entitled *The Occident and the Orient* could discuss only Arabs, Turks, and Indians in the latter category.[27] But in the course of the 1800s, according to Raymond Schwab,[28] India gradually displaced the Levant as the primary subject of Orientalist research, with China beginning to emerge clearly on the map as well. Encompassing such a vast zone into a single regional category was seldom questioned. While many scholars differentiated the "hither" (more familiar) East of Southwest Asia and North Africa from the "farther" (more exotic) East of India and China, all such distinctions remained secondary to the opposition between the Orient as a whole and the dynamic, restless West.

Throughout these permutations, the scholarly Orient was never coincident with the Asia of conventional geography texts. Where the latter category was a continental entity, the former was always defined in cultural terms. As such, the Orient could encompass North Africa, a zone that had never been considered part of Asia; even southeastern and southern Europe could be identified as having certain Oriental traits. At the same time, large expanses of Asia—most notably Siberia[29]—could be excluded from the scholar's Orient. Since Orientalists of the classical mode were fundamentally concerned with texts, their mental maps made no room for places lacking an indigenous literary tradition. The Orient was, by general consensus, limited to lands of non-Western "civilization."

But if scholars originally mapped their spurious historical Orient as somewhat offset from the questionable geographic entity called Asia, the two concepts have undergone a parallel eastward shift in the popular imagination until they now refer to virtually the same area. By the middle of the twentieth century, Western scholars were increasingly inclined to detach Southwest Asia and North Africa from the "Orient"; as early as 1930, René Guénon concluded that the Arabs were not fully Oriental but rather *"intermédiaires naturels"* between the Occident and the true (farther) Orient.[30] Six decades later, at least in the popular imagination, it would seem that the Sinic realm has displaced the Islamic as the quintessential Orient. For most college students, as indeed for the majority of Americans, the term *Orient* now conjures up primarily visions of China, Korea, Japan, and peninsular Southeast Asia.[31] Remarkably, this major geographical reorientation in the public imagination has gone almost unremarked in scholarly works.

A possible cause of this eastward displacement of the Orient may be found in the rise of biological criteria as the basis for partitioning humanity. Most inhabitants of the eastern Mediterranean look more European than Chinese, and in the "racial science" of the early twentieth century, they were increasingly classified as Caucasians (although Turkish-speaking peoples occasionally appeared as "Mongolians"). Oriental peoples, by contrast, came to be defined by most lay observers as those with a single eye fold. It is perhaps on this grounds that Burma (Myanmar) continues to be thought of as Oriental, while India is usually excluded. This pseudo-racial Orient is now well entrenched in public perceptions. In consequence, the lingering scholarly tradition of referring to the area between Morocco and Iran as the Orient has come to seem quaintly archaic.

THE AMBIGUOUS EAST

The eastward migration of the cultural Orient has been partly counterbalanced by the complex perambulations of its successor category, the East (see map 4). The term *East* has an unusually complicated history. On the one hand, it has sometimes been used to refer to an area within Christendom: the Orthodox lands of the Byzantine and Russian churches.[32] On the other hand, and perhaps just as often, *East* has been used as a synonym for *Orient,* signifying the vast and foreign realm beyond the threshold of Europe. To complicate matters further, Cold War discourse appropriated the lexicon of *East* and *West* to demarcate zones of communist and noncommunist regimes. As a result of this complicated

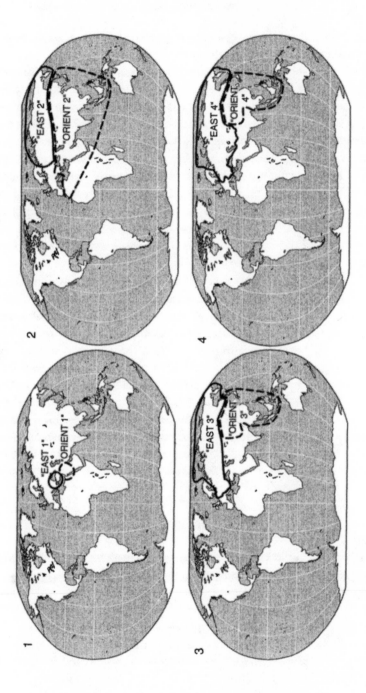

overlay of meanings, unpacking the connotations of the term in any given usage today is no easy task.

Russian-speaking lands have long been enmeshed within the larger European cultural sphere, albeit as part of a distinctive eastern variant.[33] Yet western Europeans have usually viewed Russia as only vaguely European, and certainly not Western.[34] Justifications for this exclusion are several. Some point to Russia's distinctive history of Orthodox Christianity; others highlight its periodic isolation from the main cultural currents of Western Christendom; and still others emphasize its intimate historical contact with Turkish and Mongolian peoples. Some writers go so far as to imply that several hundred years of subjugation by the Kipchak Khanate (or Golden Horde) infected Russia with a kind of Asiatic virus.[35] But whatever its justification, the result is a tool with which western Europeans can excise much of what they find objectionable in Russian history from the story of their own civilization.[36] To place Russia into a totalized East is to imply that Russia's long experiences with autocracy and totalitarianism stem from its eastern geographical position adjacent to Asia: an instance of geographical determinism — and historical denial — pushed to an extreme.

Since labeling the residents of Moscow "Asiatic" cannot be justified within the strictures of standard continental thinking, the exclusion of Russia from Europe and its classification as a portion of the East has been expressed in other ways. In the eighteenth century, northwestern Europeans began to reposition the central axis around which they conceptualized their own civilization. In earlier periods the primary division was that between

---

Map 4. *Migrations of the "East" and the "Orient."* The two terms are often used synonymously; what we have attempted to do here is to abstract the more cultural connotations of the term *Orient* from the more geopolitical connotations of the term *East*. In European historiography, the Orient begins its career in Egypt, the Levant, and adjoining areas ("Orient 1"). Subsequently, the Orient expands eastward and westward to include all non-European areas of Eurasian civilization ("Orient 2"). Finally, the Orient is pushed eastward outside of its original range altogether to encompass only East Asia, Southeast Asia, and eastern Central Asia (the "pseudo-racial Orient," or "Orient 3"). The postwar Orient ("Orient 4") is virtually identical to "Orient 3," although minor adjustments have been made to reflect Japan's territorial losses.

The East, in a more geopolitical sense, originally referred to the heartland of the Eastern Roman Empire, centered on Constantinople and identified with the Eastern Orthodox faith ("East 1"). Eventually, Russia became its focus ("East 2"). "East 3" shows the westward expansion of this conceptual space; in the interwar period it commonly encompassed all of Germany. The Cold War "communist East" is depicted as "East 4."

a refined south and a rugged north, with Muscovites classified in the latter category (along with Scandinavians and often Germans as well). During the Enlightenment, however, some writers began to distinguish the civilized West of Europe from its semibarbarous East.[37] As an "Eastern" realm, the Russian empire could be easily consigned to the category Orient. This strategy grew more popular in the nineteenth century.[38] The quintessential Victorian explorer, Sir Richard Burton, for example, claimed that the Russians were the craftiest of the Orientals (craftiness being in his book a defining Oriental quality).[39] Indeed, even Poland was sometimes put in this category, occupying as it did an intermediate position within "Europe's Orient."[40] By the early twentieth century, however, the eastward shift of the term *Orient* had largely precluded that option. Its present-day replacement is the geopolitical category of the East.

East is a broader and vaguer term than either *Asia* or *Orient,* yet one that retains the notion of critical distance from the culture of far western Europe. Like *West* (its ontological twin), *East* has been given both positive and negative readings. On the one hand, Russian nationalists themselves have adopted the term to distinguish their homeland—a realm they characterize in terms of communal solidarity and harmony—from the individualism, violence, and competition associated with the West. Yet most have been careful not to take this line of thinking too far; in the late czarist period, anti-Westernizers insisted that "the gulf separating Russia from the Occident [was] considerably less deep than that separating it from the Orient."[41] Many Russian Slavophiles, however, insisted that Slavic-speaking peoples shared a fundamentally non-Western soul;[42] a few went so far as to allege an organic affinity uniting Russia with China and even India.[43]

This kind of thinking reflected the rise of romanticism as well as nationalism in geographical discourse. As racial ideology became increasingly linked to these concerns through the nineteenth and into the twentieth century, writers from farther west, especially Germans, began likewise to expel all Slavs from the realm of Western civilization. Leopold von Ranke, for example, limited the West to the Germanic and Romance nations.[44] Shifting the East-West boundary in this way meant slighting nearly one thousand years of cultural relations in favor of a bogus appeal to national or racial spirit.[45] Nor did this go unnoticed by those Europeans who suddenly found themselves outside the pale. As Polish historian Oscar Halecki bitterly observed, "Those who call European civilization Western are inclined to decide in advance one of the most difficult and controversial questions in European history. They accept the idea of a fundamental dualism in Europe and consider only its western part really European."[46]

As the East thus advanced westward into Europe, the West correspondingly withdrew, until its core consisted of little more than Germany, France, the Low Countries, northern Italy, and Britain. Before long, some scholars had pushed the boundary still farther, dissociating even Germany from the West. Strikingly, German nationalists took the lead in this movement. Incensed by their humiliation in the First World War at the hands of Britain, France, and the United States, a number of Germans began to declare that Germany was not really a Western nation after all. In 1918 Thomas Mann argued that the war represented a continuation of the ancient struggle between Germany and an overly civilized Western realm that was originally centered on Rome and later identified primarily with France.[47] A few years later the infamous geopolitician Karl Haushofer pronounced that Germany occupied a strategic place in world history as the link between East and West—and that its mission for the future lay in "educating" the Orientals into modernity. About the same time, Asiatic affinities were discovered in German philosophy and spirituality: "Haushofer's disciples attest[ed] that German Romanticism [was] closely akin to the culture of Russia and even . . . to that of China and India."[48]

This message was brought to the American public during the height of the war by the German refugee Hans Weigert, a staunch critic of geopolitics. Weigert inverted the moral signs of Haushofer's vision without altering its substance. As he saw it, "[I]n its depths the German soul remained closely related to the East, especially among the leading groups in which the idea of Prussianism and service to the State were a strong living force. The national socialism that the Asiatics Lenin and Stalin had brought about in Russia—it could be understood by those Germans to whom the idea of Prussian socialism was more than an adroit catchword."[49]

By the mid–twentieth century, historians across Europe were echoing the refrain that Germany was—in its "soul"—a non-Western country. Like many of his contemporaries, Douglas Jerrold (Arnold Toynbee's nemesis) viewed the two world wars, not as contests taking place primarily within a single cultural zone, but as heroic struggles in which the West (Britain, France, and the United States) rose up to defend itself against what was essentially an outside power (Germany).[50] As recently as 1987, Theodore Von Laue similarly contended that Nazism represented an external attack on the West, a region that he generally limited to France and Britain.[51] Hans Kohn, being more attentive to historical niceties, argued that Germany had at one time been "of the West," but that it had become so "alienated" that it was no longer part of that geocultural formation.

Tracing this alienation as far back as the decline of the Swabian Hohen-staufen dynasty in the thirteenth century and the subsequent rise of a "semi-German" Prussia, Kohn saw the nation's non-Westernness mani-fested above all in the "deviat[ion] from the main lines of western devel-opment" that marked German intellectual life during the nineteenth cen-tury.[52] No one would write a history of Western music that excluded German contributions, but many are evidently happy to redraw the boundary when it comes to the history of Western politics, especially be-tween the rise of the Second Reich and the fall of the Third.[53]

After the Second World War, the conventional East-West division across Europe was once again reoriented, this time to correspond to Cold War geography. On the new map, East became synonymous with com-munism, West with capitalist democracy. Such a division proved highly useful for tidying up old ambiguities. All Slavic countries could now be cleanly assigned to the (communist) East (with Yugoslavia being allowed to form a kind of bridge), as could most areas of Orthodox Christian her-itage; so too the "Asiatic" Magyars could, at long last, be unambiguously excluded from the realm of Western civilization.[54] Even the arbitrary di-vision of Germany had its appeal. The Rhinelands could now be safely claimed for the West, while Brandenburg-Prussia — supposed locus of the dark side of a divided German spirit[55] — was consigned to the East. In this view, only upper Saxony and Thuringia presented any real inconvenience in being cast into the Eastern zone.

Such a readjustment of the East-West dividing line served to bring long-standing metageographic structures into agreement with the map of Cold War geopolitics. In the process, however, a false concordance was sug-gested between historical culture areas and contemporary geopolitical pat-terns. Nor has this vision disappeared with time. As recently as 1994, Philip Longworth argued that the Iron Curtain merely formalized a centuries-old division — the same one that once separated "Charlemagne's Europe from the barbarian east."[56] The "civilized Europe" of the West, in a word, could be tendentiously defined as the noncommunist realm (although in practice examinations of it rarely extended much beyond Britain, France, and West Germany).[57] The implication was, of course, that communism was in essence an anti-European philosophy[58] to which the quasi-Orien-tal East had been geographically "predisposed."[59]

In the wake of the 1989 revolutions, this particular yardstick has lost most of its salience, and the postwar division between East and West has become problematic once again.[60] As yet, it is unclear what new concep-tion will displace it. One popular gambit in recent years has been to fudge the issue by reviving the concept of a Central or Middle Europe. Not surprisingly, this concept has proven to be both spatially slippery and ideologically charged. While the geographical core of Central Europe is usually seen as including Germany, Austria, the Czech Republic, Hungary, and Poland, the actual boundaries of the region vary tremendously from one context to the next.[61] Moreover, some writers question whether this old idea retains currency in the postwar era, since it was primarily the presence of German and Jewish populations that gave the region its distinctive character in the first place. When those minorities were all but routed from Poland, Hungary, and Bohemia, many argue, Central Europe (particularly its eastern half) lost its regional identity.[62]

Compounding such problems is the fact that the concept of Central Europe remains offensive to many residents of the alleged region. Some fear that the concept of a *Mitteleuropa* might again become a pretext for German domination of Hungary and the western Slavic lands.[63] Others reject the implication that Central Europeans are somehow less than full participants in Western civilization. "The present government in Prague," Eric Hobsbawm tells us, "does not wish to be called Central-European for fear of being contaminated by contact with the East. It insists that it belongs exclusively to the West."[64] In other places and contexts, however, eastern Europeans find the concept useful. For example, many Bulgarian scholars—whose homeland is seldom classified within Central Europe by outsiders—embrace the concept as a means of reassuming an unambiguously European identity.[65] Writers from the former Eastern bloc who have adopted the Central European label often still insist that Russia does not belong to Europe at all, giving rise to "the bizarre situation . . . of a continent with a west and a centre, but no east."[66]

Perhaps most striking in this brief account of the shifting East-West boundary across Europe is the credulity with which an evanescent political feature, the Iron Curtain, was assumed to reflect long-standing cultural divisions. To view a region like Bohemia or upper Saxony as part of "Eastern Europe" in anything but a narrow political-economic sense was naive enough. Viewing it as part of a grand and mythologized East, a zone of inherent despotism and oppression, was never anything but self-serving. Yet on the very eve of the 1989 revolution, a prominent scholar could still contend that the geopolitically defined region of Eastern Europe—including areas of both Catholic and Orthodox heritage—formed a coherent, historically constituted entity, replete with its own "tradition and value system," and identifiable as the transitional zone between a Western realm of democracy and progress and an Eastern zone of stagnation and despotism.[67] The time has come to abandon such notions for good. Let us hope that James Joll was not entirely mistaken in his 1980 prediction that "if the Iron Curtain were to be torn down, then we would again begin to realize how much eastern Europe, and even Russia itself shared a common European tradition and how it does not really make sense, historically at least, to talk of a Europe which does not include Königsberg and Cracow, Breslau and Budapest, or even for that matter Goethe's Weimar."[68]

Between East and West: The Birth of the Middle East

As the foregoing discussion shows, the concept of the East has taken two simultaneous, yet contradictory, historical trajectories. In the realm of cultural discourse, the East has migrated steadily eastward, from the Levant to India and now to China. When an American speaks today of Eastern cultures, Eastern societies, or Eastern peoples, the referent is usually monsoon Asia, if not (still more narrowly) Asia's far eastern fringe.[69] Meanwhile, however, the East of political economy has migrated westward, to encompass Russia, eastern Europe, and even (in some incarnations) part or all of Germany. As a result, the East-West divide has acquired a peculiar duality. In some contexts, East still means Asia; in other contexts, it is virtually synonymous with Russia. Thus an East-West Trade Center might be concerned either with United States–Japan exchanges or with trade between western and eastern Europe.[70] While the "West" has at least kept a stable core, these are very different Easts indeed.

These opposite conceptual displacements in the end serve complementary functions. On the one hand, the eastward march of the cultural East has allowed maximal difference to be maintained between Europe and the Orient's supposed core, now located in the far east of the Eurasian landmass. This preserves the sense of "Western" European distinctiveness from a safely distant, enticingly exotic Other. At the same time, however, the advance of a geopolitical East into eastern and central Europe has allowed Europeans west of that boundary to disown the uglier episodes in what is in fact a shared political history. Just as Spain in its predemocratic days could be dismissed as lying south of the real Europe (the French often quipping that Africa begins at the Pyrenees),[71] so the totalitarian Soviet regime and its satellites could be conveniently dismissed as lying outside Europe to the east. By drawing a major global boundary between their own culture and that which gave birth to Lenin, Stalin, and even Hitler, and by insisting that this boundary is not merely an artifact of modern political life but a deep historical rift, members of the new Atlantic-centered West have effectively insulated themselves from the doubts that would otherwise arise about the perverse potentials within their own ("Western") civilization.

Yet in solving one problem, the new geographical categories created another: what was to be done with the original Orient, Southwest Asia? As the land of Islam, it had to be kept at a conceptual distance from the West, yet as part of the "Caucasian" world it could not be included in the new racially defined Orient. So how exactly were the societies of the eastern and southern Mediterranean to be accommodated in the new global map? The most popular solution has been to designate a new entity, the Middle East, and to give it quasi-continental status as an interstitial area linking Europe, Asia, and Africa.[72]

A CONTRADICTORY LOCATION

The notion that the area bracketed by Persia and Egypt constitutes a transitional zone between East and West goes back many years, to the beginning of European explorations in the Indian Ocean. By the early nineteenth century, as reflected in the geographical writings of Georg Wilhelm Friedrich Hegel, it was well established. Hegel divided the Oriental realm into four distinct worlds: China and India, which together formed the "farther" East, and Persia and Egypt, together comprising the "hither" East. While he regarded all four civilizations as Eastern in their lack of freedom and fundamental stagnation, he nonetheless considered societies of the hither East to approach European conditions far more closely than did their counterparts in the farther East. In Hegel's reification of direction, the farther east one traveled across Eurasia, the more fully one would encounter an Oriental essence. "The European who goes from Persia to India," he wrote, will observe "a prodigious contrast. Whereas in the former country he finds himself still somewhat at home, and meets with European dispositions, human virtues and human passions—as soon as he crosses the Indus . . . he encounters the most repellent characteristics, pervading every single feature of society."[73] This contrast was reflected in Hegel's conception of the history of the two realms as well: "With the Persian Empire we first enter on continuous History. . . . While China and India remained stationary, and perpetuate a natural vegetative existence even to the present time, this land has been subject to those developments and revolutions, which alone manifest a historical condition."[74]

Already in Hegel we thus find an important inversion of traditional European conceptions of Southwest Asia. Whereas this area had once been Europe's primary oppositional cultural sphere, it was now recast as an interstitial zone, mediating the differences between Europe and the still more exotic reaches of Asia (India and the "quite peculiarly oriental" China).[75]

The Middle East of our contemporary geographical imagination retains this liminal position, along with a certain ambivalence about the proper taxonomic level to which this region belongs. On the one hand, as suggested by the label *Southwest Asia*, it is sometimes seen as one among several Asian macroregions, on a par with East, South, or Southeast Asia (see chapter 6). On the other hand, it is sometimes seen as a "culture continent," more distinct and in some ways more important than these other Asian regions. In other words, there is a tendency to put the Middle East *as a cultural zone* on a par with Europe and Asia. As conceptual space, then, the Middle East occupies two different positions: at once a subset of Asia (with a North African annex) and a semiautonomous transitional zone on the Europe-Asia frontier. While this ambiguous position allows the region to be either included in Asia or excluded from it at will,[76] it is almost always seen as more Asian than European.

Indeed, by most textbook definitions, the Middle East lies essentially within Asia, with the important exception that it also includes a portion of northeastern Africa (Egypt and Libya). (Indeed, Egypt for some writers is virtually synonymous with the Middle East.)[77] Lands to the west of Libya in northwest Africa (historically and indigenously designated the *Mahgreb*), by contrast, are only informally linked to the region. According to Islamicists Bernard Lewis and P. M. Holt, "as used at the present time, the term Middle East includes Turkey, Persia, and perhaps Afghanistan, the Fertile Crescent, Arabia and north-east Africa, with a rather vague extension southwards and westwards from Egypt into Arabic-speaking Africa."[78]

The exclusion of the Mahgreb from the Middle East is logical in strict locational terms, since the phrase suggests that the region ought to lie somewhere east of Western Europe. Since northwest Africa lies *south* of Europe but not east (as noted, all of Morocco is west of London), it hardly fits this spatial conception. For just this reason, prominent British and American geographers of the 1940s inveighed against the careless and "uninformed" application of the term *Middle East* to denote lands west of Egypt.[79] Yet as we have seen, geographical labels have a tendency to slide about the map, whether following sociocultural groupings, political boundaries, or simple prejudice. In practice, the designation *Middle East* is not always confined to the territories identified by Lewis and Holt; it is also used in a broader sense to denote a significantly larger cultural region. Just as the Orient at one time encompassed Morocco because of that region's historical and cultural ties to the Levant, so the Middle East of popular usage has been readily extended westward as far as the Atlantic.

MIDDLE EAST AND NEAR EAST

In contrast to both of its modern connotations, however, this metageographical concept formerly had a different meaning altogether. As originally conceived, the Middle East was not a cultural or historical region but an arena of strategic operations. The term itself was coined in 1902 by the military theorist Alfred Thayer Mahan to refer to the area surrounding the Persian Gulf.[80] To the geopolitical strategist, the problems of this zone were distinct from those of the Near East (the Mediterranean borderlands), as well as those of the Far East (an area stretching in the imagination of the time from India to Japan). Other writers soon began to use the term in similar contexts, although the area they had in mind did not remain constant.[81] Valentine Chirol, for example, conceived of the Middle East in terms of India's vulnerable borderlands, focusing his attention on Persia but including Nepal and Tibet as well.[82] Indeed, growing anxiety over their South Asian empire prompted some British strategists by 1914 to shift the label *Middle East* to India and its adjacent territories.[83] While this usage never gained acceptance, even today a few writers conceptualize the Middle East primarily in terms of Britain's earlier defense of its South Asian empire.[84]

World War II brought a major reconfiguration of the conceptual Middle East, in which once again military considerations were paramount. A key change had occurred in 1932, when the Royal Air Force merged two formerly separate commands, the Middle Eastern (headquartered in Iraq) and the Near Eastern (in Egypt), referring to both henceforth as Middle Eastern. By 1939 the British army followed suit.[85] The newly enlarged Middle East to which this gave rise included sizable areas of tropical Africa, both in southern Sudan and in Somalia.[86] By the end of the war, popular imagination in both Britain and the United States had seized on this new use of the term and now regarded the Middle East as a zone extending from Libya to Afghanistan. After the war, this same area was recast as a cultural and historical region, although this again required some mutations in its borders. Somalia, for example, was now usually deleted (to be appended instead to sub-Saharan Africa), while Morocco, Algeria, and Tunisia (regarded in 1947 by the geographer W. B. Fisher as a "purely African unit")[87] were often grafted on.[88]

This reformulated Middle East was partly a replacement for another recent coinage, "Near East." The notion of a distinct Near East had emerged in diplomatic circles only in the late nineteenth century.[89] By the early twentieth century, however, the term had gained wide currency, usually designating the lands of the Ottoman Empire in both southwestern Asia and southeastern Europe.[90] One plausible reason for the popularity of this category was that it conceptually Easternized Greece and the Balkans. Western chauvinists had long held that both had been corrupted centuries ago by Oriental influences (a trend some traced as far back as Alexander's days), and by placing these lands within the same region as Syria and Egypt, they could effectively excise them from Europe altogether.[91]

With the demise of the Ottoman empire, however, this broad usage of the term *Near East* began to disappear from the public vocabulary.[92] Certain geographers were happy to see it go, taking up the new category of Middle East in its stead. W. B. Fisher, for example, objected to the idea of a Near East both because it lacked environmental unity and because "too much confusion now attaches to the term."[93] As it happened, Fisher was not able to demonstrate any greater environmental unity in the Middle East—a region whose delineation likewise created considerable confusion. But while a few scholars attempted to revive the old distinction between the Near and Middle East in the 1950s, most acquiesced in the new formulation, either in deference to military usage and popular opinion or perhaps because they simply dismissed the issue of nomenclature as trivial.[94]

The term *Near East* nonetheless persists in three related areas of scholarly endeavor: philology, ancient history, and antiquarianism. The farther back in time the period under consideration (and the more recondite the study), the more likely the geographical frame will be labeled the Near rather than the Middle East. A recent Yale University Press catalog, for example, contains one section on "The Middle East" and another called "Ancient Near East"; the area in question is essentially the same, only the temporal boundaries have changed.[95] Similarly, several Ivy League universities retain departments of Near Eastern Studies, which focus on the same territory now commonly called the Middle East but concern themselves primarily with ancient texts and archaeological excavations. It might be asked why the term *Near East* continues to be employed in this way. The answer, presumably, is that in the ancient world, the West was synonymous with Greece—prior, that is, to its corrupting "Orientalization" during the Hellenistic period, when the torch of Western civilization supposedly passed to Rome. From the standpoint of Athens (if not that of London), Asia Minor and the Levant can be nothing but the "near" East.

Even from a contemporary perspective, Near East would seem a more accurate spatial designation for Southwest Asia than Middle East. If France occupies the center of the West and coastal China the heart of the East, then the Fertile Crescent is surely in the near portion of the Oriental zone to a European (and of course the European vantage point is the only one from which official definitions of Eastern lands have ever been made). By this sort of objective locational logic, the Middle East would presumably encompass such areas as India, Tibet, the Tarim Basin, the Tien Shan Mountains, and the Yenisey River Valley. In practice, however, the term *Middle East* has only rarely been extended east of Afghanistan.[96]

IMPLICATIONS OF AN
INTERMEDIATE EAST-WEST ZONE

If there is one constant principle in our narrative so far, it is that viewing the world from the perspective of Europe's Atlantic seaboard results in severe spatial distortions. In the myopic worldview we have inherited, the farther east one looks, the less significant geographical divisions become, and the more readily disparate areas may be conveniently lumped together. The continuing use of the terms *Near, Middle,* and *Far East* only highlights the intellectual inertia in our metageographies. All three are well ensconced in the American imagination, and few scholars or lay people have given them much thought. Like continents themselves, the division of the Afro-Eurasian landmass into relative degrees of Easternness has generally been taken for granted, viewed as merely a matter of convenient designation rather than an issue of any intellectual import.

We would argue, by contrast, that these habits of global regionalization subconsciously guide our geographical imagination along some rather twisted pathways. Consider carefully the exact geographical denotations of the terms used. The term *Middle East* may well be appealing because it deemphasizes proximity to the West in favor of distance, suggesting that the region lies at the middle, rather than the edge, of the non-European world. If, from the standpoint of London, Turkey[97] lies in the middle portion of the East, then where might one find its "near" extent? The only possible answer is southeastern Europe. Although this region is seldom called the Near East in popular American publications today, its designation under that term was formalized in the 1920s by no less an organization than Britain's Royal Geographical Society, which resolved that "henceforth, the 'Near East' should denote only the Balkans."[98] Such a move proved convenient for a Far West anxious to dissociate itself from any parts of Europe not as prosperous as itself. While few openly agreed with Metternich that Asia begins at the gates of Vienna, many were only too happy to declare that all regions ever ruled by the Turks had thereby become more Eastern than Western.

This "de-Europification" of southeastern Europe is not merely an old gambit of Western jingoists, but one that continues to be deployed with vigor. As in the case of Russia, the Balkans' Eastern heritage can be blamed for any problems in the region. A classic recent statement of this view can be found in George Kennan's 1993 article on the "Balkan crisis," published in the *New York Review of Books*: "What we are up against is the sad fact that developments of those earlier ages, not only those of the Turkish domination but of earlier ones as well, had the effect of *thrusting into the southeastern reaches of the European continent a salient of non-European civilization* which has continued to the present day to preserve many of its non-European characteristics, including some that fit even less with the world of today than they did with the world of eighty years ago."[99]

Although Turkish and Islamic influences are certainly powerful in many Balkan areas, it is a serious geographical blunder to imply that a country like Serbia has more in common, culturally and historically, with Eastern regions, be they near, middle, or far, than it does with the rest of Europe. For Croatia, a country of Latin Christendom that was never fully in Ottoman hands, the notion is more absurd yet. More disturbing than its sheer inaccuracy, however, is the way this geographical error rewrites Europe's history. To view such modern horrors as "ethnic cleansing" as a product of alien Eastern influences violating an otherwise virtuous Europe requires a tremendously selective memory of the recent European past. The game Kennan is playing is the same as that involved in turning Nazism and Stalinism into Oriental creeds, and it leads to a comparable denial of historical responsibility. Our flawed metageography has become a vehicle for displacing the sins of Western civilization onto an intrusive non-European Other in our midst.

Appropriations of East-West Rhetoric Outside Europe

As the foregoing discussion shows, the binary division of the world into an immense Eastern and an exiguous Western section is an anachronistic product of European cultural egotism. To be sure, a glance at the metageographies of other peoples soon suggests that self-glorifying bifurcations of the globe have hardly been unique to Europe. Premodern Muslim geographers, for instance, divided the globe into the "land of submission [to God]" (*Dar al-Islam*) and the "land of war" (*dar al-harb*). According to their most common metageographical framework, the realm of Islam occupied the central portion of the world, with the lands of the various nonbelievers located around the fringe.[100] In South Asia, Brahmans similarly saw their own homeland as both central and uniquely holy (all exterior regions being places of pollution for members of the high castes). As late as the early modern period, Indian world maps typically showed South Asia as forming the great bulk of the planet's land-mass, with one Indian cartographer depicting Europe in a few marginal circles labeled "England, France, [and] other hat-wearing islands."[101] The Chinese too usually conceived of their civilization as the center of the world, as reflected in the term *Middle Kingdom*.[102] Indeed, only the Japanese in all Eurasia historically thought of themselves as Eastern—a notion rooted not in Japan's relationship to Europe, of course, but in its intellectual subservience to China.[103]

Yet whatever their indigenous notions of world order,[104] all Asian intellectuals in modern times have had to wrestle with European geographical conceptions. In some cases, the result has been a profound ambivalence about membership in the supposed Asian fraternity. This ambivalence has been especially marked in Turkey and Japan, the two countries that occupy the physical as well as conceptual borderlands of Europe's cultural East. Many Turkish intellectuals, for their part, have asserted that their civilization's real roots lie in the West. The Ottoman sultans long considered themselves to be the true heirs of the Roman Empire as well as of Greek civilization, claiming Alexander the Great as one of their own.[105] Since the late nineteenth century and especially after 1922, a continuing identification with the West has been manifested in such developments as the formation of a resolutely secular nation-state, membership in the North Atlantic Treaty Organization, attempts to secure membership in the European Union, and continued insistence by some Turkish intellectuals that their country fully shares a Western heritage. Radical Islamicists active in Turkey today, however, fervently oppose any further linkages with Europe, justifying their attempt to refashion Turkey into an Islamic state in part by recourse to a different rhetoric that positions Turkey within the East, or at least within *Dar al-Islam*. Other Turkish conservatives prefer to link their country with the rest of the Turkish-

speaking world, a vast zone that stretches to western China, and potentially to central and northern Siberia as well.[106]

Like their Turkish counterparts, Japanese thinkers have debated their
country's proper position in the framework of East and West. Both in
the early decades of the twentieth century and in the immediate postwar
period, many hoped to join Japan firmly to the West, seeing there the only
avenue to power or even independence. *Datsu-a nyu-o* — "leave Asia and
join Europe" — was at both times a powerful slogan in the discourse of
modernization.[107] Popularized by Fukuzawa Yukichi, the foremost public intellectual of the Meiji period, this phrase has sometimes been interpreted as embodying a hope to turn Japan into something of a European
exclave, but Fukuzawa's position was more subtle than that. Loosely following Herbert Spencer, Fukuzawa saw the modern civilization that Japan
must adopt as Western, but only "fortuitously and temporarily so";[108] the
future, he thought, would bring about a global cultural convergence that
would render such categories as East and West meaningless.

But identifying with the West has been the exception rather than the
rule in recent Asian history. Typically, the practice of bifurcating the globe
along an East-West axis has been embraced by non-European elites for
their own purposes. Understandably, leaders in those parts of the world
that have shared a common subjugation or vulnerability to Western power
have found it useful to seek deeper connections among themselves. Having been told by Europeans that they all belonged to an essentialized Orient, the various peoples of Asia have often seen some merit in emphasizing their commonalities. Not that this has precluded a considerable
variation of emphasis on what Asianness meant. "Each Asian Orien-
tophile," as Stephen Hay explains, has "entertained a somewhat different
notion of the essential features of this civilization, his image of the East
consisting usually of an expanded version of those particular traditions
he most wished to revitalize."[109] Nor has the adoption of European metageographical categories precluded energetic jostling for position within
Asia. In India, China, and Japan as well, nationalist intellectuals have
found ways to privilege their own country while articulating a basis for
pan-Asian unity.[110]

Indian thinkers who embraced the notion of a fundamental East-West
divide insisted loudly that India was the focal point of the East (although
they were equally anxious to establish connections with China, Japan, and
other "Oriental" countries).[111] Such renowned Indian nationalists as Ra-
bindranath Tagore framed the longitudinal global divide around the distinction between a spiritualistic East and a materialistic West, arguing that

the reconnection of these two halves was necessary to make a global whole—and that India was uniquely positioned to facilitate such a fusion. China's first modern geographer, Hsu Chi-yu (Xu Jiyu), likewise argued for Chinese superiority within a European-inspired metageographical framework. Hsu expressly endorsed the then-standard fourfold continental system, dividing the world into Asia, Europe, Africa, and America; he even accepted the Ural Mountains as the boundary between Europe and Asia.[112] But in emphasizing China's role as "lord" of the most extensive continent, he was able to elevate the position of his own country above that assigned it by Western scholars. China's importance was further strengthened in his scheme by an insistence that Europe had originally been civilized through a process of Easternization.[113]

Hsu's geohistorical vision was echoed at the turn of the twentieth century by Liang Qichao, the leading Chinese theorist of modernity. Liang was instrumental in dismantling the Confucian Sinocentric model of world geography, arguing that the position of China must be understood primarily within the context of Western imperialism.[114] At the same time, however, he contended that all of the world's early civilizations (including the Assyrian and Jewish) had originated in Asia, the source from which civilizing influences had radiated westward to a benighted Europe. Moreover, in defiance of Hegel, Liang located Asia's (and ultimately China's) superiority not only in the past but also in the future, holding that the center of progress and innovation, far from making a steady westward march, was subject to constant shifts in its geographical position.[115]

The place where national chauvinism has been most forcefully articulated within a pan-Asian framework, however, is Japan. Japanese scholars of the early twentieth century generally accepted the notion of an Orient stretching from the Levant to their own homeland. The artist and ideologue Okakura Kakuzo in particular argued with eloquence that "Asia is one."[116] One major response was to develop a distinctly Japanese school of "Oriental studies," modeled in part on Western Orientalism but transposing the spirit of progress from Europe to Japan.[117] After Japan's victory over the Russians in 1905, Japanese pan-Asianism took on a politically charged character, with the Japanese touting themselves as a model for successful resistance to European rule. Residents of other "Asian" nations, anxious for liberation from European colonialism, originally supported this idea with enthusiasm. But by the 1930s, it was clear that Asian unity on Japanese terms was merely another form of colonial oppression, and Japanese arguments over the spatial dimensions of Asian culture tended to recede before political expedience. Ultranationalist ideologues increasingly framed their country's imperium as the only Asiatic entity that could create an appropriate political space for Asian peoples. It is worth noting that their Asia was the maximal Asia of Western geography. Only thus could Japanese intervention in India be justified—an important precondition for the Japanese government's arming of the Indian National Army to oppose the British Raj.[118]

Including the entire purported continent within their purview, the would-be architects of a Japanese-led Asian resurgence necessarily had to contend with Islam. This was not entirely an academic issue at the time, since Japanese military authorities suddenly found tens of millions of Muslims under their rule in Indonesia, Malaysia, and elsewhere. As a result of both their grandiose designs and their immediate exigencies, the Japanese scholarly community began to develop an intense interest in Islam. Defining it as an Asian religion, nationalistic scholars saw in Islam a potential ally in the struggle against the West, and they quickly spread this message to military leaders. How well Japanese officials understood Islam is another question. Allegedly, several high-placed military officers planned at one time to "relocate" Mecca to Singapore, in order to make this Japanese-occupied city the center of the Islamic world.[119]

As these examples illustrate, yesteryear's global geographies have impressive staying power. Their moral implications may shift over time, but metageographical conceptions rooted in the European intellectual tradition are still very much with us, both at home and in the rest of the world. Nonetheless, as this chapter has attempted to demonstrate, the boundaries of both West and East have been, and still are, remarkably flexible. Their geographical referents have proven highly variable, changing drastically from one context to another. What remains to be shown is how the "contents" of these great geographical divisions—the features of society and culture that supposedly distinguish East from West—are no less problematic. It is to this thorny set of issues that the next chapter turns.

CHAPTER 3

The Cultural Constructs of Orient and Occident, East and West

Europeans have long been wont to define their own psychosociological qualities in contrast to those of their Eastern neighbors, a habit of thought that shows no sign of decline. The continuing salience of East and West as foundational concepts in our understanding of the world is manifest daily in the popular press, where a cluster of related mental attributes is consistently identified with the European approach to life and contrasted with an opposite set of traits that is said to characterize the East.

The key components of the Western cluster comprise a familiar list. European civilization is said to be characterized by a compulsion to control and manipulate nature; a tendency to regard the self as an autonomous agent in competition with others; a restless desire for growth and development; a keen appreciation of personal freedom; a hunger for material wealth; a practical, this-worldly orientation that seeks social betterment through technological means; and perhaps above all, a commitment to rational inquiry. The Eastern mind has been defined in opposite terms. Put simply, the essence of the East is seen as manifest in communitarian, aesthetic, and other-worldly values, extolling the submission of the individual to a timeless, mystical whole.[1] Variations on this theme have been elaborated at great length by generations of Western theorists, beginning with Herodotus and Hippocrates.

To maximize the rhetorical potency of this East-West distinction, it has been convenient to define the essence of each in abstract, spiritual terms rather than identifying them with specific social practices or institutions.

Consider, for instance, the handling of the Christian legacy. While many writers have been inclined to identify the West with the realm of historical Christendom, most have been uncomfortable embracing "quasi-Oriental" societies of Orthodox creed. More fundamentally, such a definition of Westernness might have difficulty acknowledging Christianity's own roots in the East. It is possible, however, historically speaking, to isolate a civilization coincident with the realm of *Western* Christianity (Catholic and later Protestant as well): a civilization that gave rise to a host of social institutions and practices that guided its subsequent development and that distinguished it from other portions of the world.[2] Yet as we saw in chapter 2, the West has never consistently been defined in such a manner. Rather than highlighting anything as specific as the heritage of (Western) Christian beliefs or the institutions of the church, those who have created the myth of special Western distinction have been most comfortable focusing on ineffable traits, such as a supposed passion for freedom and rationality.

Efforts to articulate Eastern and Western essences seem if anything to have intensified through the middle decades of the twentieth century, as European and American scholars attempted to comprehend the dismantling of European empires and the rise of Japan. By now that fad has largely passed in academia, and few contemporary scholars are so bold as to pontificate on the essential differences of the two putative regions. Yet underlying notions of Eastern and Western essences persist. In the hands of tolerant, liberal writers, such notions are revealed in casual statements that associate Western culture primarily with the skeptical tradition of the Enlightenment—rather than, say, with the fanaticism that has always infected a sizable proportion of the Christian community.[3] In the hands of others, the binarism is cruder: the West is simply identified with civility, the non-West with barbarism. Nor have such moralizing terms been rendered extinct by the recent attacks on Eurocentrism. Consider, for example, the words chosen by the syndicated columnist Charles Krauthammer to analyze the recent debacle in Somalia: "As in Vietnam, we are dealing with people who do not play by Western rules, yet we are shocked to discover that they engage in barbarism."[4] Born in a mood of Western chauvinism and embraced in earlier years with little critical reflection, such ideas continue to be unreflectingly accepted and promulgated—even by many who now champion the non-Western world and disparage Euro-American hegemony.

It is the aim of this chapter to analyze the central features commonly attributed to Eastern and Western civilization. Whereas the previous chapter focused on the cartographic slipperiness of these purported regions, attempting to follow the convoluted migrations of the geographical boundary between them, the discussion here turns to the more abstract question of how each has been culturally defined. It should be noted at the outset that, in contrast to the relatively understudied mapping of East-West divisions that has been our subject heretofore, the interrogation of these loaded terms as cultural concepts is by now a well-advanced intellectual project. A seminal early contribution was Edward Said's *Orientalism,* an indispensable touchstone for subsequent work in this area. The polemic by geographer J. M. Blaut, *The Colonizer's Model of the World,* represents a more recent meditation on related issues.[5] Many of Said's and Blaut's arguments anticipate our own, and their book-length works on this topic develop a number of themes in far greater detail than is possible here. Our primary purpose is a relatively narrow one: to review the historical-geographic evidence to see what merit there is in a handful of the dominant East-West characterizations sketched above.

We preface our empirical investigation by tracing the twin inflections of East-West polarity: the dominant version, which takes a positive view of the West, and the minority stance, which applauds the civilization of the East. Juxtaposing passages that celebrate Western civilization's achievements with those that denigrate them reveals an important symmetry, inasmuch as both parties to this debate often rely on essentializing rhetoric to advance their arguments. But the similarity goes farther. Not only do the West's defenders and its attackers share a dependence on East-West comparisons, but more particularly there is a remarkable congruence in the contours of their respective cultural stereotypes. Their values may be strongly opposed, but partisans and critics of the West alike tend to portray East and West in essentially the same set of terms.

Having established the consistency and durability of the basic polarity, we proceed to a selective analysis of its contents. We begin by considering the dominant idealist or intellectual element in the received definition of the West: a purported penchant for rationality. Western rationalism has been said to manifest itself both in the abstract realm of thought and in the practical realm of social institutions. Both claims are explored at some length and weighed against empirical evidence of rationality's historical career on both sides of the East-West divide. The second section of the chapter takes on the most important political-economic characterizations of the divide: namely, the notion that the West is the land of democracy and economic dynamism, while the East is historically the land of despots and of a uniquely stultifying "Asiatic Mode of Production." Once again, our approach is to survey the empirical evidence, asking whether either concept offers a valid model of political-economic development in historical Eurasia. In all cases, our reading of the evidence challenges the notion that East and West are meaningful geohistorical categories.

For and Against the West: Two Versions of a Persistent Polarity

The positive vision of what it means to be Western took shape during the early modern era, a by-product of the explosion of global knowledge during the age of exploration and conquest. During the Middle Ages, most Europeans viewed the East—and especially the Far East of India—as a quasi-mythological land of riches, marvels, and Pagan idolaters (as well as sagacious if misguided Brahmans).[6] For a brief period during the eighteenth-century Enlightenment, moreover, attitudes toward the East were sometimes remarkably positive (as discussed below). But by the nineteenth century, all literate Europeans would have been familiar with a bipolar representation of the globe that contrasted a progressive and enlightened West to a stagnant and obscurantist East. As Ronald Inden notes, the former image was contingent on the latter: "Without the dark rock of Indian tradition under its feet, European rationality would not have seemed so bright and light."[7]

Assessments of the East grew steadily more negative as European powers advanced into the region and began to subjugate its peoples. Non-Western cultures came to be dismissed as entirely stagnant, if not barbaric, while racism came to be cloaked with a new intellectual respectability. To be sure, as Raymond Schwab demonstrates, one strain of Romanticism sought wisdom in the East (particularly India).[8] But the dominant trend in early-nineteenth-century Europe was toward increasing ethnocentrism. Johann Gottfried von Herder and especially Georg Wilhelm Friedrich Hegel were both contemptuous of Asiatic civilization; one recent author has not unfairly described Hegel's detailed description of India as "the most withering hymn of hatred and contempt sung about the Orient in occidental social philosophy."[9] And nineteenth-century British utilitarian writings on the Orient, especially those on India, show an unbridled arrogance that had been largely lacking during the Enlightenment. James Mill wrote that in Asia "darkness had always prevailed";[10] his son John Stuart Mill informed his approving readers that "the greater part of the world has, properly speaking, no history, because the despotism of Custom is complete. This is the case over the whole East."[11]

These ideas had a profound effect on early-twentieth-century American scholarship. A psychological variation of the "East versus West" dynamic was offered in the writings of Meredith Townsend, a scholar typical of those who published at the turn of the century. Writing in 1901, Townsend opined that "the European is essentially secular, that is, intent on securing the objects he can see; and the Asiatic essentially religious, that is, intent on obedience to powers which he cannot see but can imagine."[12] Townsend further informed his readers that Asiatic peoples are habitually willing to submit and in fact desire to be governed by an "absolute will."[13] Midcentury writers were often more sympathetic toward Eastern thought, hoping to foster mutual understanding between East and West, yet most were informed by similar views. F. S. C. Northrop, Sterling professor of philosophy and law at Yale University, concluded his massively involuted inquiry into global differences with the assertion that "the meaning of Eastern civilization in its relation to the meaning of Western civilization is as follows: *The Orient, for the most part, has investigated things in their aesthetic component; the Occident has investigated these things in their theoretic component.*"[14] Northrop also emphasized the "passive, receptive attitude" of the East,[15] and he repeatedly located the spirit of the West in its scientific achievements. In a related vein, the German Japanologist Kurt Singer, writing in the 1940s, confidently contended that the Japanese were by nature irrational. Singer argued that while "the Western mind is eminently architectonic," the "Japanese mind is . . . un-technical. It does not believe in the efficiency of rational patterns; and what it achieves is more often than not incapable of generalization and defies rational analysis."[16]

Similar themes have remained pervasive in the postwar era, if often in muted form. Indicative of the new generation, with its greater avowed respect for Eastern peoples, is C. Northcote Parkinson. Writing in the early 1960s, Parkinson argued against a program of Westernizing Asian peoples by contending that "the world's most brilliant ideas have been the result of friction between East and West."[17] Yet the same author absolutized the differences between these two "halves" of the Eurasian landmass as much as anyone, holding that world history is only explicable if viewed as powered by a sort of alternating current between the Orient and the Occident. As it turns out, the polarity that supposedly drove this engine is the same one we have already encountered: in the West, individuality, love of freedom, and technological progress; in the East, the inverse of these qualities. Parkinson also had moments of remarkably transparent

Western chauvinism, as when he averred that humor is a characteristically Western trait, having been for all intents and purposes invented by the Greeks. Among the Asiatics, in his reading, life was far too dreary for mirth to thrive:

> For the vast majority of Asian peasants, in lands where peasantry comprised over 90% of the population, life was appallingly dull. It was one thing to cultivate vines or olives on a hillside in Delos with a distant glimpse of the wine-dark sea and an awareness of market day tomorrow. It is quite another to grow rice around Patna or Hanoi.[18]

Such baldly dismissive sentiments have largely disappeared from academic writings over the past few decades. Yet the old stereotype occasionally resurfaces in surprisingly stark forms, as in this passage from a book published by a conservative think tank in Los Angeles in 1971:

> The Great Asiatic civilizations developed in a pre-logical era; the mind groped for truth through intuition, symbol, magic, and mysticism. It was irrational. It refused to see the external world as an autonomous reality capable of being shaped and adapted through an understanding of its laws.
> The West, thanks to Greek genius, succeeded in rising to the level of rational thought, founded on respect for a principle of no concern to the Oriental mind. . . . [19]

An even more recent example of such thinking can be found in the widely proclaimed book *In Defense of Elitism,* written by the Pulitzer Prize winner William A. Henry, III. In this 1994 work, Henry links European civilization not only to the concept of individual rights but also to the possibility of economic opportunity and even to the notion of charity.[20] His basic thesis seems to be that "might makes right," and he commends all instances of European imperialism for "dispersing administratively and technologically superior cultures and compelling inferior ones to adapt."[21]

Similar prejudices persist in subtler forms in the academy as well, deforming both pedagogical practice and research agendas. As Gerald James Larson writes in regard to his own discipline, "Philosophy as a field still harbors many, possibly a majority, who think that serious philosophizing has occurred only in the West, and who dismiss non-European modes of thinking . . . as abstract nonsense."[22]

INVERTING THE PARADIGM

Alongside the dominant rhetoric of Western self-congratulation runs a strong countercurrent of self-criticism, denouncing Western ethnocentrism and calling all of Europe's modern achievements into question. Critics of European civilization in the twentieth century have been able to draw on long-standing tropes that associate darkness with the West. In the ancient Mediterranean world, as in traditional China, the West was often linked to death through the metaphor of the setting sun.[23] Similarly, in the earliest Christian tradition, the East—the locus of Eden—was usually seen as "symbolizing all the light-giving spiritual qualities," while the West was slighted as the land of "impenetrable darkness and materiality."[24] To be sure, such views were submerged during the Middle Ages, displaced in the main by a Greek-derived denigration of Eastern mysticism and slavish disposition. But a minority tradition in European intellectual life has continued to uphold the East as a land of superior wisdom.[25]

Perhaps surprisingly, one of the main repositories of such thought can be found in the writings of Enlightenment thinkers. The philosophes of eighteenth-century France showed an unprecedented enthusiasm for the civilizations of Persia, the Arabic-speaking lands, and China.[26] To appreciate the magnitude of this intellectual revolution, it is worth recalling the extent to which earlier Western works were marred by "the humanists' long prohibition against looking beyond Greece for fear of running into barbarism, and the clerics' against looking beyond Judea for fear of running into idolatry."[27] All of this was to change in the eighteenth century. Voltaire scoffed at earlier "universal histories" for their ethnocentric biases, which he largely corrected in his own history of the world; according to Peter Gay, "defective as his chapters on the Orient may be, Voltaire wrenched the center of history away from the Christian or the European world."[28] Even Islam became the target of appreciative revisionism during these years. "The image of Mohammed as a wise, tolerant, unmystical and undogmatic ruler," writes Bernard Lewis, "became widespread in the period of the Enlightenment."[29] And the Chinese sage Confucius was so widely revered as to have been called "the patron saint of the eighteenth-century Enlightenment."[30] One might also note the respect granted to Indic culture by such Enlightenment savants as Henry Colebrooke, Charles Wilkins, and William Jones.[31]

In fact, one could almost say that the high Enlightenment in France was characterized by contempt for the historical institutions and values of Western civilization. A passage usually ascribed to Voltaire tells us: "As soon as India became known to the barbarians of the West . . . she became the object of their cupidity. This was all the more so when these barbarians, becoming civilized and industrious, created new needs for themselves. . . . The Portuguese and their successors were only able to provide pepper and cloth to Europe by means of slaughter."[32] In a similar vein, Denis Diderot—first among the philosophes—denigrated Christianity as the most "absurd," "atrocious," "puerile," and "intolerant" of all religions.[33] The same thinker showed a spirit of profound admiration when writing about the religious and philosophical systems of other parts of Eurasia,[34] and he rejected the mapping of moral values onto the global grid of continental distinctions.[35]

Not all Enlightenment thinkers, of course, were so enlightened. Some (notably Montesquieu[36] and Condorcet[37]) were virulent European chauvinists; many were anti-African racists; almost all disdained the Turks.[38] Moreover, Enlightenment thinkers saw what they wanted to see, and what their cultural conditioning allowed them to discover, in such heroic figures as Mohammed and Confucius.[39] But as self-interested and limited as it may have been, their celebration of other cultures' religious and intellectual traditions had less in common with the dominant Eurocentrism that both preceded and followed them than with the attitude of today's cultural Left.

There was, however, one important difference. As the preceding passages suggest, what Voltaire and his cohorts found in East Asia were precisely the values that we now habitually associate with the West: rationality, moderation, and "unmystical, undogmatic rule." In our own day, by contrast, those who find Western civilization wanting have usually combined the philosophes' contempt for Western violence and greed with a *denigration* of precisely these values, so that rationality becomes not an Eastern virtue but a Western vice. In effect, modern celebrations of the East borrow their repertoire of East-West clichés not from the philosophes but from the older tradition of Western chauvinism.

Several recent examples of this tendency can be seen in the polemical writings of North America's radical environmentalist movement. "Deep ecologists" typically lay the blame for the earth's impending ecological catastrophe squarely at the feet of Western civilization. As Jim Mason puts it, in an analysis typical of the genre: "[I]t is the Western tradition that is the primary culprit in the destruction of the natural world."[40] Some eco-radicals locate the roots of Western rapine in the Judeo-Christian heritage; others attribute more importance to the rise of science and the rationalistic spirit in the early modern period. But most seem to concur that the ecological problems of the contemporary world are entirely Western in origin and have become globalized only through a process of rampant Westernization.[41]

In explaining why this is so, deep-ecological rhetoric repeats most of the conventional stereotypes of East and West. Traditional Eastern cultures are pictured as existing within a changeless, harmonious equilibrium with the natural world: a harmony fostered by the religious traditions of Buddhism and Taoism, which are celebrated for extolling a receptive approach to life, discouraging individuality, and stressing social harmony.[42] Western culture, conversely, is indicted for encouraging precisely the same traits that its defenders claim for it: competition, individualism, and rationalistic inquiry. Oriented toward money and technological prowess, Westerners are depicted as pillagers of the planet, obsessed with satisfying their own greed and with fulfilling their religion's malignant mandate to control the earth.

Several other branches of contemporary social thought concerned with challenging Eurocentrism likewise retain the premise that reason and progress are defining characteristics of the West. The Afrocentrist Clinton Jean, for instance, pinpoints rationality as the West's key failing and contends in a familiar key that Asiatic societies have "enjoyed a *stability* that Western history has not been able to achieve," making possible a kind of "spiritual security" that has been noticeably absent in the West.[43] To Jean and like-minded critics, socioeconomic development is the most notorious by-product of Western rationality. It is also the main tool of Western objectification and social destruction, "articulat[ing] the state with profits, patriarchy and objectifying science and technology."[44]

Similar sentiments animate much deconstructive work in the humanities, despite the ritual denunciation of binary thinking in critical theory. Consider, for example, a recent statement from a theorist of Japanese and Western cinema: "There is no need to remind ourselves that the West and the non-West do not voluntarily engage in cross-cultural exchange. The relationship between the two has always taken the form of political, economic, and cultural domination of the non-West by the West."[45] In this formulation, European global hegemony—a historical product of the past three centuries—is totalized to cover the entire span of human history. At the same time, the "West" and the "non-West" are treated as actors possessing their own volition.

Such a tendency to reify vast portions of the world may be extreme, but it is not unusual. As Aijaz Ahmad cogently argues, Edward Said himself constructed a reified and timeless West, a region that is essentially defined by its "unitary will to inferiorize and vanquish non-Europe."[46] Meera Nanda shows how postcolonial theorists, along with the leading practitioners of the sociology of science, are dedicated to the proposition that scientific rationality is uniquely Western and antithetical to the indigenous systems of knowledge characteristic of "Third World" societies. (She also demonstrates how this plays into the hands of right-wing populists: "The sad irony is that we now have the most radical, cutting-edge thinkers in the west giving intellectual ammunition to the worst kinds of nativist ideologues.")[47] Finally, as Zhang Longxi contends, even the guiding lights of contemporary cultural radicalism—including such renowned thinkers as Roland Barthes, Jacques Derrida, and Michel Foucault—have promulgated wildly distorted views of "Eastern" cultures; some have even deliberately fashioned a made-up East that "just serves the purpose of being a foil or contrast to the Western self." The task ahead, Longxi argues, is not only to "demythologize the Other," but to dismantle completely the "false polarity between East and West."[48]

Collapsing the East-West duality is no easy matter. Although enlightened thinkers such as Marshall Hodgson and S. Radhakrishnan[49] showed the necessity of such a move more than thirty years ago (anticipating the views of many contemporary social theorists), the habits of totalizing distinction are deeply ingrained. To transcend the myth of special Western virtue (or vice) would require a wholesale reappraisal of the spurious differentiation between Western and Eastern cultures. Yet merely cataloguing every distinction between the East and the West ever proposed would be a lifelong project.[50] What we have assayed here is merely one step toward that larger effort: an attempt to challenge the conventional mapping of a few key "Western" and "Eastern" traits. Our focus in the remainder of this chapter will be trained on two narrow but central sets of concerns: first, the question of whether rationality has a lopsided geographical distribution, and second, the much-debated issue of whether Western democracy and progress are legitimately contrasted to Eastern despotism and stagnation.[51]

The Rationality Question

Few would argue that the industrial, political, and intellectual revolutions of the eighteenth century were so profound as to create an entirely new civilization. Modern Europe's cultural continuity with its medieval past is too obvious to be denied; even those who downplay classical antecedents see Western culture as having crystallized during the Middle Ages. But most thinking on the rise of Western civilization is in fact premised on the notion of deep historical roots extending as far back as ancient Greece. In particular, we are often told that it was the Greek "invention" of rationality that made Western progress possible in the first place. The gestation period may have been centuries long, but the baby was eventually delivered. In this account, the seed of Western civilization is said to lie in a peculiarly Greek spirit of rationality. But *has* there been some kind of vital rational force animating Western history—a force that has been largely absent from the East?

Before attempting to answer this question, it is essential to define more precisely just what is to be meant by such loaded terms as *reason, rationalism,* and *rationality.* All three are among the more complex words in the English language, meaning distinctly different things for different writers and when used in different contexts. In casual speech, *rationalism* is often employed simply to mean logically consistent thought; among philosophers, it usually connotes a distinct theory of knowledge. Even in the latter, narrowly epistemological sense, its meaning may vary from strict deductionism to a broader positivism that attempts to integrate deductive and inductive modes of inquiry (the hypothetico-deductive approach). In a different arena altogether, social organizations may also be held to embody rationality, a rational organization traditionally being defined as one that is run efficiently through hierarchical, rule-bound, and bureaucratic structures.[52] Finally, in economics, rationality often implies decision-making that maximizes personal or social satisfactions over the long term.[53]

While rationality can thus refer either to modes of thinking or to social structure, it is often assumed that where reason is found in one realm, it will also be found in others. In other words, a society whose members examine the world rationally is expected to make its economic and political organizations function in a rational manner. Moreover, rationality is believed to have similar manifestations in all areas of life. In both thought and action, rationality is associated with laws and regularities; rational thought discovers natural laws, while rational organizations enforce consistent social laws, often ones that build on those of nature. Likewise, rationality of both the epistemological and the practical varieties is expected to be associated with a discounting of emotion in favor of dispassionate analysis or impartial enforcement of law, as the case may be.

In the historical record, however, these presumed correspondences prove elusive. It is difficult to document the claim that epistemological rationality (of whatever variety) parallels or precedes the development of organizational rationality. Max Weber himself acknowledged this, concluding that rational economic practices emerged in Europe precisely within the context of an epistemologically antirational religious tradition. In Weber's paradoxical formulation, systematic rationality came about irrationally, with great cunning.[54] Nor have all societies that valued deductive thinking necessarily created efficient, rule-bound, bureaucratic institutions. The birth of rationalist philosophy in ancient Greece, after all, did not lead to a push for "rational" social reform. If anything, a strict epistemological rationalist would have to regard the very notion of rational social organization as a contradiction in terms, since rationality in this context cannot be ascertained without recourse to value judgments. Notions of practical rationality can only be justified by appealing to historically grounded notions of reason and justice, not to abstract laws or universal principles alone.[55]

These niceties have seldom been systematically thought through by those seeking to differentiate West from East. As a result, sustained efforts to locate the rational in Western society have given rise to monstrously involuted logic. Max Weber's writings offer a prime example. Weber came to the curious conclusion that whereas the development of "salvation religion" proved a cul-de-sac in the East, leading only to the dead-end of contemplation, in the West the same phenomenon was transformed, first by way of monastic and then by worldly asceticism, into the progressive pursuit of virtue and ultimately "the maximization of rational action."[56] An important "rational purpose" of Western monks, he argued, was nothing less than the "supervision of heretics."[57] But this account effectively twists the meaning of rationalism out of all recognition, conflating it with the dogmatic enforcement of one specific version of revealed truth at the expense of all other creeds—including, as the record shows, competing epistemologies based purely on reason.

The latter form of rationalism—strict epistemological rationalism, in the narrow philosophical sense of the term—did originate in, and has historically been closely linked to, the Western philosophical tradition. But rationalism in its purest form was never more than one contender among many in the unsettled field of Western epistemology, and it has never been widely accepted.[58] While strict deduction may be appropriate, indeed necessary, for guiding inquiry into principles of logic or mathematical systems, analysis of most phenomena demands a significantly more flexible set of procedures. Accordingly, most Western "rationalists" have embraced a broader version of epistemological rationality, synthesizing deductive (rationalist) and inductive (empiricist) approaches.

This tradition—which emerged during the Enlightenment[59] and reached its apogee in twentieth-century logical positivism—may with more merit be designated a specifically Western approach to knowledge. According to Stephen Toulmin, such a qualified rationalism—dedicated to certainty, simplicity, universality, reductionism, and total comprehension—dominated the mainstream of Western philosophy from the early seventeenth century to approximately 1965: precisely the period of Europe's global ascendancy (as well as the period when European investigators laid the foundations of modern science).[60] For the West's champions and detractors alike, it is this intellectual tradition that is most often singled out as the key to Europe's power.

But even rationality of this broader type does not adequately represent the "Western tradition." While modern European culture certainly shows a strong rationalistic impulse, this impulse by no means accounts for all or even most of Western thought. In the early modern era, it would be more accurate to characterize rationalism as a persecuted minority view in western Europe. In the seventeenth century—a period of marked intolerance—rationalist thinkers had either to make peace with religious orthodoxy (partitioning the universe between reason and revelation, as it were) or to flee to one of a few small refuges, like the Netherlands, where heretical thought was tolerated. Most thinkers of the period chose the former course, seeing no incompatibility between reason, selectively applied, and belief in scriptural authority. (The writings of Isaac Newton, for instance, demonstrate the conviction with which many scientists expected to affirm revelation by showing the orderliness of creation.)[61] As John Steadman concludes, a simple identification of European thought with rationality "ignores Western history." He points out: "Just as the rational and scientific aspects of eastern societies have been underestimated, so have the mystical and religious traditions of the West. [One must] consider the dual tendency to overstress the mystical aspect of the Orient at the expense of its rationalism and to exaggerate the rational aspect of the West at the expense of its mysticism."[62]

It is true that entire scholarly communities did free themselves, slowly and painfully, from the dictates of religious dogma during the Enlightenment and especially after the Darwinian revolution. But the rationalist creed did not thereby reign unchallenged in Western thought. New alternatives to rationality have continually emerged, many of them explicitly irrationalist in orientation. The romantic reaction against the Enlightenment introduced the first post-Christian challenge to rationalism in Western culture, but hardly the last. In the words of Ernest Gellner (a proponent of rationality as both an epistemology and a "way of life"), threats to reason have always been ubiquitous in Western culture, where

"the irrationalist waxes lyrical in praise of the deep wisdom of the community or the tradition, of blood or soil or class, or the vibrant vigor of
the dark inner forces of the psyche."[63] In short, it is less rationality per se
than the recurring clash between rationalist and nonrationalist schools of
thought that may be said to characterize Western civilization.[64] To quote
historian William McNeill: "The striking transformation of the culture
of the Roman world from the naturalism and rationalism of Hellenism
to the transcendentalism and mysticism of the fifth and sixth centuries
had no real parallels elsewhere in the [Afro-Eurasian] ecumene. . . . Unusual instability, arising out of a violent oscillation from one extreme to
another, may in fact be the most distinctive and fateful characteristic of
the European style of civilization."[65]

In this view, the recent upsurge of antirationalist views may mark less
a rejection of the Western tradition than yet another "violent oscillation"
within it. During the mid–twentieth century, rationalism ruled, especially
in the newly created "social sciences." Political scientists, sociologists, and
economists created a discourse of positivism, based on a few simple postulates of purported universality and scarcely allowing for the existence
of alternative interpretations. But the hegemony of positivism was short-
lived. The subsequent popularity of antirational theories, including radical postmodernism, is in this sense less a revolution than a revival. While
denouncing "Western thought," contemporary social theory in fact man-
ifests tendencies that have always been part of the intellectual repertoire
of European civilization.[66]

REASON'S HISTORICAL
GEOGRAPHICAL COORDINATES

This brief overview suggests that we would do well to
adopt the broadest possible definition of "reason." For our purposes, reason (or reasonability) might best be conceived as a state of mind marked
by (1) suspicion toward received authority, (2) a commitment to continually refining one's own understanding, (3) a receptivity toward both new
evidence and alternative explanatory schemes, and (4) a dedication to logical consistency. The reasonable person, by this definition, is one who believes that human beings can arrive at a powerful, if partial, understanding
of the natural world through observation, deduction, and experimenta-
tion, yet who remains skeptical of any theories that overstep the boundaries of evidence (especially those purporting to offer total explanations
of the universe's mysteries). Since knowledge (outside of certain limited,

self-contained logical systems) is always uncertain, partial, and subject to continual modification, toleration for competing ideas is a general hallmark of such a stance. Finally, a commitment to reason means recognizing that different issues require different approaches.

At long last, then, we are ready to ask whether rationality thus defined is unique to the West. The answer is a resounding "no." All known cultural systems have manifested both rational and irrational impulses. At certain times and in certain contexts, to be sure, one or the other may prevail *for a time*—with the result that reason has enjoyed brief periods of ascendancy over several areas of the earth. Prior to Europe's own brilliant, flawed, and in many ways aborted Enlightenment, the greatest success of reason may well have been seen not in classical Greece but in Sung China.[67] Its basis was a remarkably empiricist turn of thought, described by Jacques Gernet in the following terms:

> The immeasurable times and spaces, the intermingling of living beings —demons, animals, infernal beings, men, and gods, through their transmigrations—this whole cosmic phantasmagoria disappeared, leaving in its place only the visible world. Man became man again in a limited, comprehensible universe which he had only to examine if he wished to understand it. . . . What is strikingly manifest is the advent of a practical rationalism based on experiment, the putting of inventions, ideas, and theories to the test. We find curiosity at work in every realm of knowledge—arts, technology, natural sciences, mathematics, society, institutions, politics.[68]

Nor was Sung China the only civilization to cultivate reason. Rational inquiry also thrived in particular times and places in Islamdom. Traditional Western historiography recognizes this, but only to a very limited extent. The usual account holds that the Islamic world kept the flame of Greek rationality alive while the West descended into barbarism, but that after passing the torch back to the West during the Renaissance, it lapsed into slumber, with reactionary Muslim clerics denouncing all knowledge that contradicted sacred dogma. Marshall Hodgson issued a bracing corrective to this view several decades ago. In particular, Hodgson documented an interest in scientific exploration that persisted in the "central Islamicate areas" well into the early modern period, although for several centuries "little [was] done to advance it."[69] The "crucial blocking of creativity" that arose in the seventeenth century, he insisted, came about not so much from internal reaction as from "competition with the newly transformed West."[70] Meanwhile, as late as the seventeenth century, rational sciences were flourishing in Islamic north India—bolstered by the arrival of scholars from Central Asia fleeing a rigidly orthodox Uzbek state.[71]

As this scenario suggests, the survival and development of rational inquiry within the Islamic realm do not mean that all Islamic societies have encouraged rationalism with equal vigor. Yet exactly the same qualifications must be made for Christendom. Catholic leaders attempted to do in Europe essentially what their Uzbek counterparts did in Central Asia.[72] Such lapses suggest that, in the West as in the East, it is more plausible to attribute the periodic ascendancy of rationalism to historical contingencies, including the press of economic and political forces, than to an intrinsic cultural aptitude for or against rationality. If anything, the religious heritage of Southwest Asia and North Africa may be more overtly congenial to reason than that of Europe.[73]

Even India, the cradle of much "Eastern" religion, was not isolated from the spirit of rationality. Despite Hinduism's many mystical claims, reason continued to thrive in historical South Asia. This is especially evident in regard to mathematics, the clearest form of pure reason (and the grounds on which the ancient Greeks staked their own claims of being a rational people). During the early medieval era, Indian mathematics was far more sophisticated than Western mathematics, which is one reason why Lynda Shaffer argues that an initial "Southernization" of global thought and technology set the foundations for later "Westernization."[74] Such South Asian precocity continued into early modern times. "An untutored Kerala mathematician named Madhava," for example, "developed his own system of calculus based on his knowledge of trigonometry around A.D. 1500, more than a century before either Newton or Leibnitz."[75] Importantly, the undogmatic nature of Hinduism and the modular character of South Asian society—in which separate social groupings could pursue their own agendas with little interference from others (provided each maintained its proper station)—opened a space for rational thought such as emerged in more westerly lands only through protracted social struggle. While the notion of India's profound spirituality as antithetical to the rationality and materialism of the West has often been embraced by Indian thinkers, others have offered strong dissent. V. R. Narla, for example, argues that South Asia might just as well be regarded as the birthplace of materialism, and he calls on his compatriots to "stop all twaddle about Eastern spiritualism."[76]

In sum, the idea that the West has been historically differentiated from the rest of the world by virtue of its deeper affinity for reason fails all reasonable tests. For the contemporary period, the notion is equally invalid. "Revealed truth" remains a vibrant intellectual force in the United States today, even if it has lost its place in the academy. And if many believers combine their faith with an abiding respect for rigorous thought, tens of millions reject reason outright, contending that scripture contains the literal and absolute truth. In the summer of 1993, massive flooding across the American Midwest was interpreted in pulpits across the land as divine retribution against a sinful nation. Even in Western academia, which has long since abandoned the received authority of traditional religion, such essentially antirational doctrines as deep ecology and Freudianism retain scholarly appeal.[77]

THE HISTORICAL GEOGRAPHY
OF ORGANIZATIONAL RATIONALITY

The cursory review of the literature sketched above suggests that no form of epistemological rationality can securely be held to distinguish Western civilization. This leaves us to contend with "practical rationality" of the organizational variety, which Max Weber and his followers have perceived in formal bureaucracies. As we have seen, the West's champions and detractors alike have frequently singled out the modern state as the culmination of a uniquely Western-rationalist developmental trajectory.[78]

This old notion is called into question by two separate lines of evidence. The first concerns the extent to which the process of bureaucratization (and the routinization of social functions more generally) is inherently rational at all. A closer reading of the historical record suggests that while these processes do create impressive organizational forms, the resulting organizations do not necessarily function efficiently, especially over the long term. The final replacement of a patchwork of feudal organizations by centralized modern bureaucracies in nineteenth-century Europe may have embodied a triumph of certain "rational" principles of governance, but it requires a dubious leap of logic to conclude that rationality therefore inheres in the bureaus. Weber may have seen modernity as nothing less than the "continual spread of bureaucratic administration,"[79] but today some analysts who retain Weber's instrumental interest in efficiency find a higher rationality in bureaucratic dissolution. Indeed, it is now common to regard bureaucracies, governmental or corporate, as hidebound repositories of traditional behavior that perpetuate irrational responses to contemporary problems. Certainly in the business world, the contemporary trend is to break down bureaucracies in order to create "virtual corporations" characterized by transitory forms, embodying elaborate horizontal structures, and in some cases based on the elaboration of personal relationships.[80]

In any case, a separate problem vexes those who would persist in regarding classical bureaucracy as the last word in organizational rationality: that it is simply not possible to maintain that the West has ever held any sort of historical monopoly on large-scale, centralized, hierarchical administrative organs. In their modern form, such institutions are a global phenomenon, representing a response to shared conditions of the early modern world. With the rapid rise in the cost of warfare following the "gunpowder revolution" of the fifteenth century, efficient revenue extraction became a necessity for state survival throughout the world—such that administrative "rationalization" became a hallmark of seventeenth- and eighteenth-century development from one end of Eurasia to the other.[81] Indeed, one reason why the savants of the French Enlightenment were so enthusiastic about China was that they considered its (supposedly) meritocratic mandarinate bureaucracy to be a model of efficiency from which the quasi-feudal *ancien régime* might learn. In a similar vein, commentators on both sides of the Pacific can be heard today arguing that Confucianism predisposes the Chinese cultural sphere to practical rationality and economic success more effectively than any Western creed.[82]

But the clearest challenge to the classical linkage between organizational rationality and Western culture, which Weber saw as uniquely mediated by radical Protestantism, is encountered in the annals of Islam. Here one can do no better than to cite a thought experiment proposed by Ernest Gellner, who otherwise accepts much of Weber's analysis:

> I like to imagine what would have happened had the Arabs . . . conquere[d] and Islamicize[d] Europe. No doubt we would all be admiring Ibn Weber's *The Kharejite Ethic and the Spirit of Capitalism* which would conclusively demonstrate how the modern rational spirit and its expression in business and bureaucratic organization could only have arisen in consequence of the sixteenth century neo-Kharejite puritanism in northern Europe. In particular, the work would demonstrate how modern economic and organizational rationality could never have arisen if Europe had stayed Christian, given the inveterate proclivity of that faith to a baroque, manipulative, patronage-ridden, quasi-animistic, and disorderly vision of the world. A faith given to seeing the cosmic order as bribable by pious works and donations could never have taught its adherents to rely on faith alone and to produce and accumulate in an orderly, systematic and unwavering manner.[83]

As Gellner goes on to explain, "By various obvious criteria—universalism, scripturalism, spiritual egalitarianism, the extension of full participation in the sacred community not to one, or some, but to *all,* and the rational systematization of social life—Islam is, of the three great Western monotheisms, the closest to modernity."[84]

The Question of Political Economy

The attribution of organizational rationality to the West has historically been closely linked with the advocacy of "free choice" in the political and economic spheres. According to the standard Anglo-American view, democracy is a quintessentially Western phenomenon: one that arose among the ancient Greeks, slumbered for a spell during the Middle Ages, reemerged in the modern epoch to become the guiding light of Western civilization, and finally was graciously offered as a gift to the rest of the world. Such a view, admittedly caricatured here, is little more than self-flattery. As is well known, ancient Greek voting was a highly restricted privilege, confined to a small group of male citizens; more important, democracy was but one of many political forms found in the Hellenic world, and it thrived only in very particular economic and military situations. The specific material coordinates of Greek democracy can be identified with some precision: (1) a highly profitable agrarian export economy, based on wine and olive oil, which allowed substantial leisure time for the small-scale cultivator; and (2) a military system founded on the infantry phalanx and the skill-demanding trireme, which provided an important role for every male citizen.[85] The convergence of these circumstances was highly limited in the Hellenic world, both spatially and temporally, and where they did not obtain democracy was absent, regardless of any broader Greek proclivities for freedom.

Nor has post-Renaissance Europe shown a notable penchant for popular participation in government. In modern Europe before the end of the Second World War, democracy enjoyed limited success. Only where economic circumstances allowed a sizable portion of the populace to acquire some measure of prosperity was any real progress made in a democratic direction. And even where and when widespread prosperity was achieved, powerful antidemocratic constituencies ensured that popular political participation would be tightly circumscribed. In a word, economic forces and the variable outcomes of political struggles would appear to have guided the geographical distribution of democratic forms more than any hypothesized Western spirit.[86] The same holds true for the contemporary non-West as well: economic development and the spread of literacy, more than any sort of cultural "Westernization," is what has generated pressures for public freedoms and democratic reforms in places like South Korea, Taiwan, and Thailand.[87] While certain Asian leaders, most notably Singapore's Lee Kuan Yew and Malaysia's Mahathir Mohamad, reject liberal democracy because of its Western associations, such reasoning amounts to little more than a self-interested attempt to justify policies of oppression.

One way to neutralize these recent developments is to argue that economic success is in itself proof of Westernization. If one starts from the notion that the East is characterized by stasis and the West by progress, then any indication of economic growth in the East can be taken as evidence of the diffusion of Western values and mores. But such an assumption as well appears to be unwarranted in a broader historical view. Recent scholarship suggests that all urbanized portions of the Afro-Eurasian ecumene maintained roughly equivalent levels of technology and economic activity until at least the seventeenth and probably well into the eighteenth century, undermining the Western dynamism–Eastern stasis formulation.[88] As late as 1683, Eurasia was home to at least three dynamic, non-European empires: the Ottomans were battering the walls of Vienna and threatening all of Christendom, the Moguls in India were unassailable by Western forces, and the Manchus in China were expanding vigorously, not threatened in the least by Europe's pale and hairy barbarians. Prior to the Industrial Revolution, slow and fitful economic and technological progress was the rule everywhere. Certainly there were episodes of intensive economic development, marked by rapid change and rising living standards. But these occurred in the East as often as in the West, with Sung China and Tokugawa Japan being among the prime examples prior to 1700.[89] Most regimes, by contrast, were profoundly conservative, and those innovations that were accepted tended to diffuse rapidly throughout the urbanized portions of the Eastern Hemisphere. In short, if rapid rates of economic growth and technological prowess are taken to be the definitive Western qualities, one can only conclude that the West did not exist until the high modern era.

This problem becomes all the more acute for those who insist that laissez-faire capitalism embodies a uniquely Western spirit. Rhetorical support for this notion comes from an associative logic whereby the same spirit of freedom represented in democracy is also located in "free-market" economics. But the empirical evidence undermines the association completely. On the one hand, there are problems in equating capitalism with the West. It is undeniable that capitalists came to control production in western Europe before they arose to dominance elsewhere. But it is equally undeniable that this achievement was predicated in no small part on high levels of economic sophistication and proto-capitalist activity throughout the Eurasian ecumene.[90] Any number of non-Western economies were organized on quasi-capitalist lines, and given time and autonomy, others could well have developed into "the genuine article."[91] In fact, there is now evidence that full-fledged capitalist relations of production did arise in at least one sector of the premodern Japanese economy, independent of European influence.[92]

One can, of course, locate certain political and economic structures in medieval and early modern Europe that were unique to the region. Moreover, some of these specifically European attributes undoubtedly contributed to Europe's later rise to global predominance. Carlo Cipolla, for example, powerfully argues that the tradition of urban self-government, based essentially on merchant and craft-guild oligarchies, was uniquely European, and, more important, that its emergence in the High Middle Ages represented nothing less than "the turning point of world history."[93] Cipolla also maintains that Europe was unique in the high social position it afforded members of the medical, legal, and notarial professions.[94] But while these are important points, Cipolla probably grants them more significance than they deserve. Moreover, he adds to his list a host of other "uniquely Western" characteristics that were most assuredly *not* limited to Europe in the preindustrial period.[95]

Particularly intractable problems beset any claims that laissez-faire policies are intrinsic to capitalist success, European or otherwise. In early modern Europe, mercantilism was complexly intertwined with capitalism everywhere; in later Prussia and prerevolutionary Russia, state initiative was instrumental in the emergence not just of industrialism but of capitalism itself. Likewise, some of the most successful forms of capitalism in recent decades have been the "bureaucratic authoritarian" variants of Japan, Singapore, and South Korea. And even the West continues to embrace multiple economic models. Significant disparities exist between the capitalisms of Britain and Italy; in terms of corporate structure, Italy actually has more in common with Taiwan than it does with the United States and the United Kingdom.[96] In short, the asserted linkages between free-market policies, economic success, and Western civilization prove elusive at every turn.

MODELING THE EAST: ORIENTAL DESPOTISM
AND THE ASIATIC MODE OF PRODUCTION

The same formulas that identify freedom with the West attribute despotism to the East. The idea that despotic power is indigenous to Asia—and a root cause of Asian "stagnation"—has been a staple of European Orientalism for centuries. Originating with Herodotus (if not earlier), it was given formal expression by Aristotle, and was revived by one of the more ethnocentric thinkers of the Enlightenment, Montesquieu. Montesquieu's motive for denouncing Asian despotism was not a democrat's concern for the oppression of the populace, but rather a noble's fear of granting the sovereign too much power over the aristocracy.[97] His work on the subject can effectively be read as a warning against the rising absolutism of the French monarchy; as Brendan O'Leary notes, "The cry against despotism was evidently no more than a slogan employed by the declining feudal nobility to defend its interests."[98]

In the centuries after Montesquieu, the thesis that Asiatic societies are inherently despotic (and therefore stagnant) was revived by Hegelians in Germany and utilitarians in Britain. Both schools of thought conceived of Oriental society as composed of an archipelago of self-contained, egalitarian village communities, served poorly by a small and perennially insecure merchant class, and ruled firmly by an autocrat who possessed all the land and whose will was carried out by a parasitic, urban-based administrative elite.[99] Unlike the West, it was a land utterly without *civil society*.[100] Borrowing from both Hegel and the utilitarians, Marx and Engels refined this concept into the notion of an "Asiatic Mode of Production." Their suggestion was that Asiatic societies were held in thrall by a despotic ruling clique, residing in central cities and directly expropriating surplus from largely autarkic and generally undifferentiated village communities. Other Marxist writers later "took the concept on a tour through world history and found that it applied almost everywhere—except to Western Europe."[101]

Like John Stuart Mill before him, Marx had linked Asian despotism (in a few minor passages at any rate) to the need for centralized control over irrigation facilities and other waterworks. This idea was seized upon in the middle decades of the twentieth century by the German historian Karl Wittfogel, who turned it into the cornerstone of a veritable "science of human freedom." A onetime Marxist transformed into a crusading Cold Warrior, Wittfogel retained a profoundly Marxian conception of historical geography. His major intellectual project was to develop a three-stage thesis regarding Asiatic despots: (1) that large-scale hydraulic systems were essential for agrarian production in arid, semiarid, or monsoon conditions; (2) that the building and maintenance of such waterworks gave rise to monolithic bureaucracies that eventually acquired total social control; and (3) that the concentration of political power in the hands of those who controlled these key structures effectively stultified social evolution. Throughout the East (as well as in parts of Africa and the Americas), Wittfogel saw remarkably similar "organizational methods of hydraulic despotism" that included "record-keeping, census taking, centralized armies, [and] a state system of post and intelligence."[102] In Western areas of rain-fed agriculture, on the other hand, such institutions did not have the material base on which to rise, with the result that the West developed polycentric states where numerous parties shared power. While bureaucracies did eventually arise in the postfeudal West, they did so in a more decentered context, where "Big Government, Big Business, Big Agriculture, and Big Labor" functioned to "check each other."[103] Thus was the Occidental experience cleanly and absolutely distinguished from that of the Orient.

This radically bilinear vision of human history was grounded in a thoroughgoing environmental determinism. In Wittfogel's view, it was the seasonal drought of Eastern lands, combined with the presence of large plains suitable for massive irrigation works, that gave rise to a social order marked by oppression and stagnation. Conversely, it was the relatively even distribution of precipitation in the West, combined with Europe's more broken terrain, that had created room for freedom and progress. Not surprisingly, Wittfogel encountered some striking geographical inconsistencies in working out his evolutionary scenario. Certain Asian regions that he considered fully subject to despotism, for example, relied primarily on rain-fed agriculture. Japan showed the reverse problem; its agriculture was founded on irrigation, yet it developed a feudal system strikingly similar to that of the West. The most intractable difficulty, however, was posed by Russia, a land of rain-dependent cropping that seemed nonetheless to exemplify the central flaws of Oriental despotism. This problem was particularly acute, as the avowed subtext of Wittfogel's work was to show how the Western creed of Marxian communism had been transformed, on Russian soil, into a dangerous Asiatic tyranny.

Wittfogel solved these problems by devising an elaborate taxonomy of hydraulic forms, by adding nuances to his theory of environmental determinism, and by abandoning his core logic altogether when necessary. Russia thus became a "semi-Asiatic" state, while Japan was relegated to the "sub-margin" of the hydraulic realm (along with Venice).[104] Zones of rain-fed cropping in Asia were likewise brushed over, in this case because Wittfogel believed that nearby irrigated districts, even if "inferior in acreage and yield to the remaining arable land," could "nevertheless be sufficient to stimulate despotic patterns of . . . government."[105] When considering Russia and other supposedly despotic areas that lacked irrigation altogether, Wittfogel suggested a process of despotic transference, arguing that totalitarian bureaucracy may spread from hydraulic to nonhydraulic and even pastoral societies. In the process, he revived the idea of the infamous "Tartar yoke" : the Mongols, having assimilated the Chinese system of governance, dragged Russia into the murk of despotism, a position from which it would eventually Orientalize Marxism and then threaten the entire world.[106]

Wittfogel's *Oriental Despotism* received scathing reviews from most Asian specialists at the time of its publication. It has also been vehemently attacked by most Marxian theorists, not only because of its overt Cold War posture, but also because its environmental determinism allowed little room for progress. Yet the book was tremendously influential in its day and still commands respect in certain academic quarters. An important 1991 book called *The Geography of Science,* for example, purports to explain the historical geographical distribution of scientific efforts in Wittfogelian terms.[107] An eminent environmental historian has recently argued that water development projects in the American West are best understood in the context of hydraulic despotism.[108] A prominent Marxist geographer, Richard Peet, has also attempted a partial rehabilitation of the theory, arguing that "profound insights into differential societal development can be derived from Wittfogel's ideas."[109] While eschewing some of Wittfogel's more straightforward environmental correlations, Peet countenances the basic scheme, arguing that "south-west Asia, Egypt, the Indus Valley, China, Peru and Meso-America" not only gave rise to early "state-dominated societies," but did so in part because they possessed "similar . . . physical environments."[110] More important, Peet retains Wittfogel's (and Marx's) belief that Asiatic history can be characterized by cyclical dynamics, whereas feudal Europe instead maintained "the potential for transformation."[111]

While few contemporary Marxists would countenance Peet's one-armed embrace of Wittfogel, the general notion of an Asiatic Mode of Production continues to occupy a prominent niche in Marxian theory.[112] In the words of one critic, "like a spectre, the Asiatic Mode of Production haunts Marxist historians, even those dedicated to its excommunication from the discourse."[113] Nor is twentieth-century Marxism the only school of social theory haunted by nineteenth-century notions of Oriental despotism. This dubious legacy of Hegel and the British utilitarians continues to inform most competing schools of analysis as well. The neo-Weberian political theorist Paul Kennedy, for example, writes in his

recent study, *The Rise and Fall of the Great Powers,* that "however impos-
ing and organized some of those oriental empires appeared by comparison with Europe, they all suffered from the consequences of having a centralized authority which insisted upon a uniformity of belief and practice,
not only in official state religion but also in such areas as commercial development and weapons development."[114] Had he done his homework,
Kennedy would know otherwise. Like the notion that European culture
has shown a unique affinity for democracy and economic dynamism, the
theories of Oriental despotism and stagnation are simply not supported
by the historical record.

DESPOTISM RECONSIDERED

The most intractable problem with the Oriental despotism
thesis is the faulty logic of environmental determinism on which it depends. Wittfogel's elaborate theory, like Marx's work on the Asiatic Mode
of Production, ultimately hinges on the presumption of fundamental geographical concordance between physical environments and developmental pathways. Such a presumption represents a fundamental falsification of historical geography, not least because the diversity *within* the
various environments of "West" and "East" is greater than any general-
izable differences *between* them. The Po Valley of Italy, to take but one example, has more in common (in terms of climate, soils, landforms, and
ultimately, hydraulic potentiality) with the river valleys of central China
than with the Scottish uplands. Likewise, central China's valleys show far
closer environmental affinities with northern Italy than with the tropical,
semiarid basalt tablelands of India's Deccan. While the theory of Oriental despotism is predicated on the notion that Southwest Asia, Egypt, the
Indus Valley, and China (as well as, in Wittfogel's and Peet's view, Peru
and Mesoamerica) are characterized by "similar environments," on closer
inspection each of these geographical entities encompasses fantastic geographical diversity. Mesoamerica alone encompasses lowland tropical
rainforests, cloud forests, pine savannas, grasslands, and scrub; its "despotic" pre-Columbian societies thrived in such diverse environments as
seasonally inundated alluvial lowlands, lowland karst platforms marked
by serious water scarcity, undrained highland basins characterized by salt
accumulation, and upland slopes of fertile volcanic soils experiencing ample seasonal rains. In short, there is simply no empirical support for the
claim that Mesoamerica has a singular environment—much less that its
environment is in any way similar to that of Egypt.

In addition to this general logical fallacy, a host of specific empirical weaknesses in Wittfogel's thesis have been identified by regional specialists. In East Asia, Andrew March finds that only a small proportion of the cultivated land in the North China Plain, the supposed core of hydraulic despotism, has ever been irrigated. Such irrigated land as did exist here has usually been watered by small-scale, locally controlled wells. In fact, after extended consideration, March finds Wittfogel's evidence so unconvincing as to conclude that "there is at work in his thought an overriding . . . belief that reality finally lies in the theoretical model rather than in the concrete facts of social geography."[115] Similarly, Kenneth Hall finds that in the premodern Southeast Asian states of Java and Cambodia, where hydraulic agriculture was certainly important, centralized bureaucracies did not develop.[116] Ancient Egypt—one of Wittfogel's keystone cases—turns out not to have possessed a centralized irrigation bureaucracy at all.[117] And in the case of India, Brendan O'Leary marshals an impressive array of evidence to demonstrate the baselessness of the despotism thesis. Large-scale hydraulic agriculture was never vital anywhere in the subcontinent, and irrigation bureaucracies were never important.[118] Finally, one might also note that the Netherlands—an area that could well be considered the heart of the West—is itself characterized by old and extensive hydraulic works (albeit for drainage rather than irrigation). While an earlier generation of scholars usually attributed the initiation of such works to the counts of Holland, the current consensus is that they were developed through a more cooperative, "bottom-up" process—as is true in virtually all of the other hydraulic zones of the world.[119]

Brendan O'Leary's masterful exposé of the Oriental despotism thesis in the Indian context demolishes not just the hydraulic actuator but the entire concept of a distinctive Asiatic Mode of Production. Neither the monarchs nor the bureaucracies of indigenous Indian states were ever particularly strong, even if they did claim absolute power. O'Leary actually speculates that the Mogul Empire may, in the end, have been inadequately bureaucratic and despotic to defend itself. As for Hindu polities in South Asia, O'Leary claims that the state was "mainly an irrelevance."[120] Likewise, a substantial body of research now highlights the historical depth and continuing vitality of the South Asian mercantile tradition, further undercutting the notion of an Asiatic Mode of Production. The idea that traditional Asian village communities were essentially undifferentiated and autarkic, with trade largely limited to luxury trinkets for the urban elite, has been conclusively disproved.[121]

The same fate has befallen another key component of the Asiatic despotism thesis: the belief that, unlike their counterparts in Europe, Asian governments monopolized landholding and thus prevented the rise of locally based, semiautonomous aristocracies that might have offered more scope for innovation. This idea is a variation on one of the oldest East-West archetypes, namely the tenet that Europe's rise was made possible in part by its uniquely fragmented or "parcelized" political landscape. Broken up as it was into small and competing political territories, this view holds, feudal Europe offered numerous refuges (especially chartered towns) where progressive forces (merchants) could gain a foothold. Parcelized sovereignty thus indirectly helped to lay the foundation for subsequent economic transformation. In Asiatic states, by contrast, monolithic governments supposedly maintained such total control as to preclude the existence of such interstitial refuges.[122]

How well does this thesis fare in the face of the evidence? To be sure, many "Eastern" rulers attempted to forestall the development of independent, land-based aristocracies. It was to this end, for instance, that Islamic states from Anatolia to Mogul India periodically transferred their local representatives from one region to another and forbade subalterns from passing power on to their heirs. But such measures did not always succeed, and private land ownership and land-based aristocracies did emerge — even in the Ottoman Empire, the most bureaucratic and centralized of the Islamic regimes.[123] In India the fragmentation of power was even more pronounced. While the Mogul Empire may have been characterized (as John Richards would have it) by a certain "autocratic centralism,"[124] André Wink insists that "the old notion of despotism or unmitigated sovereignty is of no value to understand the mechanics of Mogul politics and its incorporative, assimilative methods of expansion."[125] With regard to the Hindu Marathas, Wink's judgment is even firmer: "Instead of a centralized despotism . . . we found a territory which was parcelled out amongst co-sharers with vested rights, partly defined as 'offices' and partly as 'chiefdoms' but everywhere at odds with arbitrary rule."[126]

In short, wherever one cares to look in the early modern world, state governments had to come to terms with local power holders. All premodern regimes were, to one degree or another, "segmentary" in structure, even where official ideology told a different story. And while all major Eurasian states in the early modern period can be characterized as "agrarian-bureaucratic,"[127] central bureaucracies remained too small and too ineffectual to spread their authority evenly throughout the countryside, and all had to work with and through local notables. As Michael

Mann puts it, before the nineteenth century "autonomous 'despotic' powers exercised by a centralized political elite . . . [were] precarious and temporary." Wittfogel's notion of Oriental despotism "attributes to ancient
states powers of social control that were simply unavailable to any . . .
historical states. . . . [Such states] could do virtually nothing to influence
social life beyond the ninety-kilometer striking range of their army without going through intermediary, autonomous power groups."[128]
Parcelized sovereignty, in other words, may well have played an important role in European development, but it was not a uniquely European
phenomenon.[129]

The centralized power that Wittfogel and others thought they saw in
the traditional Orient is only possible once "modern" apparatuses of control are established. Simple bureaucratic mechanisms like record keeping
or census taking (both of which, incidentally, developed independently
in nonhydraulic thirteenth-century Europe as well as in Wittfogel's Orient)[130] are by no means sufficient to give rise to total social control. And
even in regard to modern totalitarianism, the notion of total power is still
overdrawn. Owing in part to their Wittfogelian heritage, American So-
vietologists typically viewed the Soviet Union as completely centralized,
often not even bothering to examine "what took place beyond the Krem-
lin walls."[131] The events of the early 1990s quickly showed how misin-
formed this interpretation was.

In short, the attempt to sever the connections between European civilization and Soviet totalitarianism was not sound scholarship but ideological exorcism. Karl Wittfogel strove to deny that the Soviet system of
his day had any roots in the historical West. But like Nazism, fascism, and
the divine right of kings, Stalinism was grounded in an undeniably European intellectual and political tradition. The Marxist doctrine on which
it was partially based was articulated by men of Rhenish extraction, imbued with Hegelianism, and founded on the works of Ricardo, and its
variants remain recognizably Western, even if non-Western Marxisms such
as Maoism have incorporated elements of other intellectual heritages. Both
the defenders and the detractors of the "Western tradition" delude themselves in thinking otherwise.

WESTERNIZATION AND MODERNIZATION REVISITED

While debates on Oriental despotism are of a rather recondite nature, the belief that Westernization is a necessary concomitant
of modernization has immediacy in the popular imagination. The as-

sumption is still often made that all of the desirable features of modernity are inextricably bound to the West. In a recent survey of Japan in the influential journal the *Economist,* for example, we are told that "Japan is changing; . . . modernization is indeed turning out to mean westernization."[132] Westernization, in turn, is taken here to mean nothing less than individualism, freedom of expression, tolerance, and democratization.

The formula "modernization = westernization" assumes a priority of origin over process, of geography over history. It holds, in essence, that modernization represents the cultural essence of Western Europe, because Western Europe is (supposedly) where it all began. The present work posits a different metageography of modernity. For one thing, we would challenge the claim that individualism, democracy, secularism, and the like reveal anything *essential* or transhistorical about Western culture.[133] Before the high modern era began, the West was not particularly blessed with any of these attributes. The foundational institution of the Occidental cultural region—the Roman Catholic Church—did everything it could to oppose the growth of individual freedoms, modern science, democracy, market culture, and, of course, secularism, and today it finds itself uneasily allied with radical Islam in an attempt to maintain "traditional" family structures. In fact, all the familiar "isms" of modernization were resisted by important elements of the establishment in Western Europe. Moreover, all were driven by processes that were in important ways global from the start, and all have proven both incomplete and contingent, even in the West. As a result, it is no more tenable to equate this roster of developments with modernity than it is to equate it with Western culture.[134] A recognition of contingency in history discredits our simple glosses of both Westernization and modernity, as well as sundering the presumed identity between the two.[135]

ORIENTALISM RECONSIDERED

The habits of mind that we have been exploring throughout this chapter fit generally under the rubric of "Orientalism," as defined by Edward Said in his seminal work of the same title. According to Said, Orientalism is "a style of thought based upon an ontological and epistemological distinction between 'the Orient' and . . . 'the Occident.'"[136] Said contends that this discourse was constructed by Western scholars, explorers, novelists, and administrators to help them gain authority over the East. In the writings of these "experts," the Orient was conceived in Platonic and antiquarian terms as a changeless land, supine in the face of

Occidental progress. The essence of Orientalism was nothing less than the "ineradicable distinction between Western superiority and Oriental inferiority."[137] Its legacy, he further claims, has contaminated almost all attempts made by Europeans and Americans to understand Asia. Wallowing in the totality of the received vision, scholars visualize the Orient as "absolutely different . . . from the West." In such an endeavor, empirical information is of little account; "what matters and is decisive is . . . the Orientalist vision" itself.[138]

While a number of trenchant criticisms have been aimed at Said's analysis by regional specialists,[139] *Orientalism* has rightly made a profound impression on European and American scholarship in the decades since its publication. Said has convinced virtually an entire generation of scholars to reassess their attitudes toward the East and toward their own scholarly traditions. Indeed, if there is a single project shared by contemporary radical social theorists, it might be that of deconstructing "the concept, the authority, and the assumed primacy of the category of 'the West,' " as Robert Young contends.[140]

This book has been informed by a similar goal. But its approach differs in two important respects from that of Said and like-minded historians of discourse. First, while we find much of Said's critique of traditional scholarship on the Orient compelling, we do not concur with the alternative path he proposes. In the few passages where he issues programmatic statements on this score, Said calls for a globalist perspective that eschews reliance on any kind of large-scale, cultural-geographical categories. "Can one divide human reality," he asks, "into clearly different cultures, histories, traditions, societies, even races, and survive the consequences humanly?"[141] For Said, the answer is no. He appears to find the very business of organizing knowledge into regional frameworks distasteful and to be skeptical of the chance that interesting work will ever be produced when the units of analysis are delimited in geographical terms. A Western scholar might study "discrete and concrete" problems of specific Islamic societies, in his view, but the attempt to grasp the world of Islam as a whole invites contamination by the "doctrines of Orientalism."[142]

We disagree. It remains our conviction that human diversity is geographically structured and that *some* if not all of the relevant geographical patterns precede the development of discourses about them. While the Orient of European imagination is a fictitious entity, an area like Southwest Asia and North Africa (Said's effective Orient) is in significant respects a real region, the home of a historically constituted civilization. Like all regions, it is fuzzy and ill defined. Nor is it a totalizable entity, imbued with some sort of Platonic essence. Yet it remains an indispensable unit of historical and cultural analysis. Denying the existence of geographically definable arenas of human affinity is, to our minds, as great an obstacle to global understanding as is objectifying them into timeless natural essences.

This disagreement in turn reflects a deeper difference between our approach and that represented by Said. Like many scholars of our generation, Said appears to have concluded that the Enlightenment's central attempt—which was, according to Alasdair MacIntyre, "to provide neutral, impersonal, tradition-independent standards of rational judgement"[143]—was fundamentally misguided. In a recent essay, he explicitly disassociates himself from Enlightenment universalism, advocating instead the fragmentation and disassociation favored by postmodernists.[144] We are led to draw different conclusions. If certain Enlightenment figures made unyielding claims to the universality of all of their values, we believe that this demonstrates the hubris of *totalizing* rationality, not the failure of reason itself. If claims of universalism, humanism, and rationalism have been evoked to justify unprecedented violence against colonized peoples, we believe this only means that the Enlightenment was flawed and incomplete. And if the rigid taxonomic grids of eighteenth-century geography betray important misconceptions, we believe the answer is to rectify those misconceptions by recourse to the Enlightenment's own avowed procedures, not to abandon the effort to map the globe altogether. Coming to grips with world geography requires recognizing the existence of human universals as well as of human diversity.[145] To avoid a "pernicious occidentalism" that rejects "everything to do with the West" requires salvaging the original intent of the Enlightenment project.[146]

For this reason, we do not feel compelled to indict the entire heritage of European thought or to target Cartesian epistemology in particular as the primary source of intercultural violence in the modern world.[147] Besides misrepresenting the pluralism of European intellectual life, such an analytical strategy inevitably perpetuates the myth of Europe's *most special* position in global history.[148] Far from serving to deconstruct the West, rhetoric of this sort only reinscribes it as a foundational sociogeographical category.

CHAPTER 4

Eurocentrism and Afrocentrism

As the preceding chapters have shown, the assumption of European centrality in the human past is a pervasive feature of Western thought, resurfacing even where it is most loudly denounced. In recent years, of course, this worldview has come under sharp attack: not only as a symptom of cultural arrogance, but as a distortion of the empirical record. Samir Amin is one of many scholars to articulate a compelling critique of Eurocentrism on historical grounds. In positing a direct connection between ancient Greece and modern Europe, Amin argues, the doctrine mistakenly discounts important developments in the intervening millennia.[1]

We have argued that the time has come to mount a comparable critique on geographical grounds. For undergirding the creed of European priority in history is the mistaken notion that Europe constitutes a unique geographical unit, not only as a continent on a par with Asia or Africa, but as the first among the continents.[2] The debate over the status of European civilization in the college curriculum has so far been hobbled by this pervasive misperception. In the frontal assault on the doctrine of Eurocentrism, as much as in the trenchant defense of the Western tradition, simplistic geographical notions continue to be deployed in unexamined ways. In fact, one corner of academia from which some of the sharpest attacks against Eurocentrism have been launched—namely, radical Afrocentrism—has also resurrected the worst fallacies of its ethnocentric logic. The main difference is that these discredited rhetorical tools are now being used to elevate a different landmass—and a long-denigrated one—to a position of geohistorical priority.

In noting this structural parallel between the metageographical frameworks of Eurocentrism and Afrocentrism, we do not by any means intend to suggest that these are in all ways parallel doctrines. Afrocentrism is an oppositional vision, formulated in reaction to a hegemonic worldview that has been profoundly dismissive of African culture and history. Moreover, while it has gained institutional backing on a few campuses, it enjoys nowhere near the same institutional force as its Eurocentric counterpart, nor has it played anything like the same historical role. If we nonetheless insist that this doctrine deserves critical scrutiny, it is because we take it seriously as an intellectual construct. While it may not share the widespread ideological power of Eurocentrism, radical Afrocentrism does make similar truth-claims, based on similarly faulty modes of geographical reasoning. An honest project of critical metageography must challenge such faulty reasoning in all its incarnations.

Reflecting their different institutional and intellectual histories, however, our discussions of Eurocentrism and Afrocentrism follow very different paths. In the case of Eurocentrism, whose geographical logic has been critiqued at length in preceding chapters, we concentrate on surveying its residual institutional manifestations. After tracing briefly how the idea of European primacy has dominated in historical and geographical thought, we devote most of this section to exposing the residues of that idea in contemporary academic structures, course offerings, and textbooks. We also look briefly at the "inverted Eurocentrisms" of critical social theory, noting some ways in which the agenda of the cultural Left paradoxically perpetuates the predominance of Europe.

In contrast to this essentially institutional analysis of the Eurocentric mainstream, our treatment of radical Afrocentrism necessarily proceeds in a different way. Here we begin by limning the contours of a spatial imaginary developed in a handful of openly polemical texts. Our discussion focuses on two major variants of the Afrocentric world map: one that might be called a continental vision, ascribing an essentially African identity (and virtue) to all peoples who have resided in the space of the continent, and the other a more narrowly racialist variety, identifying Africanness (and virtue) exclusively with the continent's black inhabitants. Having exposed the fallacies of continental thinking in a preceding chapter, we concentrate our present critique on the problems of racialist geography. In either form, we conclude, radical Afrocentrism is as unable to withstand sustained geohistorical scrutiny as is Eurocentrism. Whether based on geophysical or racialist criteria, and whether focused on Europe or Africa, the notion that one landmass holds special primacy in the history of the world simply does not hold up.

Eurocentrism

EUROPEAN HISTORY AS WORLD HISTORY

By the early 1800s, most Western historians had convinced themselves that only Europeans could really be said to possess history. The rest of the world was divided into two broad categories: a zone of Eastern despotism (Asia), which had once been progressive but was no longer so, and a realm of savagery and barbarism (sub-Saharan Africa, pre-Columbian America, Oceania), which had always been bereft of history. Asia's allotted role in world history was thus that of a dim precursor; civilization may have begun there, but as social development had shown an inexorable tendency to move from east to west, the progressive center of the civilized world had firmly lodged itself on Eurasia's western margin.[3]

The notion that the dynamic center of human history had migrated westward over the years, following the course of the sun, has a long history in European thought. Loren Baritz traces this idea to Virgil and other Roman poets who saw the torch of civilization passing westward from Troy to Rome; Jan Willem Schulte Nordholt links it to Orosius, the author of the first Christian world history.[4] In the Middle Ages, Geoffrey of Monmouth extended the same movement from Rome to England, and by the eighteenth century several writers were already looking to extend it yet again across the Atlantic.[5] But it was Hegel who gave the idea philosophical respectability, stating categorically that "the history of the World travels from East to West." According to Hegel, sub-Saharan Africa had no part in world history, being "still in the condition of mere nature." Asia, on the other hand, was an area in which "spirit [had] not yet attained subjectivity" and lived "only in an outward movement [of history] which becomes in the end an elemental fury and desolation."[6]

The nineteenth-century German geographer Carl Ritter took up this idea, contending that "Europe may be called the Face of the Old World, out of which the soul of humanity could look more clearly into the great and promising future."[7] But it was the noted American geographer Ellen Churchill Semple, writing in 1911, who gave this metageographical fiction its most elaborate (and lyrical) expression. "The Atlantic face of the Americas," she wrote, formed "the drowsy unstirred Orient of the inhabited world, which westward developed growing activity — dreaming a civilization in Mexico and Peru, roused to artistic and maritime achievement in Oceania and the Malay archipelago, to permanent state making and real cultural development in Asia, and attaining the highest civilization at last in western Europe."[8] Not surprisingly, this great American scholar went on to suggest that the locus of history was continuing its westward push—back to Atlantic America.

Hegel's spatial vision of history was tremendously influential through the mid–twentieth century. (Indeed, its continuing influence today is reflected in the extraordinary popularity of Francis Fukuyama's Hegelian meditation, *The End of History and the Last Man*.)[9] But Hegel was by no means the only philosopher to locate the spirit of history exclusively in Europe. Consider, for example, Johann Gottfried von Herder, whose views on this subject presented in many respects a distinct alternative to those of Hegel. Herder has recently been praised for recognizing "that it was the 'civilized' European's illusion that he was at the center of things."[10] Yet on geographical grounds, the similarity in their conceptions is significant. While concurring with Hegel that Asia was originally "the mother of all mental illumination," Herder insisted that "the Philosophy of History looks upon Greece as her birthplace." The reason: Asia long ago fell into a stupor by imbibing too much "tradition," a "pleasant poison" and the "true narcotic of the mind."[11]

These brief passages, while implicating both Herder and Hegel for limiting "true" history to Europe, also suggest that both men believed Asiatic societies were at least worthy of study. Such was not necessarily the case several decades later. By the mid–nineteenth century, Eurocentrism had so intensified that it was common for world historians simply to brush away the rest of the world in a few opening passages. A typical "outline of universal history," written in German and published in translation in the United States in 1853, opened by dismissing Eastern civilization in no uncertain terms: "By dint of their intellectual capacity [the Asiatics] quickly attained to a certain grade of civilization, but afterwards gave themselves up to an unenterprising pursuit of pleasure, until they gradually sank into sloth and effeminacy." The brief discussion of Asian history that followed began by repeating the now familiar Hegelian trope: "As the progress of the human race has in general followed the course of the sun, it will be most advisable to commence its history with the tribes of the extreme East. In the vast empire of China has lived, since the earliest period, a race of Mongolian origin, which has preserved unchanged for years the same culture and institutions."[12]

In keeping with these uncharitable views, the author dispensed with Chinese history in a single page. India, which was deemed somewhat more important (both because of its more westerly location, and by virtue of being peopled by an "Indo-European race"), merited all of two and a half pages. After a few more pages of text were given over to the "Near East," the remainder of this "universal" history was devoted to Europe.

By the early twentieth century, the equation of world history with European history had become normative in Western scholarship. Even those who questioned the direction of Western development did so within this framework. Karl Marx may have inverted Hegel's idealism, but he never fundamentally questioned Hegel's geographical vision of history—leading Samir Amin a century later to lament that Eurocentrism had become the bane of Western socialism.[13] Those influenced by Saint-Simon saw the world in a similar light, and August Comte especially "was a convinced, almost mystical, champion of *européocentrisme* who believed that "Europe was in the end to become identical with humanity."[14] Leopold von Ranke, in some ways the antithesis of Hegel as a historian,[15] nonetheless joined Hegel in celebrating European genius and world domination.[16] Jacob Burckhardt likewise viewed the universal as reducible to the European.[17]

That nineteenth-century European historians would equate their small corner of Eurasia with the universe is not particularly surprising.[18] More remarkable is the extent to which this practice has persisted into the present century.[19] Will Durant opened the popular *Story of Civilization* with a trenchant criticism of conventional historiographic practice on this point: "[T]he provincialism of our traditional histories, which began with Greece and summed up Asia in a line, has become no mere academic error, but possibly a fatal failure of perspective and intelligence. The future faces into the Pacific, and understanding must follow it there."[20] Yet even Durant proceeded to dispense with Asia in a single volume (under the telling rubric of *Our Oriental Heritage*), allotting virtually all space in the remaining ten volumes of the *Story of Civilization* to Europe. (In keeping with historical convention, the Middle East and North Africa were later reintroduced for their role as a repository of the "Western tradition" while Europe slumbered through the early Middle Ages.)

To be sure, a handful of renowned world historians of the early and middle twentieth century—notably H. G. Wells, Arnold Toynbee, and William McNeill—endeavored to transcend European parochialism. But most textbooks and popular works have stubbornly held to an older vision of European primacy. A book with the sweeping title *A History of the Modern World,* published for student use by Alfred Knopf in 1971, considers the non-Western portions of the world only to the extent that they were dominated by Europe.[21] A more recent work, purporting to elucidate the "mainstream of civilization," takes a more catholic approach and does examine Asia in its own right, yet maintains it in a distinctly subordinate position.[22]

Here "ancient India and China" are rushed through in one combined chapter, while the Roman Empire is given three chapters of its own; for the medieval period, the Latin portion of Europe rates six chapters, while the realm of Islam is combined with the Byzantine Empire(!), the two being discussed in a scant one and a half chapters. If this is the mainstream of civilization, Asia would appear to have been watered by a meager trickle, while sub-Saharan Africa has evidently been a parched wasteland.

Certainly strenuous efforts are now under way to correct the European bias of world history texts. Most significant here is the recently issued *National Standards for World History*.[23] Yet it is notable that this document has prompted a howl of outrage from politically powerful conservatives, even though it still retains a certain emphasis on the Western world.

EUROCENTRISM IN THE ACADEMIC CURRICULUM

The pervasive provincialism of world history texts is replicated in university course offerings. Although the critiques recently launched against academic Eurocentrism have again resulted in some initial reform efforts, the curriculum as a whole remains overwhelmingly concerned with Europe and North America. Fields such as philosophy, politics, sociology, religion, art history, and music by and large retain their traditional love affair with the products of Western civilization. While some required introductory humanities classes have recently been reframed in a more cosmopolitan spirit, upper-level courses have been little affected. In some philosophy departments, for example, an occasional course might be offered under the dubious heading of "Asian philosophy," but most departments remain exclusively concerned with the European tradition.

History tends to do a better job than most disciplines of representing humanity's ecumenical heritage. Major American history departments are committed to covering the entire earth, or at least those portions of it possessing a literary tradition. Modern history in particular is typically understood as a global category, and all reputable departments acknowledge at some level the importance of premodern Asian history. Yet the bias toward Europe and the United States persists. Most history faculties informally divide themselves into three clusters: those scholars concerned with the United States, those whose focus is Europe, and those who work in "the rest of the world." Not untypically the first group is most numerous and most powerful, while the third group is the least numerous and the least powerful. As a result, while Asia is not ignored altogether, it is often accorded so few faculty positions that large portions of Eurasia slip off the map altogether.

This particular manifestation of Eurocentric bias is clearly revealed in the course catalogs of North America's major universities, circa 1990.[24] Stanford University, for example, listed thirteen history classes on Europe as a whole, with another nineteen on specific European countries or regions (four each on the British Isles and France, three on Germany, seven on Russia, one on Italy, and four on eastern Europe). East Asia was relatively well represented with thirteen courses (six on China, five on Japan, and two on Korea), but South Asia was represented by only a single course, and Southeast Asia did not make the roster at all. Stanford's emphasis on Europe in its history offerings was even more pronounced when one included the classics department; as in other schools, "classics" refers to Europe, with a slight nod to the "Near East." (Classical China and Mesoamerica, for instance, are never recognized under this rubric.) Nor was Stanford at all atypical. Harvard's history department offered over fifty courses on Europe, in contrast to less than ten for all of East, South, and Southeast Asia. Columbia included as many history courses on Iberia as it did on all of South and Southeast Asia combined. Cornell was unusual in providing four courses on South Asia, yet these offerings were overshadowed by eleven classes on Britain and thirty-nine on Europe as a whole. Chicago's history department offered as many courses on France alone as it did on all of East and Southeast Asia.

The Eurocentric bias in the university history curriculum is balanced somewhat by historically oriented courses on the non-Western world that are offered in separate regional studies departments and programs. Harvard's history department may have ignored South Asia, for instance, but the university offered two courses each on the histories of India and Tibet through its program in Sanskritic and Indian Studies. Similarly, Cornell skipped Southwest Asia and North Africa in its history department, only to cover both in its Near Eastern Studies program. Regional studies centers of this kind offer undeniable advantages for American scholars focusing on understudied areas of the globe. But to segregate Asian history into separate programs in this way also reinforces the worldview of Eurocentrism. The implicit message is that while places like India may have civilizations, they do not really possess *history* (in the sense of a dynamic, self-generated transformational force). Responsible scholars have actively rejected such claims for decades. Yet the notion that only the West has history lives on at the heart of the American intellectual enterprise, encoded into the very structure of academic disciplines.

Universities do not necessarily organize their history offerings in this way because of active ideological Eurocentrism on the part of the faculty. More often, departmental structures reflect the inertia of their institutional setting. Departments of East Asian or Near Eastern Studies were created in the heyday of Orientalism, when scholarly consensus did regard Asian societies as largely ahistorical. Once established, they acquired a life of their own. Perhaps as a result, challenges to such deep-seated organizational structures are rare, even among the most ardent critics of academic Eurocentrism. Instead the emphasis is typically put on calls for additional classes, for the creation of centers devoted to the study of Asian-, Hispanic-, or African-American culture, and for the strengthening of area studies faculties.

The perpetuation of archaic institutional categories may indeed seem innocuous. What does it matter where non-European history is taught, it might be asked, so long as it is available somewhere on campus? But the continued isolation of college courses on China in a department of East Asian Studies, or of those on Iran in a department of Near Eastern Studies, has consequences for scholarship as well as for pedagogy. Such institutional arrangements remove scholars of these areas from the diverse intellectual milieus of traditional disciplines and instead segregate them into what can become regional ghettos. Can this be unrelated to the frequently heard complaint that Asian Studies tends to lag behind explorations of Western history in the development of theoretical perspectives? Certainly other factors contribute to marginalizing Asia in the scholarly imagination as well, not least the daunting effort required to conduct research in non-Western-language materials. Such an intrinsic problem is perhaps exacerbated, however, by pulling scholars outside of a disciplinary community and clustering them instead in marginalized area-studies complexes. At the very least, such institutional arrangements would seem to deserve more serious scholarly scrutiny than they typically receive.

THE HISTORICAL
AND GEOGRAPHICAL IMAGINATIONS

At one level, the priority given to Europe and especially the United States in most American universities reflects a thoroughly understandable interest in local conditions. The educational system of every country in the world endeavors to teach students about their own society; history as taught in Tokyo is Japanocentric, just as that offered in Riga is Latviacentric.[25] It is only to be expected that American history should take primacy in American schools, and since most Americans institutions do have European roots, a major focus on Europe is to be expected as well. Nor is it our intent to disparage national and local history. Learning about one's own society and its past is an essential foundation for responsible citizenship. But in the drive to understand the lineaments of American society, vast areas of the world have largely been neglected. National and local history cannot be taught accurately, nor can they serve their proper function, except when set in the context of a global and thoroughly ecumenical approach. It is this balance that we believe is lacking in American educational institutions today. Unfortunately, student demand contributes to this imbalance in the curriculum. Where existing courses on African or South Asian history do not fill, it is unlikely that extra efforts will be made to hire additional Africanists or South Asianists.

Perhaps the primary justification for continuing to slight the history of the non-Western world is that historians have traditionally sought out the "main lines" of development, concentrating their attention on a handful of places that have proven themselves to be important centers of innovation and power. Even within Europe, this has meant a heavy focus on a small set of countries: Britain, France, Germany, Russia, and Italy (the latter highlighted mainly for its role in the Renaissance). The rest of Europe has usually been viewed as marginal and is consequently marginalized. Similarly, in American history the regional history of the South and the West has at best been accorded a second-rate status. The "mainstream" of American history is either located at the national level or in the Northeast.

Such a view of history does offer certain advantages, but it is also significantly limiting, and it risks impoverishing our understanding of the world. To suggest that one area's history is vastly more important than that of the rest of the world, as our current curriculum does, is to lull students into a dangerous form of parochialism. Quite aside from the moral questionability of deeming certain places unworthy of study because they have not been globally dominant in the past, this view of history raises practical problems of second-guessing the future as well. To take one striking example, before World War II Japan was often viewed as an intriguing but rather marginal place, and only a handful of Americans possessed linguistic competence in Japanese. Today, by contrast, Japan is recognized as a historically significant nation indeed, and American universities are in hot competition for qualified Japan specialists. Clearly, past power is not necessarily a reliable guide to future importance. To map out zones of historical significance and insignificance is to presume knowledge of the unknowable.

The geographer's perspective may offer a useful corrective here. Unlike historians, geographers have traditionally been concerned with landscapes as witnesses to the diversity and variability of human experience. Rather than ranking countries according to their historical importance, they have tended to explore and celebrate whatever distinguishes one place from another.[26] Indeed, it is the very uniqueness of a place that has most often caught the geographer's fancy. The result is at best a sort of cartographic democracy, where all portions of the earth are viewed as equally interesting because each has its own combination of characteristics.[27] The geographical imagination may be stunted in other respects, but this fundamentally balanced perspective on place offers an important contribution toward a truly global, ecumenical study of our species and its place on the planet.

If the geographical imagination offers a potential avenue of escape from Eurocentrism, however, geography as a field has hardly been immune from the syndrome.[28] On the contrary, geography perhaps more clearly than any other discipline testifies to the failure of recent theoretical reorientations to offer any real alternatives to the Eurocentric approaches of the past. Not a few practitioners of the more avant-garde schools of geography actually argue for overthrowing the field's long-standing commitment to global understanding in favor of a renewed focus on the West.

The clearest sign of a retreat from cosmopolitanism in geography is the steady decline in overseas field research.[29] The turning away from work in non-Western settings began in the 1950s and 1960s, when young scholars sought to transform geography into a universalistic spatial science.[30]As long as "laws" discovered in Iowa had explanatory power as well for the landscapes of Nigeria, it was only logical to conduct research close to home.[31] The 1970s and 1980s saw the rise of several competing theoretical orientations in the discipline that reasserted the importance of local particularities. Yet in all of these movements — especially in geographical humanism — research efforts remained largely focused on Western societies.[32]

Nowhere is the retreat from international topics more clearly evident than in regional and cultural geography, the two subdisciplines previously devoted to a globalist approach. In the new regional geography, attention is typically focused either on Anglo-American regions or on the spaceless, placeless, regionless flux that is said to characterize the postmodern condition.[33] Most practitioners of the "new cultural geography," meanwhile, are primarily concerned with urban settings in the English-speaking world, seeking either to explicate the locational structures of distinct cultural (or subcultural) groupings, or to explore ideological reflections found in the modern urban landscape.[34] There is nothing wrong with such studies in their own right. What is troubling is the accompanying dismissal of "old" cultural geography — the kind that entails fieldwork in remote and often rural areas—as "a celebration of the parochial" and a "contemplation of the bizarre."[35] Such dismissals imply that certain places (especially postmodern cities like Los Angeles)[36] are fundamentally more important than others, and that the particularities of remote or distinctive landscapes may be regarded as somehow trivial. For those attempting to forge a globalist perspective, such disciplinary trends offer little comfort.

INVERTED EUROCENTRISMS IN RADICAL THEORY

As we have seen, Eurocentric thinking has been so ubiquitous that no field of study has been able to avoid it completely. In the same way, a world vision that places Europe at the center of analysis informs the works of numerous scholars all along the continuum of political beliefs. Even many writers who claim to be making bold breaks from Eurocentrism often continue unwittingly to lend support to key elements of the doctrine.

Neo-Marxist writing, in such fields as world-systems and dependency theory, asserts the centrality of the West in explicitly spatial terms, labeling Europe and North America (and sometimes Japan as well) as the global core. Such analyses take an important step away from traditional Eurocentrism by historicizing this structure, arguing that the primacy of the core stems more from its exploitation of the rest of the world than from its intrinsic progressiveness. But there remain inherent problems in the spatial model of core and periphery. "Africans and Asians," writes the perceptive C. A. Bayly, "are again in danger of dropping out of the picture, while terms like 'peripheral' and 'semi-peripheral zones' have replaced the evocative 'Dark Continents' of the Victorians in contemporary debate."[37]

Newer voices in the academic Left denounce Eurocentrism in all forms, hidden as well as overt. The vanguard of this movement consists of scholars associated with postmodern theoretical approaches. By adopting a relativistic viewpoint that stresses diversity of experience and multiplicity of meaning, such critics hope to demolish the monolithic ideological structures of the past. The more extreme postmodernists go further, questioning whether human communities around the globe share *any* basic cognitive and experiential ground. To the extent that each group is believed to have its own reality—one that is largely inaccessible to outsiders—an authentic interpretation can come only from within the cultural context in question. In this view, the very process of studying other peoples is suspect. In presuming to speak for and about foreign peoples, the Western observer can cast them only into the role of an objectified, voiceless

"Other." The triumph of such a scientistic (and masculine) mode of see-ing (or "gaze") — i.e., the continuing value placed on knowledge gained through invasive scrutiny by expert outsiders — is widely denounced to-day as a pernicious legacy of imperialism.[38]

This perspective offers powerful tools for analyzing the intellectual dis-courses and colonial mentalities that have long sustained Eurocentrism.[39] But it also has troubling implications for scholarly practice. As Samir Amin points out, to claim "that only Europeans can understand Europe, Chi-nese China, Christians Christianity, and Moslems Islam" is effectively to surrender to an ethnocentric world, where "the Eurocentrism of one group is completed by the inverted Eurocentrism of others."[40]

Afrocentrism

Within the United States, perhaps the most prominent ex-ample of the "inverted Eurocentrism" that Amin laments is to be found in radical Afrocentrism.[41] Afrocentrism is a diverse philosophy, but the core concern of most theorists is to illustrate the profound creativity, ci-vility, and historicity of African peoples.[42] Particular attention is paid to ancient Egypt, which is seen both as having been fundamentally African (in cultural attributes as well as racial affiliation) and as having been the ultimate font of Mediterranean civilization.

The evidence presented to substantiate this view is sometimes com-pelling. Many ancient Egyptians — elites and commoners alike — might well be identified by today's criteria as black Africans rather than as Cau-casians or as members of any putative Mediterranean race.[43] Moreover, Egyptian influences were of some importance in the development of Greek culture — a relationship recognized by the ancient Greeks themselves and not denied in Europe until the romantic reaction of the nineteenth century focused scholarly attention on overtly racist doctrines.[44] In stark contrast to modern times, the "racial" attributes of different peoples were evidently rarely considered significant in the classical era and were seldom employed as a measure of civilization.[45]

This argument is forcefully advanced by Martin Bernal, author of the widely cited book *Black Athena*. Bernal acknowledges the polemical na-ture of his work, contending that the "political purpose of *Black Athena* is, of course, to lessen European cultural arrogance."[46] Nonetheless, *Black Athena* is probably the most deeply researched contribution to the

"African origins of civilization" school. Its first volume in particular represents a substantial, if controversial, intellectual achievement. Here Bernal offers a major new reading, not only of ancient civilization but of modern intellectual history as well. Particularly compelling is the way he documents the origins in nineteenth-century Germany of the now-conventional myth that ancient Greek culture was formed de novo, in contrast to an earlier understanding that Greece owed much to Egypt and other eastern Mediterranean civilizations.

Supported by a variety of evidence and possessing a profound moral authority, the critique of Eurocentrism articulated in *Black Athena* and elsewhere in the Afrocentric corpus is both pointed and persuasive. But some proponents of Afrocentrism, including Bernal, go beyond rejecting the notion of Western priority to make untenable claims on behalf of Africa, upholding it as the unique locus of innovation or virtue. In so doing, radical Afrocentrists embrace the same faulty geographical thinking they so effectively expose. While cutting Europe down to size constitutes an essential metageographical advance, substituting Africa for Europe in a revised ethnocentric scheme merely perpetuates the myth of continents in a new guise.

"Africa," however, does not always denote the same thing in this literature. Molefi Asante, representing one camp, insists that African culture is coterminous with the conventionally defined continent. Along with Mark Mattson, Asante straightforwardly contends that peoples throughout the continent share essential cultural traits: "There is an African culture, just as one might speak of an Asian or a European culture."[47] He accordingly finds the term *sub-Saharan* suspect.[48] North African peoples, he insists, are just as African as those who happen to reside south of the great desert. Even groups who migrated to north Africa from other continents in historical times—the Carthaginians no less than the Arabs—should be considered part of the same cultural sphere as sub-Saharan groups like the Yoruba or the Xhosa. In Asante's view, such famous Arabic writers as Ibn Khaldun and Ibn Battuta demonstrate the superior qualities of a distinctly African civilization.[49]

In contrast to scholarly claims concerning the importance of ancient Egypt, this sweeping assertion of pan-African similarities flies in the face of historical evidence. While northern Africa in the prehistorical and earliest historical eras may indeed have been culturally linked to the sub-Saharan zone,[50] such relationships had deteriorated by Roman times. After the Arab migrations, which began in the seventh century, the notion of a fundamental unity of northern and sub-Saharan Africa cannot be seriously maintained. Even the notion of a unified sub-Saharan cultural zone is difficult to substantiate; linkages between western Africa and southern Africa were at most vague and indirect, while highland Ethiopia, the Khoisan-speaking areas of the southwest, and Madagascar are perhaps best separated from the sub-Saharan zone altogether. Several portions of sub-Saharan Africa were also closely connected to other portions of the world: the eastern littoral, for example, to Oman and the rest of southern Arabia, and highland Ethiopia to Yemen in earlier millennia and to India in more recent centuries. And while one might agree with Patrick Manning that the modern slave trade in some ways created "one Africa out of many,"[51] one would still have to limit this latter-day unified Africa to the area south of the Sahara.

There is also irony in Asante's sanctification of Arabic writers such as Ibn Khaldun as exemplars of uniquely African sensibilities. Ibn Khaldun, a Tunisian of Andalusian descent, was undoubtedly one of the greatest historical philosophers of all time. But he was also heir to a tradition of bigotry toward sub-Saharan Africa, albeit one based on cultural and environmental, rather than racial, reasoning.[52] Peoples living in the southern climatic zones, he unabashedly wrote, have "qualities of character [that are] close to those of dumb animals."[53] He further argued, in true environmental determinist fashion, that the "stupidity," "levity," and "emotionalism" of black Africans was the result of climatic influences.[54] To be sure, his view of northeastern Europeans was just as dismissive: "It has even been reported that most of the Negroes of the first zone dwell in caves and thickets, eat herbs, live in savage isolation and do not congregate, and eat each other. The same applies to the Slavs. The reason for this is that their remoteness from being temperate produces in them a disposition and character similar to those of dumb animals."[55]

Ibn Battuta, another prominent scholar from the Maghreb, entertained similarly disparaging ideas. This fourteenth-century geographer was probably the world's greatest premodern traveler, but he was not at all impressed by the sub-Saharan kingdom of Mali. The many un-Islamic practices of Mali's residents appalled him (as did the stinginess of its sovereign).[56] To say that Ibn Battuta was more at home in Arabia than in Mali would be an understatement of the highest order. Like Ibn Khaldun, Ibn Battuta ill suits the role of pan-African cultural hero.

RACIAL ESSENTIALISM
AND ENVIRONMENTAL DETERMINISM

For these and related reasons, a separate strain of Afrocentric theory challenges Asante's notion that North Africa's Arabic and Berber speakers should be considered carriers of a distinctly African tradition. Opuko Agyeman, for example, views Arabs as historical enemies of the African people and disparages Islam as a devastating foreign religion. "Afrocentricity," Agyeman argues, refers "exclusively to the Africa of the Africans, of black people, and decidedly not to any continental mystique of a geographical area which includes Africa's invaders—whether they be Arabs who set foot a thousand years ago, or the Dutch who made their incursion some five hundred years ago."[57] But while he rejects the notion that the continent imparts some kind of mystical bond to all who dwell in Africa, Agyeman does identify precisely such a primordial glue in race. Moreover, he goes on to claim: "The other side of this medal of racial consanguinity—of the corporate essence of the racial family—is the concept of the indivisibility of [black] African destiny."[58]

The racial essentialism evident in Agyeman's writing is widespread in extremist Afrocentric thought (despite Asante's protestations that "the concept of race has no biological or anthropological basis").[59] To the racial essentialist, races constitute discrete subspecies of *Homo sapiens* and are characterized by inherently different modes of life and patterns of thinking. One is black or white not merely in skin tone, but in one's very essence. Moreover, central to this version of racial essentialism is a belief in fundamental geographical concordance: a belief, in other words, that a wide variety of different features will automatically have identical geographical distributional patterns. Agyeman, for example, implicitly posits the existence of a basic divisional line across the southern Sahara: to the north of this line, one finds white peoples and non-African ways of thinking; to the south, one finds the black race and African ways of thinking.

The logical linkage between region and race is supplied by environmental determinism. This kind of thinking looms everywhere behind doctrinaire Afrocentrism, but it is articulated most clearly in the work of the movement's founder, Cheikh Anta Diop. Diop argues for the existence of two fundamental races, the black and the white, each associated with a distinct portion of the earth, the South and the North.[60] Each of these human subspecies, in turn, owes its fundamental character to the environment in which it developed. The black race, encompassing both Africans and other southern peoples (such as the Dravidians of India), originated in a realm of abundance and agricultural productivity, which engendered "a gentle, idealistic, peaceful nature, endowed with a spirit of justice and gaiety."[61] The environmental determinants of the white race, however, were of a different order: "[T]he ferocity of nature in the Eurasian steppes, the barrenness of those regions, the overall circumstances of the material conditions, were to create instincts necessary for survival in such an environment. . . . In the unrewarding activity that the physical environment imposed on man, there was already implied materialism, anthropomorphism, . . . and the secular spirit. . . . Man in those regions long remained a nomad. He was cruel."[62]

Nor is Diop alone in these beliefs. Judging from recent publications, it would appear that the linkage of race, place, and personality has numerous supporters. Clinton Jean, author of a recent book called *Behind the Eurocentric Veils,* echoes Diop's climatic determinism in explaining why, unlike the Eurasians, African peoples are inherently peaceful, humanistic, and nonsexist.[63] On their way to discovering agriculture and inventing civilization, Jean tells us, Africans evolved for millennia in a state of bliss, which was only disrupted when violent barbarians invaded from the Eurasian outback.[64] This vision of global history is almost identical to that of early-twentieth-century European and American racists. The narrative is the same; only the moral signs have been reversed.[65]

Strikingly, the most prominent proponent of African racial *cum* geographical essentialism is Martin Bernal, whose *Black Athena* was singled out for praise above. In the more abstruse second volume of this work, where he lays out the archaeological and philological evidence for his thesis, Bernal adopts a strikingly reactionary posture, embracing both racial essentialism and environmental determinism with an enthusiasm rarely seen in American social science since the 1920s. On the former score, Bernal occasionally approaches absurdity, as in his convoluted explanation of an exiguous modern black population in Abkhazia (part of [former Soviet] Georgia).[66] The book's claims about climate and culture also take some bizarre turns, including two instances where specific ideological and artistic developments are explained as responses to volcanic eruptions.[67] It is his more mundane evocations of environmental determinism, however, that do the most harm to the book's credibility.[68]

Besides evoking an environmental determinism that would make most geographers blush, Bernal also forwards an extreme form of cultural diffusionism, implicitly denying that "uncivilized" peoples had any creativity by tracing all progress to one or two cultural hearths. The notion that "civilization" had a single, racially determined point of origin—from which it spread out to enlighten the rest of the world—was widespread a century ago, embraced not only by white racists, but by certain Hindu chauvinists in India.[69] In particular, Western scholars were inclined for many years to attribute anything laudable in sub-Saharan Africa to southward diffusion from Egypt.[70] Bernal retains this vision, with the single (if signal) difference that he reorients the arrows from south to north.

The problem of substituting one chauvinism for another in this way has not gone unrecognized in African studies. As V. Y. Mudimbe cogently writes: "Modern African thought . . . is at the crossroads of Western epistemological filiation and African ethnocentrism. Moreover, many concepts and categories underpinning this ethnocentrism are inventions of the West."[71]

Bernal himself admits as much, in a backhanded way. He concludes the second volume of *Black Athena* by acknowledging its "outrageous" character, noting that many of its outrages stem from a revival of scholarly beliefs that have been dismissed since the early twentieth century.[72] In our view, those beliefs were dismissed for good reason. If researchers in the last sixty years have abandoned geographical determinism and strict diffusionism, it is because their premises have been thoroughly discredited.

THE GEOGRAPHY OF RACE REVISITED

So, too, with the basic categories of racialist discourse. As noted above, race has always been an inescapably geographical concept. In earlier years, many scholars assumed a simple continental correspondence, associating each landmass with a discrete race readily identifiable by skin color. While this model was abandoned long ago as both geographically and biologically unsupportable, the general notion of correspondence persisted. In theory, different races have always been tied, at least in their origins, to distinct areas of the world.

It does not require an especially discerning eye to realize that there is nothing red about indigenous Americans or yellow about East Asians[73] — or that blacks are not really black and whites are far from white.[74] Accordingly, in the heyday of "racial science" in the late nineteenth and early twentieth centuries, Western scholars devised ever more elaborate racial classificatory schemes — and drew correspondingly intricate maps to show their distributional patterns.[75] While the ideological underpinnings of this endeavor began to be rejected in the United States after the 1930s,[76] scholars of race continued to readjust their grand taxonomies through the 1960s.

A major impulse behind this project was to enlarge the realm of the "Caucasian race," largely at the expense of the "Negro race" in northeastern Africa. Such a gambit actually dates back to the late nineteenth century, following Ethiopia's defeat of an invading Italian army. "Since racism did not permit Westerners to acknowledge that black men could vanquish whites, Europeans suddenly discovered that Ethiopians were Caucasians darkened by exposure to the equatorial sun."[77] St. Clair Drake nicely illustrates the ironic aspects of this maneuver: "As late as the sixties, in some areas of the [United States] a person was considered 'black' by law or custom if he had, or was suspected as having, *any* African ancestry, however remote and small in degree. At the same time, in dealing with Africa some anthropologists refused to classify any Africans as 'Negro' or 'black' if they were known to have, or were assumed to have, any white Caucasoid ancestry! They were dubbed 'Hamites,' not *true* Negroes.'"[78]

The residue of the "Hamitic thesis" remains in contemporary maps of racial distribution, which often persist in classifying the residents of Ethiopia as Caucasians. (One geography textbook published in 1990 actually extends the realm of Europeans as far as eastern Kenya!)[79] The most egregious example of such cartographic arrogance can be seen in a world atlas devised by scholars in the U.S. War Department in 1946, which extends the "white race" to encompass all of Uganda as well as the northeastern section of Zaire. This same map, incidentally, portrays Hungary and most of Finland as occupied by members of the "yellow" race—presumably by virtue of the fact that Hungarians and Finns speak Uralic, rather than Indo-European, languages.[80] This is by no means the only occasion on which race has been conflated with language, but rarely has the exercise yielded such counterintuitive categorizations.

In attempting to depict the complexities of racial affiliation that they believed they were discovering, physical anthropologists of the 1950s and 1960s delineated some rather bizarre distributional patterns. In 1954, for example, Carleton Coon discerned a "Negroid race" in most of sub-Saharan Africa, Melanesia, remote reaches of the Philippines and Malaysia, as well as the entire (modern) state of Andra Prahdesh in India.[81] He also mapped an "Australian race" through most of the Australian outback, Tamil Nadu and Kerala in India, and the southwestern tip of the Arabian Peninsula. By 1966, however, Coon had substantially revised his view of racial geography.[82] He now placed the "Asian Negroids" in an "Australoid" category and rechristened the African blacks as "Congoids." His mapping of these supposed racial groups nonetheless continued to show incongruous patterns, evident in his notion that Luzon and the Malay Peninsula were occupied in 1492 by Australoids, when the rest of the Philippines and western Indonesia were supposedly inhabited by Mongoloids. He also viewed much of east central India as Australoid in 1492, and while he saw this territory as having substantially decreased by the 1960s, he also mapped another two areas of Australoid occupancy that had inexplicably emerged in the meantime in western India.[83]

In the process of translation from specialized works to popular texts, the contradictions of race as a spatial category have, if anything, been amplified. A geography textbook from 1978, for example, maps southern India, an area inhabited by very dark-skinned peoples, as "Caucasoid"—a category said to be "indexed by skin color."[84] Another geography text from 1990 depicts the same area simply as "White," and it further informs the reader that eastern New Guinea is "Australian" in race, while western New Guinea is "Black" except for the "Pygmies" inhabiting the island's central mountains.[85] Here we see the myth of the nation-state, compounded by the myth of continents, being further contaminated by the myth of race. The resulting concoction is offered as fact in an introductory world geography course for university students.

Maps of "race" no longer have any legitimate place in global geography texts. For even as Carleton Coon was elaborating his contorted racial taxonomy, Ashley Montagu was demolishing the biological basis of race as a concept.[86] Montagu and like-minded scholars advanced three powerful arguments. First, all identified racial characteristics are biologically superficial, and in no way challenge the fundamental biological unity of humankind. Second, any given racial attribute tends to mutate gradually as one traverses the landscape; distinct transitions from one "racial group" to another are almost never encountered. (While racialist theory ascribes this phenomenon to "racial mixing," "pure races" have never been isolated.) Third, and most compelling from the geographic point of view, each racial characteristic has its own distributional patterns. The global map of skin color, for example, bears little resemblance to the map of hair form or to the map of head shape. One can thus map races only if one selects one particular trait as more essential than others. (The 1963 *Encyclopedia Britannica,* for example, regards hair texture as the prime racial determinant.)[87] Yet no real evidence can be offered to show why the selection of such an isolated trait is not entirely arbitrary.

Nor is any firmer basis for racial categories to be found under the skin. The most rigorous investigation to date of global genetic patterns—L. Luca Cavalli-Sforza, Paulo Menozzi, and Alberto Piazza's massive *History and Geography of Human Genes*—finds that racial categories are genetically meaningless. The authors conclude that while one can identify " 'clusters of populations' exhibiting genetic similarities," such clusters simply cannot be "identified with races."[88] In fact, their painstaking genetic mapping and multivariate analysis fairly demolish the familiar "races of humanity." For example, the northern Chinese are shown to be more closely related to northern Europeans than they are to southern Chinese.[89] And Africans, far from forming a uniform race, actually show more genetic diversity than do all of the world's non-African peoples put together.

In a word, while race is indisputably an important cultural construct, it fails as a natural category on biogeographical grounds. The presumed correspondence between the distribution of racial traits simply does not exist. Only by abandoning the doctrine of geographical concordance— the belief that a wide array of unrelated features should correspond in their spatial patterns—can we begin to ascertain the macrogeography of human life.

In some senses, both Eurocentrism and Afrocentrism make easy targets. Both rely on simplistic spatial categories, and both invoke modes of reasoning that have been widely discredited in recent scholarship. If we have given extended coverage to these crude geographical constructs, it is out of our conviction that even the most elemental building blocks of popular macrogeography deserve careful scrutiny. This conviction animated our first three chapters as well, where we focused on the crudest of schema: continents, East-West, and Orient-Occident.

Having revealed the problems of those simple but pervasive categories, we are now ready to turn our attention to the more elaborate frameworks within which professional scholars have organized their knowledge of the world. The next chapter explores the essentially historical traditions of civilizations and world systems; the succeeding chapter takes up the essentially geographical tradition of demarcating distinct world regions. Along the way, we begin to face the challenge of going beyond all of these frameworks, to come up with a more supple and accurate map of the world.

CHAPTER 5

Global Geography
in the Historical Imagination

It might seem odd for geographers to look to historians for insight into the construction of metageographies. According to conventional wisdom, historians are concerned with time, not with space; their attention is trained on issues of periodization rather than on problems of regionalization. Such was often true in the early days of the profession, when historians typically took established national states to be their proper subject. But this assumption no longer prevails. In the twentieth century, with the rise of social history in all its guises, the subjects and scales of historical analysis have undergone enormous diversification. States may remain a convenient framework for many investigations, given the linguistic and administrative constraints governing access to documents. But both micro- and macro-level studies are now commonplace, and national boundaries are routinely transgressed and problematized in recent work.[1] In breaking free of the national grid, numerous historians have made explicit contributions to the project of rethinking metageographies.

The present chapter begins by looking at two of the most important approaches to high-level geohistorical patterns to have emerged in this century. Not coincidentally, both arose out of the literature on world history, a dynamic subfield that has been cartographically oriented since its inception.[2] While state-centered narratives can be written with only a cursory nod to geography (especially when the story of national development is told from the viewpoint of centrally located elites), and while even the histories of whole civilizations can be composed with little attention to spatial variation (as by the routine device of generalizing from a few key places), such ageographical expedients have not been readily available to those writing about long-term global processes. The imperative of discussing distant parts of the world within the same temporal framework has forced scholars in this field to pay explicit attention to issues of geographical scale and to the relationships between places over time. But just because the subfield of world history is overtly geographical does not ensure that its geography is adequate for the tasks assigned it. Attempts to write about global processes have pushed historians into a long search for appropriate regional categories.

Our goal in this chapter is accordingly to identify and critique some of the major conceptual discoveries they have made on along the way. We begin by analyzing the geographical vision of Arnold Toynbee, one of the first historians to attempt an analytic synthesis of the whole sweep of human history. Toynbee took civilizations as his operative categories, describing these geohistorical formations as quasi-isolated and essentially comparable units of analysis. As we will see, Toynbee's general reading of world geography made a significant impression. Although its credibility with professional historians has long since dimmed, the assumption that civilizations are isolable units of human association continues to inform most historical atlases and textbooks. The same notion is also becoming respectable once again in the academy, as social scientists attempt to understand the resurgence of religious rhetoric and cultural identity claims.

Meanwhile, however, a powerful alternative to Toynbee's approach has taken hold in the field: one that emphasizes integration rather than isolation. In the view of most practicing world historians today, the appropriate units of analysis are less distinct civilizations than interactive world systems, typically seen as functionally differentiated into wealthy metropoles or cores and their dependent hinterlands or peripheries. The systems map of the world has taken many forms, ranging from the "soft" variant of historians like Fernand Braudel to the more formal theoretical models of sociologist Immanuel Wallerstein. What all have in common, however, is a grounding in studies of capital and commodity exchange and an emphasis on cross-cultural connections.

Our brief reprise of the spatial conceptions that emerge from both the civilization-based and the world-systems schools of world history suggests that both have essential contributions to make to our metageographical lexicon. Neither literature alone is adequate to represent the complexities of global geography; both macrocultural regions and cross-cutting economic ties must be held simultaneously in view. In the words of William McNeill, "somehow an appreciation of the autonomy of separate civilizations . . . across the past two thousand years needs to be combined with the portrait of an emerging world system, connecting greater and greater numbers of persons across civilized boundaries."[3]

McNeill's mandate is a tall order, far exceeding our own limited geohistorical vision. Moreover, the scope of the present work does not permit us to review the enormously complex processes that have shaped economic and political space over two millennia.[4] What we do attempt, by way of concluding the present chapter, is to review what we see as the major principles of a historically informed world cultural geography. Drawing selectively on a few particularly stimulating empirical studies, we try to distill the rudiments of a method for thinking systematically about historically constituted sociospatial identity.

Civilizations

THE WORLD ACCORDING TO TOYNBEE

Until the early twentieth century, most of what passed for world history was either the story of European civilization writ large or popularizations with questionable claims to scholarly integrity. Although distinct civilizations, each associated with distinct parts of the world, had long been distinguished, these were spatially conceptualized in only the vaguest terms. Arnold Toynbee, however, raised the ambitions of the genre to a higher level. His lifework, *A Study of History,* was ecumenical, erudite, and massive, totaling twelve volumes published over twenty-seven years (1934 to 1961). *A Study of History* sought to identify universal principles at work throughout the globe's diverse societies over time. In historical-geographical terms, it had three organizing premises: (1) that a sizable but finite number of discrete, quasi-autarkic civilizations could be identified in the human past; (2) that these constituted the appropriate subjects of world-historical analysis; and (3) that all civilizations known to the historical record had experienced essentially parallel trajectories of birth, growth, decline, and fossilization. In many cases, Toynbee located the driving force behind the creation of a new civilization in the rise of a new universalistic religion (or a new branch of an existing religion).

An important question is why Toynbee chose to delimit discrete civilizations in the first place. Understanding this fundamental decision requires setting his project within the historiographical context of its day.

Above all else, what animated Toynbee was the desire to overthrow the immensely popular notion that Western civilization was the only world civilization that mattered and to dismantle the fundamental metageographical distinctions between East and West, Asia and Europe. The usual rhetorical justification for this common view was the purported principle of the "unity of History": a Hegelian idea that all human civilization constituted a singular evolutionary line, culminating in the rise of Europe.[5] Toynbee's words on this subject were unequivocal. In his view, "the survival of the misconception of 'the unity of History' [was] to be explained by the persistence of three underlying misconceptions: the egocentric illusion, the catchword of the 'unchanging East,' and the misconception of growth as a movement in a straight line."[6] While Toynbee's diagnosis was acute, his alternative created problems of its own, downplaying all ties between civilizations in order to assert the intrinsic importance of each.[7]

As much as the multiple-civilizations framework represented an advance over the dominant alternatives of its day, Toynbee's geographical vision remained flawed in three important ways. The first weakness of his scheme was its rigid demarcation between civilized and uncivilized societies, leading him to omit the latter category from his purportedly global vision. For Toynbee, a civilization was generally rooted in two features: religious commonality and the possession of textual traditions associated with one or more literary languages and scripts. The emphasis on literary texts has rightly come under fire for leading to a preoccupation with elite culture, as well as for its underlying (but unexamined) philosophical presuppositions.[8] But textualism also conspired with the emphasis on religious traditions to produce a kind of global geographical elitism. Regions lacking literary integration and without common religious institutions were in most cases simply excluded from view.

The most conspicuous exclusion from Toynbee's "world" map is sub-Saharan Africa. In Toynbee's atlas, virtually all lands south of the Sahara (excluding Ethiopia and the eastern littoral) were marked, circa A.D. 1952, simply as "primitive (still uncommitted)," a strategy that effectively denied the area any positive identity (see map 5).[9] Moreover, where he did recognize civilized peoples in Africa, as in the Sahel belt, he did not deem them to have created a civilization in their own right, appending them rather to a unified zone of Islam. This maneuver was hardly unusual—in the 1950s Western historians almost universally ignored Africa altogether[10]—but it represented an anomaly within Toynbee's system. By the criteria that he used elsewhere in his work (as when he identified no less

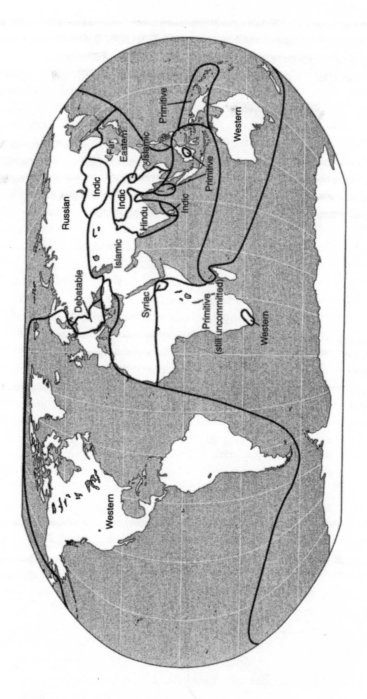

than eight distinct historical Christian civilizations), the Islamic peoples of the Sahel were sufficiently differentiated from those of North Africa to have merited a separate designation.[11]

The second weakness of Toynbee's geographical scheme stemmed from his insistence on grouping peoples strictly according to religious genealogies. This principle sometimes led him to shoehorn widely disparate social groups into a single ill-fitting category. For instance, faced with four spatially dispersed religious minorities—the extant Christian communities of the Levant, the Parsees of western India, the Christians of Ethiopia, and religious Jews all over the world—Toynbee grouped them all together as fossilized remnants of an otherwise extinct "Syriac civilization." Likewise, Tibetans, Mongolians, Burmese, Thais, and Cambodians—united in their adherence to Buddhism, but deeply divided by language, lifeways, and sect—constituted in his view the fragments of a single civilization, "Indic society" (which had earlier vanished in India itself). Categories like these proved singularly unhelpful to those interested in broader social analysis and were never widely adopted.

But the most serious weakness of Toynbee's geographical scheme (albeit one shared by most regional thinkers of his day) was more fundamental: the extent to which he stressed isolation at the expense of integration. The boundaries between civilizations were grossly overdrawn. To his credit, Toynbee took care to describe civilizations as "intelligible fields of study" rather than as objective entities that simply awaited discovery. In addition, he continually transgressed in practice his own

Map 5. *Toynbee's Civilizations of A.D. 1952.* In this modified map (Toynbee 1934–61, volume II [1959], page 93), seven civilizations are pictured as extant (in areas large enough to be depicted on a map of this scale): Western, Russian, Islamic, Hindu, Indic, Far Eastern, and Syriac (in central Ethiopia). Eastern Europe is portrayed as "debatable"—presumably between the Western and the Russian civilizations. Note that Yunnan in China is mapped as Islamic, even though Muslims are a minority in the region. In Toynbee's view, India is no longer "Indic," having given up Buddhism for Hinduism. His modern Indic civilization encompasses only those areas dominated by Theravada Buddhism (mainland Southeast Asia and Sri Lanka) and Lamaist Mahayana Buddhism (greater Tibet and greater Mongolia). Note that Turkey is counted as part of the West, as is the Philippines. Most of sub-Saharan Africa, as well as Melanesia, central Borneo, and the Andaman and Nicobar Islands, is depicted as still existing outside of the framework of civilization (although the label suggests that these "primitive" areas are bound to "commit" eventually to one of the existing civilizations—presumably the Western). For Toynbee's view of the West as dynamically expanding into all areas of non-Western civilization, see map 3, zone 7.

methodological pronouncements about the importance of viewing each civilization as self-contained,[12] arguing (among other things) that universalistic religions tended to emerge precisely at the junctions of differing social spheres.[13] Yet scant effort was made to acknowledge the profound interconnections among the nineteen to thirty-four discrete civilized societies into which he parceled the world.[14]

CIVILIZATIONS IN WORLD HISTORY TEXTS

The influence of Arnold Toynbee on the spatial imagination of world history is difficult to assess. On the one hand, few historians (or scholars in any field) read *A Study in History* anymore, and many basic elements of its geohistorical vision are distinctly out of favor. Few today would overtly defend Toynbee's organic analogy of civilizations' growth and decline, and no successor has attempted to outline a system of global geography as intricate and comprehensive as his. Indeed, as we will see below, the mainstream of professional world history has gone in very different directions. Yet many widely read works in the growing field that Toynbee called "universal history" have retained essential aspects of his metageographical scheme. While the list may have been shortened, a catalog of independent civilizations has displaced the "unity of History" schema in the standard English-language reference works on the global past.

Structuring a discussion of premodern history around a handful of large-scale civilizations has much to recommend it. The high level of congruence between historians' civilizations and geographers' world regions (at least within Eurasia) suggests that universalistic religious communities have in fact demarcated areas of intense and enduring interaction (as discussed in chapter 6). The problem with most global histories is not that they adopt a civilization-centered framework. Rather, it is that they perpetuate Toynbee's tendency to overstress the internal coherence of such regions (following area specialists in downplaying interregional connections),[15] while abandoning his ecumenical reach (settling instead for a focus on a few "primary" areas). What has been lost is Toynbee's insistence on giving equal attention to *all* civilizations. As a result, the standard cartography of premodern world history has been radically contracted, simplified, and prioritized—with some disturbing ideological implications.

The best place to see this schematization at work is in general-interest books on world history. The typical single-volume work begins on the eve of European expansion (circa 1500), highlighting four cultural zones across Eurasia: the realms of Christian, Islamic, Hindu, and Confucian civilization. The rest of the world is relegated to a limbo of interstices, hinterlands, repositories of barbarism, or (in the case of the Americas) doomed areas of less-advanced civilization. A stripped-down example of this standard metageography can be seen in L. S. Stavrianos's *The World Since 1500*.[16] Stavrianos begins his work with a tour of the major world civilizations immediately prior to European expansion. Here, three principal Eurasian cultural zones—Europe, the Muslim world, and the Confucian world—are contrasted as a group with the more isolated cultural spheres of sub-Saharan Africa, the Americas, and Australasia. The exclusions created by Stavrianos's definition of the Muslim world are revealing. Effectively, it comprises three great empires: the Ottoman (in Turkey and the Balkans), the Safavid (in Persia), and the Mogul (in India). This formulation subsumes predominantly Hindu India into the culture of its ruling elite; it also places Islamic societies of the Sahel outside the Muslim world, and leaves the Islamic regions of Central and Southeast Asia—along with their Buddhist neighbors—off the map altogether. Nor is the reduction of the community of Islam during this period to the three "gunpowder empires" unusual. Such important sixteenth-century Islamic polities as the Uzbek Khanate are rarely discusssed in world history textbooks, even in the minimal sense of acknowledging their existence. The resulting image of the world in 1500 is one where a handful of powerful states are lit up on an otherwise dark canvas.

A more sophisticated regionalization scheme can be seen in a newer work, J. M. Roberts's *History of the World*.[17] Roberts's scope is impressive. The most comprehensive section of his work (Book IV, "The Age of Diverging Traditions") includes chapters on Islam and the Arab Empires, Byzantium and Russia, the Turks and the Mongols, Western Europe, India, China, Japan, the Americas, and sub-Saharan Africa. The latter may be accorded only five pages, but at least it is recognized as a civilization in its own right. Yet even Roberts omits any discussion of Southeast Asia prior to European colonialism. The conventional cellular view of Eurasian historical geography—one that discerns only Europe, the Islamic world, India, and China, perceiving all other areas as merely the fuzzy edges of these great "civilizations"—evidently still does not have room for a complex and hybrid region like Southeast Asia.[18]

HISTORICAL ATLASES AND SINOCENTRISM

If a simplified framework of civilizations remains the mainstay of world history texts, a similarly flattened geohistorical conception remains the frame of reference for most historical atlases of the world. The limitations of such a vision can be seen particularly well in the way these atlases depict eastern Asia. Paralleling the Eurocentricity of their overall conception, most historical atlases exhibit a secondary, but no less pronounced, Sinocentrism in their vision of East Asia. This perpetuates three distortions. First, a multinational world region—comparable in many respects to Europe—is represented as though it were composed of a single state comprising a unitary civilization. By focusing almost exclusively on the territorial extent of the Chinese Empire under its various dynasties, historical cartographers implicitly reduce East Asia to the scale of a single country, whereas Europe is allowed to play the role of a culture area of continental scope. In the process, they also unwittingly obscure the fact that present-day China is more an empire than a nation-state, holding in its tight embrace extensive territories in Central Asia that were conquered by the Manchus but never fully assimilated to Chinese ways of life.[19]

Second, the intimate connections between China and its closest cultural neighbors—Japan, Korea, and Vietnam—are obscured. Japan is most often pictured by itself on separate maps, a move that highlights internal Japanese developments without showing the archipelago's deep historical connections to the mainland. (The effect would be like that of excluding Britain from all maps of Europe.) Korea, by contrast, is simply ignored. In most English-language historical atlases, the peninsula is either excluded from view altogether, portrayed as a mere appendage of China, or reduced to a blank outline, as if it were bereft of history.[20] Although Korea is a country the size of Britain, with a language of its own and a long, intricate, and well-documented political, social, and cultural history, it is rarely considered significant enough to merit depiction in English-language historical atlases. (The major exception here comes in treatments of the Korean War, an era of obvious interest to American audiences.) As a result, such important Korean kingdoms as Silla, Paekche, Koguryo, and Po-Hai are effectively excluded from our historical-geographical consciousness.[21] When it appears at all, Korea is reduced in the American historical-cartographical imagination to little more than a parade ground for foreign invaders.[22]

The occlusion of Korea in Western atlases mirrors a strikingly similar effacement of Korean civilization in prewar Chinese and Japanese cartography. Here, the peninsula's historical distinctiveness was denied to justify the imperial pretensions of its neighbors. A prominent school of Japanese historical geography in the early twentieth century, for exam-

ple, deemed Korea a "virtual nonentity," a mere region—rather than a coherent nation or country—whose "natural" function was to serve as a
buffer between Japan and the rest of Asia.[23] Likewise, China's first modern global geography text, produced in the mid-1800s, flatly omitted Korea on the grounds of its similarity to China.[24] Korean nationalists, not
surprisingly, have fought back; Ch'oe Namson, for one, sees Korea not
as anyone's periphery but rather as the core of one of the world's three
great culture regions (the other two being China and "the Indo-European
realm").[25] These battles over geographical representation within East Asia
highlight the perniciousness of the prevalent Sinocentric bias in our own
historical atlases. No European country inhabited by more than sixty million persons and occupying 85,000 square miles would ever be ignored
in an American historical atlas; only by regarding China as an Asian equivalent of France can Korea be treated as an Asian Luxembourg.

A third and final manifestation of Sinocentricity can be seen in the treatment of Vietnam. Maps of early historical China often include the Red
River delta—a fitting practice, since the area that is now northern Vietnam was an integral part of the Chinese political system from the Han
dynasty until 939. In later periods, however, Vietnam is usually ignored
altogether, vanishing from maps of China and seldom appearing anywhere
else. This represents a serious oversight, since the Vietnamese were important players in the historical development of eastern Eurasia, especially
after they won their political freedom in the tenth century. While maintaining a variety of close contacts with China, the Vietnamese became aggressive colonizers of lightly populated territory in neighboring Southeast Asian states.

For this reason, it is usually in the framework of Southeast Asia that
post-tenth-century Vietnam does appear. But such sightings are rare, for
Southeast Asia receives the most cursory treatment of all major world regions in English-language historical atlases. Despite its centrality in the
history of world trade and cultural interchange, Southeast Asia is cartographically relegated to the position of an insignificant backwater, a seemingly passive recipient of Chinese, Indian, Muslim, and later European
influences.[26] As will be seen in chapter 6, Southeast Asia is not an easy
area to "place" on a clearly bounded world regional map. Historically, it
has been a Janus-faced region, with intimate connections to both South
and East Asia, further complicated in recent centuries by a mind-bend-
ing pattern of mercantile and political ties that crisscross the planet.[27] Such
complex relationships may not be easy to portray in world historical atlases, but responsible cartography requires at least making the effort. Com-

plexity does not excuse obliteration. Unfortunately, a metageographical framework based around a short list of "primary" civilizations makes such obliteration almost unavoidable.

GEOPOLITICAL REIMAGINATION
AND THE HUNTINGTON THESIS

As the foregoing discussion suggests, the conviction shared by Toynbee and others that discrete civilizations were the fundamental units of human organization has had a marked impact on world history texts and atlases. Outside of the field of history, however, this vision has been less influential. The regional structure of the world in 1500, on the eve of European expansion, is obviously not that of the late twentieth century. Especially in the Americas, the lines of cultural cleavage and the circuits of interconnection have been completely reconfigured; the pre-Columbian map is of little account in formulating policy for the contemporary United States or Brazil. Even within the "Old World" ecumene, major cultural assemblages have been substantially transformed and spatially rearranged. Indeed, the dominant social science view in the post–World War II decades held that cultures across the globe were converging toward a single Westernized modernity.

Recent events, however, have conspired to bring civilizations back into view, even in policy circles. With the breakup of the Soviet Union, the end of the Cold War, and the dissolution of the midcentury trinity of First, Second, and Third Worlds, deeply rooted divisions based on older cultural distinctions are returning to view. In the Eastern Hemisphere, the religious groupings and other large-scale social entities identified by world historians have proved remarkably persistent. Moreover, while there have been some significant spatial shifts, the core territories of each regional civilization have endured. History now appears more relevant than social scientists of the postwar period imagined.

In an effort to grasp the new lines of fracture in the volatile post-1989 world, the political theorist Samuel Huntington has proposed a remapping of potential high-level conflict zones based on the persistence of major civilizations. Huntington proposes a distinctive metageography of global warfare, which he sees as having been socially and spatially structured along different lines in four successive eras. Wars between princes in the Middle Ages gave way, in his view, to wars between modern nation-states, which in turn yielded at midcentury to conflicts between ideological blocks. In the 1990s, Huntington sees the beginning of the next era, one in which major battles—those with the potential for significant escalation—will be fought less between states or ideological alliances than between discrete civilizations.[28]

Huntington's thesis deserves serious consideration. The enduring reality of macrocultural divisions has already proven militarily significant, and may well become more so in the near future. As the recent fighting in Bosnia attests, political struggles across major civilizational boundaries have assumed a special intensity in this decade. Equally cogent is Huntington's insight that eastern and western Europe are more significantly divided by the old split between Orthodox and Latin Christianity than by the Iron Curtain, which already lies in a rusted heap. Yet Huntington errs in the same way that Toynbee did, by ignoring both the numerous crosscutting groupings and the deep internal subdivisions that fractionate each of the world's major civilizations. The European Union notwithstanding, there is no reason to believe that western Europe has irrevocably abandoned its legacy of internecine conflict. Similarly, any notion that China, the two Koreas, Vietnam, Japan, and Singapore will find grounds for a military entente in their common Confucian heritage is dubious at best. Most doubtful of all, however, is Huntington's vision of an impending Islamic-Confucian alliance threatening the West (a scenario based primarily on evidence of recent arms sales between Pakistan and China). If the development of a united Confucian block is unlikely, the emergence of a workable alliance between Islamic and Confucian civilizations is even more so. Centuries-old religious ties may indeed have a place on our contemporary world maps,[29] but it would be simplistic to ignore the competing claims of economic class, local community, transnational commercial exchange, and nationalism.

Systems

THE DEVELOPMENT
OF A WORLD-SYSTEMS PERSPECTIVE

If the geographical categories employed in world history are experiencing something of a comeback in social science, a reverse movement over the past three decades has brought a spatial vision from social science firmly into the discipline of world history. The 1970s and 1980s in particular witnessed a veritable paradigm shift in the field, as leading scholars rejected Toynbee's conception of separate-but-parallel development in favor of an emphasis on cross-cultural integration. By now—in the scholarly literature of world history, if not yet in all of its pedagogical texts—a functional terminology of core, semiperiphery, and periphery predominates over that of discrete civilizations as the spatial vocabulary of preference.

The literature from which this vocabulary has emerged might broadly be called world-systems studies. One of its lineages can be traced back to the University of Chicago, where Marshall Hodgson and William McNeill labored in the 1950s and 1960s to craft a new approach to macrohistory. McNeill in particular was inspired by Toynbee's work, though he sought to overturn its key suppositions. His *Rise of the West,* first published in 1963, has been hailed by many for vanquishing the then-prevalent model of world history as a procession of discrete civilizations. Instead, McNeill stressed the manifold connections among all of the civilized reaches of the Old World ecumene (the zone stretching from North Africa to Japan)—although he now says that he did not push this vision far enough, paying "inadequate attention to the emergence of the ecumenical world system."[30] McNeill also diverged from Toynbee in taking the emergence of Western domination within the developing global system as his central theme, as his title makes clear. On this he encountered opposition from his Islamicist colleague Marshall Hodgson, who argued instead for a less Eurocentric approach. Hodgson's own geohistorical scheme was ahead of its time, and his early death contributed to the general neglect of his ideas (except in Islamic Studies) over the next several decades.

A second and more direct antecedent of the systems geohistory approach may be found in the work of the *Annaliste* school of historical research, heir to the classical tradition of French geography. Fernand Braudel in particular developed a method of research that focused on systemic interactions that transgressed both state and civilizational boundaries. But Braudel's primary interest was Europe, and his broader geographical ideas were not carefully developed.[31] It was only at the hands of the sociologist Immanuel Wallerstein that Braudelian geohistory was transformed into the resolutely global world-systems theory.

World-systems studies today constitutes a highly diverse field, transcending disciplinary boundaries and embracing a wide variety of approaches. Systems theorists who follow Wallerstein (especially the early Wallerstein) argue that the world system can be formally modeled. They insist on its modernity (according to Wallerstein, there was no premodern "world system"), its European origins, its hierarchical structure, and its basis in economic circulation; they also ascribe specific social attributes to its core, semiperipheral, and peripheral zones. Debates over the empirical validity of these claims are rife,[32] and in certain instances Wallerstein himself clearly misconstrued premodern geohistorical formations.[33] But despite their important differences and occasional errors, all parties to the debate are united in the conviction that the wellsprings of historical change are to be found in global interactions—and in the corollary belief that isolated "civilizations" simply do not constitute appropriate subjects for world-historical analysis.

One of the most profound advantages of systems theory over earlier world history approaches was its truly global scope. Here even the so- called noncivilized areas of the world were seen as having played a vital role, especially in supplying labor and materials to the economic core. Still, the central focus of Wallerstein remained strikingly similar to that of the early McNeill. Both were concerned to explain the "rise of the West" (with the former stressing exploitative power, and the latter more benign systemic interactions). Moreover, by the 1980s many historians were growing wary of the mechanistic metaphors often employed by systems theorists, while others were impatient with the limited focus on the modern world. As a result, writers such as Andre Gunder Frank began to go back to the work of the Chicago historians, McNeill and Hodgson. In the early writings of McNeill, they found impressive historical depth and a less confining theoretical framework. In Hodgson, moreover, they discovered a scholar of profound vision and creativity who had largely transcended a Eurocentric approach. The disparate intellectual heritages of McNeill, Hodgson, Braudel, and Wallerstein have since been recombined to create a new approach that might best be styled world-systems history (as opposed to world-systems theory). Historians increasingly employ the term *world system* in a deliberately open-ended way, emphasizing relations across macrocultural boundaries without necessarily agreeing on the specific contours or dynamics of those relationships over time.[34]

PROBLEMS WITH THE SYSTEMS APPROACH

All of these intellectual breakthroughs have moved the project of metageographical conception forward in crucial ways. Taken together, work in this genre has decisively extended the purview of premodern history to the world as a whole, or at least to the "Old World ecumene" as a whole. To be sure, Europe remains the center of gravity in the original world-systems formulation of Wallerstein, making it guilty of "Orientalism all over again," according to David Washbrook.[35] But such a charge cannot be leveled against the expanded view of Frank, Janet Abu-Lughod, and others. Their work shows clearly that Western hegemony was an aberration in global history, just as it shows the way toward a less hierarchical vision of cultural interaction in the premodern past.

Nonetheless, world-systems history remains limited in several ways. A first weakness of the genre is a tendency to overemphasize structural necessity in explaining historical change.[36] Both the power and the perils of such a perspective are evident in Janet Abu-Lughod's recent work *Before European Hegemony*. Abu-Lughod presents a detailed portrayal of the Afro-Eurasian world system of the period between A.D. 1250 and 1350, when Europe lay at the periphery of a multicentric, trans-"continental" system of exchange. In its portrayal of this system's linkages and structure, *Before European Hegemony* is a brilliant success; in explaining the subsequent ascendancy of the Europeans, it is less convincing. Abu-Lughod argues that the fall of the Mongolian Empire, and particularly the Mongol loss of China in 1368, seriously disrupted the world system by severing overland trade routes, while "the withdrawal of the Chinese fleet [from Southeast Asia] after 1435, coupled with the overextension into the two eastern-most circuits of the Indian Ocean trade of the Arab and Gujarati Indian merchants, neither protected by a strong navy, left a vacuum of power in the Indian Ocean."[37] The resulting collapse of erstwhile linkages in the East, she argues, allowed the Portuguese, Dutch, and English simply to sail in and violently take possession of a world system that they had done virtually nothing to produce. It was this process, she argues—coupled with the Europeans' windfall from the Americas—that catapulted the West into the position of global hegemony.

This theory represents a piece of tight and elegant systemic reasoning, but the historical record does not always support such structural logic. It is not clear that the premodern world system was so coherently constructed as to be completely destabilized by political events in East and Central Asia. When Chinese ships withdrew from Southeast Asia, Southeast Asian junks simply took their place.[38] Similarly, the decline of Mongolian unity in the heartland did not permanently disrupt trade; land-based caravans continued to integrate the Eurasian economies even in the absence of a Central Asian empire.[39] In fact, whatever "crises" may have hit Asian trade networks prior to 1450, European hegemony did not begin to be established until 1600, and it was by no means sewn up until the late eighteenth century—leaving plenty of time to heal any breaches in the indigenous fabric of trade.[40] The origins of European hegemony must thus be sought elsewhere than in the systemic breakdown of the fourteenth and fifteenth centuries.[41]

While its structuralist bias may seldom impinge on its metageographies, the same cannot be said for the second weakness of world-systems studies: its uneven spatial coverage. Although Wallerstein's *modern* world system is pictured as encompassing, at least in its later stages, the entire inhabited planet, the emphasis on macroregional exchange means that the vision of much world-systems work in the premodern era is not truly global in scope. While Europe is sometimes displaced from center stage, one finds much less attention to the vast lands that lay outside the "Afro-Eurasian ecumene," notably central and northern Siberia, sub-Saharan Africa, the Americas, Australia, and the Pacific islands.[42] The resulting unevenness of coverage is not without its defenders. William McNeill, for instance, while regretting that his earlier work ignored sub-Saharan Africa, continues to argue that the area "never became the seat of a major civilization, and the [sub]continent therefore remained peripheral to the rest of the world."[43] In a similar vein, Andre Gunder Frank contends that "for the long period before 1492, this 'whole' world history should concentrate on the unity and historical interrelations within the Asio-Afro-European 'old' 'eastern' hemispheric *ecumene*."[44]

For their own historical purposes, McNeill's and Frank's emphasis on the Old World ecumene is indeed supportable. Scholars working on premodern Central Asia, for instance, have little reason to study developments in the remote Americas. But for those whose concern is the making of modern world geography, it is essential to appreciate the accomplishments and contributions of native American civilizations, and to understand why the peoples of the Americas were so vulnerable to decimation by European conquest and the ecumene's virulent pathogens—as McNeill himself has made so clear in other works.[45] Careful historical consideration of sub-Saharan Africa is equally vital. Not only do African traditions retain great vitality in the present, but sub-Saharan Africa's isolation in the past has been overemphasized. Modern scholarship suggests much borrowing between this area and the eastern Mediterranean, especially in the formative period of early civilization.[46] Ultimately, the broader agenda of historical geography requires attending to all manifestations of the human experience, through all reaches of the world. The fact that an area was formerly relatively disconnected from the Eurasian belt of civilizations is no reason to slight it.

Finally, there is a third weakness of much world-systems analysis that geographers must reckon with: namely, its economic bias. The formal models of world-systems theory in particular have a tendency to glide over cultural relations in preference for analyzing material issues, particularly those of circulation or exchange. As a result, Wallerstein and his followers have not unreasonably been accused of flattening world history into a series of economic transactions, denying autonomy (and historical significance) to cultural processes.[47]

This economistic bias has two cartographical consequences. One is a vagueness, if not invisibility, that afflicts the representation of cultural entities on many world-systems historians' maps. Janet Abu-Lughod, for instance, divides the realm of Hindu civilization into a series of circuits of exchange, and sometimes seems close to erasing any sign of commonality among those circuits.[48] Andre Gunder Frank takes this a step further, seeming at times to deny that localized entities like "civilizations" have anything but an evanescent and relatively inconsequential existence.[49] The bulk of his work ignores almost completely the different realms of meaning that are the primary differentiating features of those difficult-to-grasp entities.[50] While Frank has written that "we should regard the system like a three-legged stool, supported equally by its ecological/economic, political/military, and cultural/religious/ideological legs,"[51] it is rare for such balance to be struck in practice.

The second potential geographical danger of the economistic bias in world-systems analysis is a temptation to map cultural centrality directly onto economic centrality. While core-periphery notions can be usefully employed to elucidate some patterns of cultural geography,[52] cultural cores can by no means be simply read off from their economic counterparts. Consider the role of western Europe and eastern North America in the late nineteenth and early twentieth centuries. In terms of capital accumulation and raw political power, the London–New York axis constituted the unquestioned center of the world system at the time. Yet Anglo-American cultural hegemony was (and remains) far from complete. Powerful cultural forces certainly emanated from the English-speaking world, and have continued to do so, beaming their way eventually across all reaches of the planet. But the colonized portions of the world, except those that were depopulated and densely "settled" by Europeans, never surrendered their cultural autonomy. India and China simply cannot be said ever to have constituted a cultural periphery in orbit around a European core.[53] Perceiving them as such only occludes the complicated cultural transformations and reactions that colonialism set in motion.

Neither of these distortions is a necessary concomitant of world-systems analysis. Well-trained regionalists like the late Marshall Hodgson have demonstrated that it is possible to combine a broad vision of global connections with a nuanced appreciation of cultural processes. More recently, Jerry Bentley has shown how an ecumenical approach can be profitably trained on the history of cultural systems. In particular, Bentley demonstrates how the diverse ideological structures found throughout the ecumene influenced each other, often to a remarkable degree, in the premodern era. Instead of a handful of discrete and virtually isolated civilizations, Bentley presents us with a world of cultural exchange and continual boundary crossings. Such melding of ideas and cultural systems was by no means divorced from economic considerations, as merchants were often important conduits for cultural contact. Yet the cultural patterns that Bentley elucidates are not reducible to those of mercantile exchange.[54] His findings amply demonstrate the truth of Fernand Braudel's claim that "cultures . . . are ways of organizing space, just as economies are. While they may coincide with the latter . . . they may also be distinguished from them: the cultural map and the economic map cannot simply be superimposed without anomaly."[55]

Differentiation and Integration

The challenge from the historical literature, in short, is to synthesize the best of two contrasting traditions: one that recognizes the integrity and durability of cultural macroregions, and another that has developed a vocabulary for analyzing the interconnections between them.[56] What follows is a first cut at this kind of relational analysis. In our view, the best starting point for such a project is to sketch a rough map of the major religious communities established in premodern times. This might seem an old-fashioned way to proceed, since social identities are continually being reinvented and reformulated at both micro and macro scales. But we are convinced that the centuries-old cultural assemblages that have accreted around distinct religious and philosophical traditions—particularly those that are still viable in our day—are the most basic building blocks of global human geography. Where we depart from Toynbee is in seeing the boundaries between those blocks as mutable and porous. More specifically, we argue that the *degree* of boundary porosity between different "civilizations" has historically varied in consistent ways. We accordingly devote considerable attention to assessing the variable "social distance" between each neighboring pair of macrocultural cores in the premodern world. The result is a modified (if still highly simplified) base map of enduring cultural regions.

We then proceed to refine this scheme in two ways. First, we overlay a series of alternative identity maps that transcend or otherwise subvert the macrocultural boundaries just drawn. Some of these alternative groupings are larger than the conventional civilizational units; others are smaller. But all straddle the borders between cultural cores, superimposing several levels of complexity on our basic metageographical foundation. Finally, we introduce a handful of refinements to the notion of a cultural *boundary* itself. Our conceptual vocabulary here is small but powerful, focusing on four key concepts that have been articulated in recent historical works: the middle ground, the diaspora, the cultural archipelago, and the matrix. We certainly do not claim that these add up to an exhaustive list. What we do claim is that, taken together, these concepts provide critical tools for teaching, thinking, and writing about metageography with subtlety and sophistication.

It should be noted that our focus here is on deep historical patterns, i.e., those that emerged in the centuries before European imperialism. This means in effect a primary concentration on cultural relationships, although we note economic connections to the extent that they influence cultural patterns. While this creates a sample that is limited in both temporal and spatial extent, it is our hope that this brief empirical discussion will begin to illustrate the principles elucidated in the foregoing methodological critique.

CULTURES AND CONTACTS IN THE PREMODERN ERA

The first step in charting global geographical patterns in human history is to identify a *hierarchy of divisions* among the major cultural regions of the earth. Somehow, the geographer must schematically convey the fact that, while all cultures are distinctive, the social distance between them has historically varied enormously, depending on the intensity of their interactions. Rather than conceptualizing premodern civilizations as comparable social cells, separated by walls of a standard thickness, it is more accurate to follow Marshall Hodgson in seeing the cleavages between cultural groups as ranging from the superficial to the profound.

Because the earth's largest oceans posed the greatest challenges to regular communication in the premodern era, the highest-order division on the historical map is the long-recognized line separating the Eastern Hemisphere's supercontinent of Africa and Eurasia, on the one hand, from such disparate extramural regions as the Americas, Australia, New Zealand, and the Pacific islands, on the other.[57] The paucity of transoceanic contacts

before 1500 created dramatic ecological as well as cultural differences between Afro-Eurasia and the rest of the inhabited globe. Among other
things, it was the resulting vulnerability of the Americas (and other de
facto biological islands) to the supercontinent's virulent pathogens that
made European conquest there especially swift and ruthless.[58]

Within the supercontinent of Afro-Eurasia, a second-order division
would have to be inscribed through the Sahara Desert. North and east of
this hot, dry zone lay what world historians have called the ecumene, a
string of societies extending from the Straits of Gibraltar to the shores of
Hokkaido that were loosely knit together for millennia. South of the Sahara, by contrast, lay a culturally and biologically distinctive world, one
whose contact with "ecumenical civilizations" was partial. Yet even here
there was a broad zone of contact and intermelding between these two
"megaregions" in a zone extending all the way across the Sahel and down
the Indian Ocean littoral to modern Mozambique.[59] Since the societies
of this border area were themselves closely conjoined to other societies
located farther to the south and west, economic and to some extent even
cultural influences spanned the gap between sub-Saharan Africa and the
linked civilizations of the ecumene. Still, a distinctly African cultural-civ-
ilizational cast marked many lands south of the Sahara,[60] and it is reasonable to argue that internal interconnections were more profound here
than those extending across to other regions. In the far southwest of the
continent, however, the nonagricultural Khoisan peoples may be said to
have occupied a world apart, separated from their northern neighbors by
a cultural boundary almost as thick as that inscribed across the Sahara.

The Afro-Eurasian ecumene also stopped short in northern Eurasia.
The peoples of northern and central Siberia to a large extent formed their
own sociocultural universe, whose separation from the ecumene was ev-
idenced by the fact that they proved almost as vulnerable to the infectious
diseases of civilization as did the inhabitants of the Americas.[61] Nonetheless, Siberia was still linked to the southern zone by fur trade and technology transfer (especially in metallurgy), largely through Turkish and
Mongolian intermediaries prior to the coming of the Russians.[62]

Limiting our focus next to the ecumene itself, a third-order hierarchical division can be drawn separating East from West.[63] This divide does
not coincide with the conventional line between Orient and Occident;
Christendom and Islamdom are far too closely related for that.[64] Rather,
the deepest historical rift in Eurasia, as Hodgson rightly observed, runs
along the border of the "Confucian realm," distinguishing the Sinified
zone of East Asia from the rest of ecumenical civilization. If one Old

World civilization was more remote than the others, it was the "greater Chinese." The peoples who settled along the Yellow and Yangtze Rivers were separated by thousands of miles from the Nile-Tigris-Indus nexus that gave birth to the Eastern Hemisphere's other major civilizations. Behind their mountain barriers, they continued to develop along their own pathway for centuries, maintaining important but tenuous contacts with their neighbors to the west. The resulting divergence between the East and the "rest" is perhaps most clearly evident in the field of written communications. Whereas most literate societies in the supercontinent came to employ alphabetic systems, all of which can ultimately be traced back to a single center of innovation in the Levant, East Asia developed a wholly independent system of ideographic writing. It was this distinctive writing system that became the crucial vehicle for spreading Chinese notions of philosophy, cosmology, and statecraft to the neighboring peoples of Korea, Japan, and Vietnam.[65]

If the deepest division within the ecumene thus marks off East Asia from its non-Sinified neighbors, the next most important division lies a bit farther west, distinguishing what might be called Greater India from Europe, North Africa, and Southwest Asia (see map 6). In religious terms, this division follows the boundary between the interpenetrating realms of Hinduism and Theravada (as well as Tibetan) Buddhism, on the one hand, and "the people of the book" (Jews, Christians, and Muslims), on the other. While these two groups shared many cultural characteristics (especially in science and mathematics but even in certain mystical practices and religious concepts), basic social and ideological structures nonethe-

---

Map 6. *"Civilizational" Boundaries circa 1200 C.E.* The thickness of each line indicates the relative cultural distance between different "civilizations" or clusters of related cultures, circa 1200 C.E. (approximate boundaries only). Thus, within the "Afro-Eurasian ecumene" (bounded by the second thickest lines), East Asia is most distinctive, while a lesser divide separates the greater Indic world from the lands of the Abrahamic traditions to its West. (In 1200, admittedly, many northern Turkish groups were not yet converted to Islam, while the Uighurs were largely Manichaean; the Mongols, who would later opt for Buddhism, were either Nestorian Christians or shamanists at this time [see Toynbee 1934–61, volume 11 [1959], pages 146–47].) Only a thin line differentiates the realm of mainstream Christianity from that of Islam (the latter including substantial Nestorian and Monophysite Christian communities). Within Christendom, the thinnest line distinguishes Eastern (Orthodox) from Western (Catholic) variants. While this line is sometimes elevated to the position of utter demarcation between East and West, it actually divides only the west of the West into two subzones.

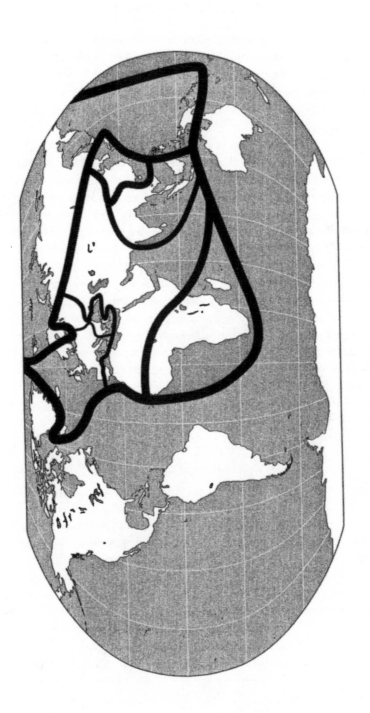

less set the Hindu-Buddhist milieu unambiguously apart from the neighboring civilizations to the west. Of course Islamic civilization began to penetrate this realm vigorously beginning in the twelfth century, but such an intervention may be regarded as an opening chapter in the story of increasing global integration—and resistance to such integration—of the *modern* world.[66]

This leaves us to conclude that the cleavage between Christendom and Islam is the shallowest of the major divisions across the ecumene in the premodern world. Professing sibling Abrahamic faiths and sharing a common Hellenic intellectual heritage, Christians and Muslims are historically united by broad social and philosophical commonalities, even if those commonalities have been resolutely denied on both sides of the Mediterranean. While Rome and Mecca represent two distinctive varieties of human culture, both belong to the same phylogenetic order and even the same family. The line dividing predominantly Christian Europe from its Islamic neighbors is properly accorded not the first-order but the fifth-order division in our metageographical taxonomy. And the opposition between Eastern and Western Christendom constitutes a still finer intrafamilial distinction. What is sometimes seized upon as the principal East-West divide on the planet is the most minor civilizational boundary of all.[67]

CROSSCUTTING REGIONAL STRUCTURES:
SUPERREGIONS AND SUBREGIONS

As important as the foregoing schema may be, it provides no more than a crude foundation for historical metageography. Taxonomy is an appropriate tool for classifying organisms, which exist as highly discrete species descended from common ancestors. As a means of classifying human communities, however, it is suggestive only. A closer look at the civilizational units classified above shows not only that they vary greatly in the depth of their differences from one another, but also that they overlap and interpenetrate in multiple directions. A plethora of crosscutting regional patterns greatly complicates the map of civilizations. Over time, such patterns have multiplied, making the geographical structures of the modern world ever more complex.

One category of crosscutting patterns might be called superregions: large, basically contiguous blocks of socially differentiated terrain that nonetheless share important if limited cultural features. Superregions are usually the legacy of conversion or conquest, processes that have allowed the members of a given group to spread their influence well beyond the boundaries of their original homeland. Some of the most easily recognizable superregions are those created by modern imperialism. Each major world empire—from the Portuguese and Spanish to the Russian, Japanese, and American—has left a certain imprint on the cultural geography of its colonized territories. And those imprints have proven remarkably durable. The division between Francophone and Anglophone Africa, for example, is now one of the most important fracture lines within the region. The Tunisian historian and philosopher of Europe and the Middle East, Hichem Djaòt, expresses this situation eloquently: "If we now turn to the parts of itself that a triumphant Europe exported, we are struck by their omnipresence. Does not every city outside of Europe have itself a European double? Every country has its own language, plus one or more European languages. Every nation, even at the heart of its historical continuity, bears an ancient past, plus a specifically colonial past."[68]

In the territories that formerly fell within the short-lived Japanese Empire, the colonial overlay is not nearly so deep. Yet not a few observers have identified in the bureaucratic states of Korea and Taiwan a legacy of the social organization established during the colonial years. In agricultural cooperatives, neighborhood self-surveillance groups, and the like, defenders and critics of Japanese imperialism alike see the enduring stamp of a "social infrastructure" imposed by the Japanese regime.[69]

If colonization has thus given rise to one class of superregions, missionary activity has created another. The realm of Buddhism represents one spatial structure that cuts across deep cultural boundaries. Bridging the profound cleavage between the Sinified peoples of East Asia and the very different civilizations of mainland Southeast Asia as well as eastern Central Asia is a common heritage of cosmology and religious expression. In fact, Arnold Toynbee and others have had solid reasons for choosing the boundary between Judeo-Hellenic and Hindu-Buddhist peoples, rather than that between the Sinified and the non-Sinified world, as the deepest cleavage in the ecumene. Yet the Buddhist superregion has only a tenuous hold over most of its vast territorial reach. Reabsorbed by Hinduism in India, displaced by Islam in most of insular Southeast Asia, and persecuted extensively centuries ago in China, Japan, and Korea, Buddhism has long since receded into the socioreligious background except in isolated pockets of its once-vast terrain (notably southern and central Sri Lanka, western mainland Southeast Asia, and eastern Central Asia). In consequence, the notion of belonging to the wider realm of Buddhism has far less significance to most persons in this part of the world than do alternative constructs for apprehending their place in the world. Only in the more abstruse philosophical, literary, architectural, and artistic dimensions is a Buddhist superregion identifiable.

A much more vibrant and coherent religious superregion can be seen in *Dar al-Islam,* the community of believers in Allah. Islam represents arguably the classic example of a cultural superregion. Its geographical structure shows a pronounced distinction between a durable core zone and a series of more recently converted peripheries, whose boundaries have fluctuated considerably over time. In the core—itself a vast area comprising most of southwestern Asia and northern Africa—Islam has predominated for many hundreds of years. Islamic institutions and practices have deeply penetrated the social formations of this area (notwithstanding the persistence of substantial non-Islamic minorities), giving substance to the notion that the "Middle East" forms a distinct world region or civilization. But even in the Islamic periphery, the implications of belonging to this ecumenical community of faith have been of profound and continuing importance. The realm of Islam may be said to constitute a distinct civilization of its own, even if, as Akbar Ahmed contends, there has never been a discrete Islamic society.[70]

In a very real sense, Islamic civilization formed the heart of the Afro-Eurasian ecumene in the period immediately prior to European expansion. Straddling important trade routes between the Mediterranean and the Indian Ocean, the Islamic core was resolutely cosmopolitan. It was also vigorously expansionary. Over the centuries, Muslim armies, traders, and mystical itinerants created a much larger community of faith extending far from the Arabian heartland. At its height around 1450, the borders of this larger Islamic world stretched from Granada in southern Spain, across the Sahara through the Sahel and into the savanna, through southeastern Europe, north to the central Volga and southwestern Siberia, east across what is now Xinjiang, deep into the heartland of Chinese civilization (especially in Gansu, Ningxia, and Yunnan), and across the whole Indian Ocean world as far as the southern Philippines. This was, without rival, the most geographically embracing of all social formations the world had ever seen.

Not surprisingly, pre-Islamic practices, institutions, and social codes remained conspicuous in its farther reaches. In many areas, Muslims even found themselves in a religious minority. Yet this wider realm of Islam constituted a coherent civilization in its way: one in which an educated traveler like Ibn Battuta could travel across dozens of cultural boundaries and always feel, in some profound sense, at home. While subsequent *re-*

conquistas at the margin have reduced its territory somewhat, *Dar al-Islam* remains vast and dynamic in our day, expanding in sub-Saharan Africa (and even, in a more discontinuous fashion, in the Americas) while gaining greater coherence and intensity through most of its extent. And it is this larger community of faith—not the region we know as the Middle East—with which most devout Muslims identify. Arabia may be the historical core, but *Dar al-Islam* is the realm.

As these examples show, macrocultural boundaries are continually transcended by more extensive political and religious entities. But crosscutting geographical formations do not always assume the colossal scale of superregions. More local communities of identity also straddle the divides between macrocultural units all over the world. Consider, for example, the composition of the Islamic core itself. This region, extending from Southwest Asia across North Africa, is divisible into at least two major subregions: an Arabic cultural zone to the west and a Persian zone to the east. But neither Arabic nor Persian culture stops at the borders of the old Islamic core. At various times, the sway of Persian civilization has extended well to the east and north of its Iranian homeland, deeply penetrating elite culture through much of South Asia, while Arabic culture has made similar, if more restricted, inroads in parts of southern India.[71] Both the Arabic and especially the Persian subregions are also intimately linked with the Turkish realm—itself an extensive geohistorical formation that at one time subsumed much of the Islamic core as well as vast reaches of inner Asia[72]—which may yet reemerge with some vigor in the near future.

Such cross-civilizational linkages can carry us further afield, in whatever direction we care to travel. Deep historical connections tie western Turkish peoples (in Anatolia, the Balkans, and the Volga) to the greater Greek and Russian cultural spheres.[73] Although a legacy of religious antipathy leads both sides to deny (often vehemently) any commonality, recent scholarship on Muslim and Christian communities in Bulgaria documents considerable cultural synthesis on the ground.[74] Likewise, the history of the Volga Tartars has been complexly intertwined with that of Orthodox Russians for hundreds of years. The same Tartars have played a key role in the intellectual life of the whole Turkish world, their educational system extending as far east as China.[75] In this way, western Turkish culture alone could be said to straddle three of the most important civilizational divides in the Old World ecumene. The eastern Turkish zone, in its turn, is historically intertwined with the Mongolian culture area, although contrasting patterns of religious conversion have led these two peoples into different "civilizational" camps;[76] by way of the Yakut, ancient patterns of Turkish culture extend through Siberia to the Arctic Ocean. In religious terms, the Mongols came to share close ties to the Tibetans, through ideological linkages which can ultimately be traced to South Asia. Yet they have also absorbed much Chinese culture, both directly and through intermediary linkages with the Tungusic peoples of Manchuria. The wider and older Tungus cultural area, for its part, crosses the boundary between northeastern China and Siberia, where centuries of Russian domination have established another set of cultural ties, oriented in a completely different direction. The resulting cultural overlays have made the cultural map of Central Asia and northeastern Asia unusually intricate, giving rise to historical polycultural states like Yehe, where a Mongol aristocracy presided over a mainly Tungusic people whose culture was essentially Chinese.[77]

The point of this meander across Eurasia is a simple one. No matter where one draws the lines between culture areas, macroregions, or civilizations of the world, alternative groupings, both large and small, can be identified that violate those boundaries. To gain a meaningful purchase on geographical structures at any scale, it is essential to identify these crosscutting cultural regions.

REEXAMINING THE BOUNDARIES:
MAPPING NEGOTIATION AND DISPERSION

So far, we have considered three kinds of geohistorical formations: civilizational regions or cores, the religious and imperial superregions that transcend them, and the smaller subregional groups that straddle their boundaries. What all of these structures have in common, at least in the abstract way they have been represented here, is a more-or-less compact, contiguous form. On closer inspection, however, almost all metageographical patterns confound this simple shape. On the one hand, the line at which two major culture regions meet often takes the form, not of a precise boundary, but of an extensive borderland with a dynamic hybrid culture of its own. On the other hand, cultures do not merely meet along their borders; they also interpenetrate each other's core spaces. Through exile, migration, trade, conquest, conversion, or other processes of cultural contact, outliers of one group routinely end up as minorities in the midst of another population's territory. In particularly cosmopolitan areas, or where successive waves of migration have deposited people from a wide range of backgrounds, it may be impossible to identify a majority group at all.

Given the ubiquity of these boundary-blurring cultural formations, the next step in constructing a sophisticated metageographical framework is to acknowledge this seeming chaos. Having painted over our boldly drawn outlines of civilizations with a series of crosscutting superregions and sub-regions, we must next refine our representation of boundaries themselves. But this calls for a finer set of brushes—a supplementary spatial lexicon capable of identifying a complex range of geographical forms. While any number of terms might prove useful here, we will focus on four key concepts: the middle ground, the diaspora, the cultural archipelago, and the matrix.

The notion of the middle ground was developed by historian Richard White to describe interactions in the Great Lakes region, during the seventeenth and eighteenth centuries, between Indian villages and French traders (or, later, their British and American successors). For most of this period, White shows, Europeans met native Americans in a political limbo where neither could dictate terms to the other. Accordingly, instead of *acculturation* (where a weaker people adopts the ways of a stronger), the rule in the Great Lakes region was *accommodation*, as both Frenchmen and Algonquians were forced "to justify their own actions in terms of what they perceived to be their partner's cultural premises."[78] In White's formulation, the "middle ground" was where this accommodation took place: a temporary, fragile, and essentially abstract cultural space. But it is also clear that this cultural space could be created only within a specific geographical place: the hinterlands of the European empire. The middle ground, he writes, was the place in between: "in between cultures, peoples, and in between empires and the nonstate world of villages. . . . It [was] the area between the historical foreground of European invasion and occupation and the background of Indian defeat and retreat."[79] The physical terrain of White's middle ground was the territory west of the Great Lakes, known to the French as the *pays d'en haut*. As early as 1650, this region lay within the ambit of imperialism, in the far periphery of an incipient world capitalist system. But as White describes it, "[T]his is an imperialism that weakens at its periphery. At the center are hands on the levers of power, but the cables have, in a sense, been badly frayed or even cut. It is a world system in which minor agents, allies, and even subjects at the periphery often guide the course of empires. This is an odd imperialism and a complicated world system."[80]

While White's conceptual vocabulary has yet to be deployed outside of American history, we believe his understanding of cultural formations in the borderland has broad salience for anthropogeographic analysis. Despite the enhanced resources of the twentieth-century state, and despite the much longer reach of modern capitalism, anthropologists looking at contemporary borderlands have repeatedly found processes that echo those of the seventeenth-century *pays d'en haut*.[81] Evidently, ours too is "an odd imperialism and a complicated world system." The concept of the middle ground can help in making sense of those interstitial places where cultures continue to negotiate on unevenly shared terrain.

If the middle ground is one indispensable conceptual tool for refining our cultural map, another is that of the diaspora. Far-flung dispersals of a given people from their original homeland are a recurrent theme in the historical record—a theme that has, if anything, been amplified by the military and economic upheavals of our own day.[82] The dramatic disruptions of Europe's Jewish communities, fleeing a succession of inquisitions, pogroms, and genocidal policies, gave rise to the archetypal historical diaspora. The outpouring of migrants from southern China, making the overseas Chinese a significant minority throughout Southeast Asia and the eastern Pacific, represents another classic diaspora.[83] A third example—and one of the hardest to trace—is that of the Roma (or Gypsies), a people without a clear homeland who have resided for centuries in small, often itinerant, and elusive communities in Europe, Southwest Asia, and North Africa. The Roma are often ignored in both historical and geographical accounts of Europe, in part because of prejudice against them, but also because of the complexity of their own social geography (compounded by their frequent recourse to concealing their identity).

More easily documented—if less widely known—is another group of dispersed societies: those that specialized in long-distance trade in medieval and early modern times. The historian Philip Curtin collectively identifies these communities by their function as "trade diaspora."[84] Before the nineteenth century, according to Curtin, a handful of cultural groups had particularly extensive outliers in "alien" territory, settlements that might be strung out across thousands of miles along major trade routes on sea and land. Maintaining a common identity over daunting distances, such transnational communities served as important conduits for cross-cultural trade. The Armenians, widely dispersed across central Eurasia, comprised one active trade diaspora; Gujarati communities, located in similarly far-flung outposts along the shores of the Indian Ocean, constituted another. These groups and others like them made their living by conveying goods across major—and sometimes militarized—cultural boundaries. Often they acted as cultural intermediaries, becoming adept at the ways of life encountered in their respective "host" cultures without abandoning their preexisting identities. That such peoples helped cre-

ate a framework for economic integration in premodern times is undeniable; that they were often instruments of cultural exchange has been
recently demonstrated as well.[85] Even the Christian community of
Ethiopia—long held up as a model of the isolated civilization—is now
known to have been closely linked with the Armenian diaspora, probably the most extensive of all.[86]

A useful metaphor for the sociospatial form engendered by diasporas
is the archipelago. In physical geography, an archipelago denotes an island
chain extending out into the sea. In cultural geography, it denotes outliers, or "exclaves," of a given culture group. Such outliers are commonplace on the map of world religions; prominent examples would include
the millions of Arabic-speaking Christians scattered through the Middle
East or Mandarin-speaking Muslims in the heart of China. But the boundary between archipelagic cultures and their surroundings is rarely so sharp
as that between land and water. Minority pockets are usually assimilated
in important ways into the surrounding society, despite retaining a distinctive repertoire of practices that sets them apart as a community. Failure to theorize this promiscuous mixing of cultural categories has weakened most regionalization schemes. Toynbee, for instance—as we have
seen—classified adherents of minority religions as though they were
wholly isolated from their social and cultural milieu, identifying them solely
with the place (and time) of their faith's origins. Only thus could he argue that European Jews were not truly European at all, but rather encap-
sulated representatives of an otherwise long-dead Syriac civilization.[87]

Toynbee to the contrary, the more common practice among dispersed
populations—even those that fiercely maintain separate identities from
their neighbors (and ties with their coreligionists back home)—is to share
in the broader social and cultural systems of the geographical milieus in
which they live. To capture this sociospatial syncretism or hybridity, the
metaphor of the archipelago must ultimately be supplemented by that of
the cultural matrix. In a matrix model, identity is a matter of one's position in a multidimensional lattice. Religious faith might be one dimension; language, a second; "lifeways" or material culture, a third. In theory,
any combination can result; in practice, the multiplicity of actual positions is truly remarkable, especially where individuals are able to select
different identities for themselves in different contexts. In Bulgaria, for
instance, most Christians speak Bulgarian and most Muslims speak Turkish (as their first language), but there are minorities of Bulgarian-speak-
ing Muslims (Pomaks) and Turkish-speaking Christians (Gagauz)—as
well as substantial numbers of Roma (Gypsies) of both faiths who

nonetheless maintain their own distinctive spiritual ideas and practices. Pomaks, moreover, identify themselves sometimes as Bulgarians (and indeed have often been forced to do so in the recent past) and sometimes as Turks, whereas Muslim Roma often seek a Turkish identity—much to the consternation of self-identified "true Turks."[88] One way to deal with these crosscutting patterns of sociospatial identity cartographically is by a combination of intermixable signs (i.e., colors and stripes), keyed to a legend that mirrors the form of the social matrix itself. Because of the often contingent and contested nature of local identity on the ground, however, even the most sophisticated matrix mapping fails to capture the full complexity of the situation.

In short, as we proceed to finer levels of resolution, cultural identity flatly eludes binary categories. It is pointless to ask whether Mandarin-speaking Muslims (the Hui)[89] belong to Chinese or to Islamic civilization, or whether sixteenth-century western European Jews belonged to Western or to Jewish civilization (the "Syriac" option can be safely dismissed as a figment of Toynbee's imagination). Such groups are by necessity of dual natures. Far from being fossilized "relics," they have become complexly embedded in their new environments, embodying at a personal level the spatial interpenetration of differing civilizational traditions. Along with the middle ground, the transnational diaspora, and the archipelago, the matrix is thus a critical concept for advanced metageography. While such a brief inventory of boundary-blurring formations remains incomplete, we believe that these concepts identify some of the recurring sociospatial patterns that complicate the global map.

POSTMODERN LANDSCAPES?

If nothing else, the foregoing examples demonstrate that global human geography is not reducible to any single one-dimensional schema. Generating a credible map of sociospatial identity at any scale is a daunting task, and contemporary developments barely touched on here greatly exacerbate the difficulty. "Globalization" is more than a catchword, but how does one get a handle on its geographical implications? Electronic media circulate massive banks of images and information to a worldwide audience, nurturing a rudimentary sort of global capitalist culture; accelerating flows of labor, capital, and commodities transform the landscape of the planet at an unprecedented pace, leaving extraordinary numbers of uprooted and dispossessed peoples in their wake.[90] On the one hand, cyberspace creates new "virtual communities" irrespective of physical distance (though not necessarily of language);[91] on the other hand, the vagaries of war, exile, and transnational migration continue to dispersed long-established cultural groups. It is sometimes maintained that the financial power of the overseas Indians now matches that of India itself,[92] and the same is said of the overseas Chinese. Islam, too, is a more globalized religion than ever, having become part of the multicultural framework of every large city in the West.[93]

Perhaps in response to this flux and fragmentation, nationalism and regionalism seem to be simultaneously on the ascendancy, and in certain parts of the world centuries-old processes of cultural-geographical hybridization are brutally erased through the mechanism of "ethnic cleansing."[94] Yet new identities continue to be created almost daily, while old ones are cast off or recombined. Trying to plot such a complex and changeable social world on a map seems frankly outlandish. But while this situation clearly puts "total cartography" beyond our reach, it makes it equally urgent not to abandon the search for sociospatial patterns altogether. True, the more closely one looks at the world, the more complicated, and the more contradictory, the cultural landscape appears. Yet meaningful patterns can still be identified, not only in the form of enduring historical legacies, but even in the seemingly chaotic mutations of the present day.

As teachers, we know from experience that some kind of elementary master system, clearly identifying the most striking, coherent, and enduring cultural aggregations, is an essential foundation for geohistorical knowledge. The kinds of crosscutting patterns discussed above are important for the advanced student of global geography, but would quickly prove deadly for the first-year undergraduate. It is for this reason that we have dwelt at length on the premodern map of civilizations or macrocultural cores. Yet the stripped-down view of civilizational realms, while truly foundational for world history, is not the best framework for contemporary global geography. It is too focused on the past, has too little to say about large areas of the globe, and tends to overstress elite cultural features. For those seeking to understand the modern world, the most comprehensive heuristic metageographical framework is that of world macroregions. Although rooted to some extent in the civilizational view of historians, the concept of world regions has its own intellectual genealogy and its own set of promises and potential pitfalls.

It is to these that we turn in our final chapter. Before taking on the world regional framework, however, we would underscore once more that all such schemes are only a rude approximation, created for heuristic convenience rather than as an end in themselves. The macrocultural map is but a stepping-stone toward a serviceable picture of the planet. To abandon the framework of civilizations and other distinct large-scale cultural entities altogether would be to deny an enduring reality. Yet to freeze our students' geographical understanding at this elementary level would be to withhold the sharper conceptual tools they will need to think their way through a complex world.

CHAPTER 6

World Regions

An Alternative Scheme

The burden of our argument to this point has been to show that received metageographical categories, from continents to civilizations, are inadequate frameworks for global human geography. This chapter explores the promise—and the problems—of what we believe is the most serviceable alternative: the framework of "world regions." Like civilizations, world regions are large sociospatial groupings delimited largely on the grounds of shared history and culture; unlike civilizations, they do not presuppose a literate "high" culture, with the result that a world regional scheme can be used to classify all portions of the globe. Although the number of regions, as well as the borders between them, vary somewhat from one map to another, most world regionalization systems arrive at essentially the same set of macrocultural zones: East Asia, Southeast Asia, South Asia, Southwest Asia and North Africa, Europe, Russia and environs, sub-Saharan Africa, Latin America, Australia and New Zealand, and the United States and Canada. Oceania and Central Asia are sometimes added to this list.

As these regional labels suggest, the world regions terminology relies heavily on that of continents, even as it attempts to displace them. But a careful look at the list reveals that its creators have made two distinct advances over the continental scheme. The first is to break up the supercontinent of Asia, elevating each of its major subdivisions (of which four to six can be identified, depending on the criteria used) to the same level as the other "continents." The second involves drawing new boundaries based on historical connections rather than landforms. Thus Africa is divided into a sub-Saharan zone and a northern, Mediterranean block (which is in turn linked with Southwest Asia); the zones of Spanish colonization in South and North America are joined together to form Latin America; and Europe's eastern edge is no longer placed along the Ural crest. In other words, transposing the continental scheme into a world regional one requires shifting boundaries as well as reshuffling hierarchical levels in the taxonomy of global geography.

What drives both of these maneuvers is a fundamental change in the underlying logic of spatial division. Where the continental scheme is based on a spurious identity between human groupings and the landmasses they inhabit, the world regional framework (at its best) attempts to delineate areas of shared ideas, related lifeways, and long-standing cultural ties. Employing a social logic of this kind for mapping out human geography makes sense in many general contexts, as demonstrated by the extent to which the world regional framework is currently used. It already forms the foundation for most college courses in world geography, serves as the organizing structure for many history programs, and is employed by the U.S. State Department to organize its global programs. Most of its units have also gained currency in the popular press.

Yet while world regions are clearly preferable to continents, their limits must be carefully specified. The new taxonomy will inevitably fall short of many readers' expectations, for the simple reason that no geographical framework can legitimately be deployed as an all-purpose global divisional scheme (as the continental system was routinely expected to be). In the case of world regions, there are specific empirical and ideological problems to be addressed—some of them traceable to the scheme's origins in military strategy. But as we shall see, other problems inherent in this framework are of a sort that would vex even the most carefully considered regionalization system.

Origins of the World Regional Framework

The world regional framework employed in American government and universities today has deep roots in European cartography. Its prototype can already be glimpsed in the informal areas delimited on Western maps hundreds of years ago to partition the globe into meaningful geographical units. But it was not until World War II that these incipient world areas were organized into a formal system of macroregional categories.

THE PREHISTORY OF WORLD REGIONS

The conception of global spatial order embodied in European cartography of the seventeenth and eighteenth centuries was emphatically hierarchical, with a clear Linnaean influence.[1] Yet a close look at the categories employed in atlases of the time reveals an incongruous mix of units at any given taxonomic level. Subcontinental entities might be defined by either political or nonpolitical criteria, with the result that kingdoms and empires were jumbled together with areas defined on the basis of language or other cultural features. Within Europe, for instance, cartographers usually distinguished Spain, France, Italy, and Hungary as units at the same taxonomic level, even though the first two were seats of dynastic empires and the latter two were not states at all at the time. The partitioning of Asia was similarly ad hoc. In a typical atlas of 1775, John Palairet identified an Asiatic Russia, Independent Tartary, Chinese Tartary, China, Turkey in Asia, Arabia, Persia, and India[2]—and Gilles and Didier Robert de Vaugondy's influential atlas of 1798 made a similar set of divisions (see map 7). Two of the units identified (Turkey and Persia) were essentially political; three others (China, Tartary, and India) denoted not states but cultural groupings. Arabia, on the other hand—drawn as it was to include the Arabian Peninsula while excluding most areas of Arabic speech—was probably conceived mostly as a physiographic unit. In other words, the primary principle of division appears to have been size. By any other criterion, the logic of the system was blithely inconsistent.

Nor were these subcontinental units spatially fixed. Regional designations such as India, Tartary, Arabia, and Persia recurred on European maps for centuries, but their geographical dimensions varied tremendously from author to author and from generation to generation. The career of the term *India* exemplifies this fluidity. As the ultimate Orient of classical European geography, "India" expanded with each new discovery throughout the Renaissance, until it encompassed, in some usages, the majority of the globe. Ortelius's 1570 map of India, for example, included all of modern-day South, East, and Southeast Asia.[3] In other cartographic representations, the Americas and even modern-day Ethiopia were placed within India's bounds.[4] Beginning in the 1700s, by contrast, the secular trend was one of gradual reduction, as India was limited first to South and Southeast Asia, and then, following the contours of British dominion, to South Asia alone.[5] Its contraction would continue in the twentieth century, when India came to refer to a single South Asian state.

The example of India represents another trend in metageographic thinking as well. Beginning in the nineteenth century, European cartog-

Map 7. *Gilles and Didier Robert de Vaugondy's Depiction of Asia, 1798.* Mainland Asia is divided here into nine units: Turkey-in-Asia, Arabia, Persia, India, China, Korea, Independent Tartary, Chinese Tartary, and Siberia. Political criteria are obviously used in the delineation of Turkey and in the separation of Independent from Chinese Tartary. Overall, however, the map is more concerned with cultural regions than with politically defined territories. (Courtesy of State Historical Society of Wisconsin.)

raphers relied increasingly on political criteria, gradually jettisoning the imprecise cultural areas of an earlier day. An atlas of 1826 by Anthony Finley, for instance, distinguished "Cabul" (Afghanistan) and Baluchistan from Persia and separated Tibet from Tartary, while replacing "China" and "Chinese Tartary" with the Chinese Empire.[6] During the mid-1800s, states and colonial territories became the primary geographical units below the continental level in all portions of the world. One result of this shift was a much finer mesh of subcontinental divisions, especially within Asia. Two decades after Finley, a new atlas by S. August Mitchell divided mainland Asia into no less than eighteen units, identifying such previously invisible territories as Herat, Khiva, and Kokand (see map 8).[7]

To organize the resulting plethora of political entities in the supercontinent of Asia, a new intermediate level of categories eventually became necessary. In the middle nineteenth century, several geographers solved this problem simply by carving Asia into locational blocks, inventing such labels as Northern Asia, Eastern Asia, Southern Asia, and Western Asia.[8] Eventually a similar "directional" division of Asia would triumph, giving us such modern regions as East, Central, South, Southeast, and Southwest Asia. But the new terminology took a long time to appear in printed atlases, and the older regional names disappeared only gradually. *India,* in particular, continued to be used in a regional, rather than narrowly political, sense well after the partition of the old British India in 1947.[9] More common in the early decades of the twentieth century was a division of Asia in accordance with European colonial rule, a procedure that was also applied at the time to Africa and Oceania.[10] In short, a set of incipient world regional categories developed in the Renaissance was largely displaced by political mapping in the era of European colonization, but was partially revived in a new set of higher-order subcontinental categories that fitfully came into use in the nineteenth century.

THE EMERGENCE OF FORMAL WORLD REGIONS

The process of organizing these inchoate categories into a formal system of world regions took place only in the mid–twentieth century. Its immediate impetus was the entry of the United States into World War II. American military personnel had never before attempted to coordinate a worldwide effort, and the ensuing search for international expertise, both for planning military strategy and for orchestrating the postwar settlement, led governmental and military planners to discover how few Americans were equipped with second-language skills and cultural training. To rectify the newly apparent knowledge gap, four organizations—the National Research Council, the American Council of Learned Societies, the Social Science Research Council, and the Smithsonian Institute—banded together in the early 1940s to create the Ethnogeographic Board. Their mission was to advise the government in matters of global geography and to investigate the current status of knowledge in American academia about diverse areas of the world.[11]

In its early days, the Ethnogeographic Board used the conventional continental framework, with the single modification of substituting Latin America for South America.[12] But the war forced its members (or more precisely, their military constituency) to devise better schemes. Simply put, the old continental architecture, anchored by the vast category of Asia, proved useless for strategic planning. At the same time, the territorial framework of European colonial empires—the other prominent template for global division—was quickly becoming irrelevant. Ultimately, the result of incremental reforms was a new system of world regions, the basis of postwar "area studies."

The new framework did not emerge overnight. The divisions employed by the Ethnogeographic Board reveal a system in transition, juxtaposing such incommensurable categories as Europe, Russia, the Near and Middle East, and the Far East (this last a massive zone that would later be broken up into South, Southeast, and East Asia).[13] A more sophisticated set of categories gradually emerged, reflecting in part a scholarly recognition of the diversity of civilizations in the eastern half of Eurasia, yet the explicitly geographical issues involved in mapping cultural macrozones were never given extended consideration. Perhaps the most important reason for this striking omission was the near-absence of professional geographers from the Ethnographic Board's deliberations.

Excluding geographers was not the intent of the board's organizers. One member had argued that "geography is basic to an area-training program . . . next in importance only to . . . language."[14] Another, Wendell Bennett, went so far as to suggest that "among the social sciences, geography is the most logical leader for the area approach."[15] But geographers at the time were primarily concerned with the physical and economic characteristics of subnational regions, often working close to home; few were evidently interested in the vast cultural "areas" of the world then being delineated,[16] and fewer still were willing to undertake the linguistic training necessary for foreign area specialization.[17] Two geographers (Robert Hall and Isaiah Bowman) did serve on the board, but it was dominated by anthropologists, whose disciplinary training emphasized high levels

Map 8. *S. August Mitchell's View of Asia,*
1849. By the mid–nineteenth century, most
European and American atlas makers divided Asia largely on the basis of politically
independent territories. Southwest Asia
and western Central Asia are thus partitioned here into numerous separate entities,
while the entire Manchu Empire (including
Korea but excluding such minor tributaries
as Nepal) is depicted as a single unit. The
unitary depiction of Arabia is vestigial,
perhaps indicating the lack of Western
knowledge of the area. Note that European
colonial possessions are shown on this map.
(Courtesy of State Historical Society of
Wisconsin.)

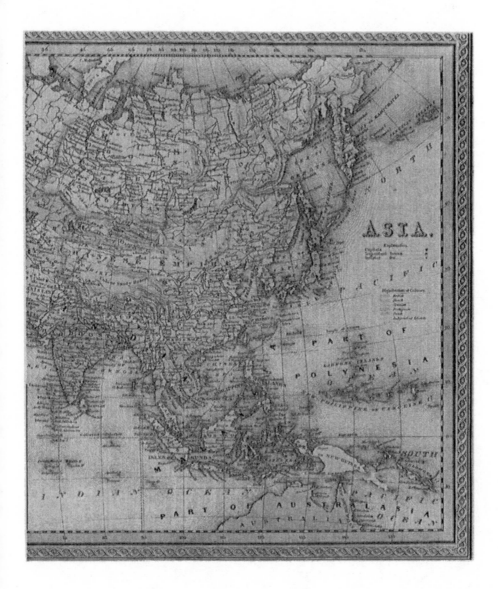

of language and cultural expertise. As a result, the world regional concept that shaped postwar area studies was essentially formulated by anthropologists, and the world regional map they posited was subject to almost no overt debate.[18]

INSTITUTIONALIZING A NEW METAGEOGRAPHY:
THE AREA STUDIES COMPLEX

Such a lacuna was in keeping with the instrumental interests that drove the whole project. For those funding the Ethnogeographic
Board, the new global framework was merely a means to an end; what
mattered was to put in place a workable structure around which to organize policy and through which to enhance international education. Governmental interest in area studies intensified in the early years of the Cold
War, with the Social Science Research Council playing a critical role. After surveying university resources pertaining to the various regions in 1947,
Robert Hall concluded that the United States was woefully ill-prepared
for operating in an increasingly integrated world system. A handful of universities offered programs that focused on Latin America or East Asia
(China and Japan), but Hall found that "very little organized or group
interest is as yet shown in any part of Europe (other than Russia) and
there is even less on such major world areas as the Near East, Africa, the
Indian World, or Southeast Asia."[19]

At the urging of the Social Science Research Council and with the as-
sistance of the Ford Foundation, the federal government soon began to
fund multidisciplinary "area studies centers" at key universities. The postwar growth and diffusion of the American area studies complex was impressively rapid. As of 1947, fourteen organized centers had been established; six were devoted to Latin America, three to Russia and Eastern
Europe, four to Japan and China, and one to the Indic realm. Five years
later, there were no less than twenty-five area and language studies programs in American universities.[20] Such programs continued to multiply
through the 1960s. Key to their growth was the passage, in 1958, of the
National Defense Education Act, whose Title VI was transferred to the
annual Higher Education Acts of subsequent years.[21] While Title VI is
labeled "language development" in the original act, its main provision established "language and area centers" that would teach "the history, economics, geography, and so on of the region."[22] By the early 1990s, American universities boasted no less than 124 national resource centers.[23]

All of this impressive growth occurred within the framework of world regions or areas. As defined under the original Title VI act, the organizational and terminological scheme of world division featured Russia, the Far East, Southeast Asia, South Asia, the Near East, Europe, Africa, and Latin America as its basic units.[24] (When India became the name of a single country, the larger region that had long been referred to by that name was simply rechristened South Asia.)[25] The only substantive changes over the next two decades involved replacing "Russia" with "the Soviet Union and Eastern Europe" and relabeling the "Near East" the "Middle East." This reorganization of the global map into world areas was readily adopted by geographers, anthropologists, and other scholars. In the 1950s, a few world geography textbooks were still organized around the architecture of continents,[26] but by the end of the decade the transition to a world area framework (usually called world regions by geographers) was essentially complete.[27]

In this standard and seemingly nonproblematic list of world regions (see map 9), serious flaws of taxonomic logic are not difficult to uncover. The contaminating influence of the old continental scheme is still apparent, for example, in the residual use of an area studies category of Africa, even though North Africa had already been unambiguously placed within the Middle East.[28] Likewise, Cold War priorities are evident in the conceptual grouping of the USSR with communist Eastern Europe and in the invisibility of Central Asia, which largely remained a "no-man's land"[29] in the area studies scheme. (By the early 1980s, only Indiana University had established an "Inner Asian" studies center with Title VI funding.) In addition—despite the prominence of Oceania in the global conceptualization of the old Ethnogeographic Board—Australia and the islands of the Pacific did not appear on the map of area studies at all. Finally, northern North America and Europe occupy ambiguous positions in the scheme. Since both regions have always been the primary focus of scholarship in the humanities and the social sciences, there was little need to fund centers to encourage their study.[30] Nonetheless, as the area studies framework gained hold, Europe was increasingly *conceptualized* as a world area, on the same level as East or South Asia.[31]

Meanwhile, the world regional grid gradually acquired a life of its own outside of American institutions. In the years since the area studies scheme emerged in the 1950s, its influence has spread across the globe, and by now its regional categories are well on their way to being indigenized, especially where they correspond at some level with local metageographical traditions.[32] Large numbers of people in South Asia and Latin America,[33] for example, identify themselves as (among other things)

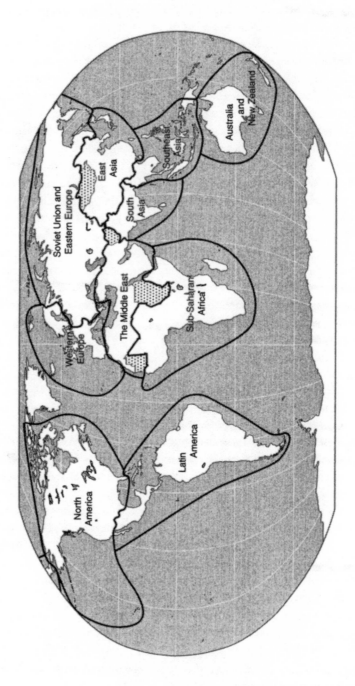

members of the same macrocultural regions through which American scholars and officials identify them.

In some cases, the appropriation of world regions at the local level has an important geopolitical dimension. The mere fact that functionaries in the U.S. State Department regard a given area as constituting a distinct world region helps to make it so, inasmuch as local elites find it expedient to follow their usage. But the same area constructs have also proved useful for organizing local economic blocks. Thus the constituent members of the Association of Southeast Asian Nations (ASEAN) — Brunei, Indonesia, Malaysia, the Philippines, Singapore, Thailand, and Vietnam — delimit their membership according to the boundary that formally, if somewhat artificially, marks the limits of Southeast Asia.[34] Tellingly, the group has not been disposed to grant membership to countries that lie outside of this region, such as Sri Lanka — which arguably has more in common with Thailand than the Philippines does. In the world regional scheme, Sri Lanka unambiguously falls within South Asia: an important justification for New Delhi's claim that it belongs within the Indian geopolitical sphere of influence.[35]

Reconceiving Asia and the Americas

For all its problems, the world regional framework represents a great improvement over continental or East-West global thinking. Its primary virtue is easily stated. By breaking up the supercontinent of Asia into four to six first-order regions,[36] it directly attacks the asymmetry of the old Europe-Asia pair. In the conventional area studies scheme, Europe is compared not to Asia as a whole, but to a unit like South Asia.[37] Certainly the precise delineation of Europe is still a vexing and ideologically charged issue.[38] Yet however Europe is defined, the European and

Map 9. *Standard World Regions circa 1975.* This map (drawn on a 1975 base map) shows the most common world regional boundaries and designations from the Cold War period. Ambiguous cases are indicated with dots: Mauritania and the Sudan (which are often placed in sub-Saharan Africa rather than the Middle East), Afghanistan (which can be located in South Asia rather than the Middle East), and Mongolia (which is sometimes placed with the Soviet Union rather than with East Asia). The unlabeled islands of the Pacific are either grouped together as Oceania, divided into three realms (Melanesia, Micronesia, and Polynesia), or simply ignored.

Indian "subcontinents" form roughly commensurable units of human geography. Both areas are of similar size and have historically maintained roughly comparable population levels.[39] Both have been historically forged into cultural communities by the sharing of religious beliefs and institutions (Christianity and Hinduism, respectively), as well as by the common use of learned languages (Greek and Latin in Europe, Sanskrit and Persian in South Asia). Finally, both areas have been sufficiently integrated to generate linguistic convergence, whereby unrelated languages came to share certain structural features (although this phenomenon seems to be more strongly developed in South Asia than in Europe).[40]

The demotion of Europe from continental to world regional status also helps rectify geographical miscues at lower levels in the spatial hierarchy. If one regards India as merely one large and important country in Asia, comparable to a relatively large and important European country like Italy, then it is only logical to regard Indian provinces as geographically equivalent to Italian provinces; Tamil Nadu comes to be placed in the same geographical order as Tuscany. If, on the other hand, South Asia is compared to Europe, then Tamil Nadu becomes logically comparable to Italy as a whole. Many South Asianists have long recognized the wisdom of this. As Bharat Bhatt notes, "[M]ore than one student of India, confronted by the variety of its regional languages and cultures, has compared the subcontinent, in this respect, to the whole of Europe."[41]

To be sure, not all "subcontinents" of Asia are so readily compared to "the whole of Europe." Three Asian world regions are relatively unproblematic: East Asia (the historical zone of Chinese influence), South Asia (the zone of Indian influence), and Southwest Asia (the historical heartland of Islam, joined with North Africa). But the interstices between these historically well-defined regions present more complications. The following section looks in depth at two such interstitial zones, Southeast Asia (once commonly referred to as Indochina) and Central Asia (the successor to Tartary). After tracing the evolution of regional nomenclature over time, we look briefly at the historical-geographical basis for the complex cultural patterns that distinguish each region today.

THE EMERGENCE OF THE CONCEPT
OF SOUTHEAST ASIA

The boundaries of Southeast Asia today are relatively uncontroversial. In scholarly usage, the region almost always encompasses ten states, in two broad subcategories: Burma (Myanmar), Thailand, Laos,

Vietnam, and Cambodia (Kampuchea) together constitute mainland or "peninsular" Southeast Asia, while Singapore, Indonesia, Malaysia, the Philippines, and Brunei make up its archepelagic or "insular" extension.[42] Yet this definition, like the area's designation as a world region on a par with South or East Asia, is quite new.

In earlier centuries, Europeans usually regarded the area we now call Southeast Asia as a mere extension of South Asia. "Farther India" was the most common designation, and the major islands are still occasionally called the East Indies.[43] French atlases of the eighteenth century often mapped a region called *L'Inde au delà du Gange* — today's Southeast Asia plus a section of modern northeast India.[44] Here Westerners were perhaps recapitulating Southwest Asian terminology, for Arab geographers had long considered this area to be the farther reaches of *al-Hind* (or India).[45] Later, as the more Sinicized northeastern reaches of the region came to be better known, *Indochina* became a common appellation for mainland Southeast Asia. The *Oxford English Dictionary*, for example, notes that "Further India, or the region between India and China, [is] sometimes called Indo-China."[46] The latter term emphasized the area's location between these two great Oriental realms, reflected in the mixture of Chinese and Indian cultural influences encountered there.[47] It also implicitly denied Southeast Asia an identity of its own. In the late nineteenth century, European conceptions of the region were further modified when cartographers began to remove the insular realm from Asia altogether, appending it instead to Oceania or "Australasia."[48]

Colonialism brought continuing adjustments to Europeans' view of this area's geographical contours. As Britain and France proceeded to divide mainland Southeast Asia between themselves, the notion of a coherent "farther India" gradually gave way to assertions of identity throughout a given metropole's colonial territories. Burma was thus to some degree conceptually joined to British India, while Indochina became virtually synonymous with France's empire in Vietnam, Cambodia, and Laos. Here the triumph of imperial boundaries over culturally constituted ones is especially evident.[49] European politics similarly reshaped the map of Southeast Asia's insular realm. While the terms *Malay Archipelago* and *Indonesia* were originally applied to the entire island chain (usually excepting the Philippines), it soon became fashionable to differentiate the "Dutch East Indies" from the British-dominated "Malaysian" territories on the peninsula and in northern Borneo. Again, imperial boundaries overrode those of language and culture; the Malay-speaking world, for example, was divided between the Netherlands (in coastal Borneo [Kalimantan] and eastern Sumatra), Britain, and independent Siam (in what is now southern Thailand). The Philippines, meanwhile, with its Spanish and later America imperial imprint, often came to be regarded as a separate geographical realm altogether.

Yet even as "farther India" was devolving into a handful of colonial possessions, the same lands were slowly being reconceptualized as "Southeast Asia" in other contexts. The term itself dates to the early nineteenth century, when its meaning was both broad and variable. In its German form, *Süd-Ost Asien,* it at one time designated the entire southeastern half of the Asian "continent"—an area now popularly viewed as comprising Asia in toto.[50] Similarly, in the first known usage of the term in English (1839), Southeast Asia was held to encompass "Hindustan, Malaya, Siam and China."[51] As recently as 1994, in the English translation of Fernand Braudel's *A History of Civilizations* (originally published in 1963), "South-East Asia" is essentially employed in this sweeping fashion.

The narrowing of "Southeast Asia" to its present boundaries was thus a gradual process. The first assertion of this interstitial area's identity as a distinctive cultural and social region is often credited to the Austrian anthropologist Robert Heine-Geldern. Heine-Geldern's Southeast Asia was not precisely that of contemporary geography, as he believed that ethnographic criteria called for the inclusion of Assam, the Andaman and Nicobar Islands, and Taiwan.[52] The ethnographic reasoning behind this boundary was generally sound.[53] But the Japanese invasion in the 1940s, and the subsequent breakup of the European colonial empires, led to a new stress on political affiliation over the cultural criteria emphasized by Heine-Geldern and his contemporaries.

In fact, the Southeast Asia of modern area studies is in many ways an artifact of military usage. It entered popular consciousness in World War II, when military strategists used it to "designate the theater of war commanded by Lord Louis Mountbatten."[54] Mountbatten's command was almost coincident with the modern region of the same name, except that it included Sri Lanka while excluding the Philippines.[55] The shift in the boundaries to their present location is often credited to the Indian historian and diplomat, K. M. Panikkar, who wrote on the region during the war years.[56] Formal legitimation came in 1945, when the U.S. State Department created a division of Southeast Asian Affairs covering the conventional modern-day region.[57] Shortly after the end of the war, world atlases similarly began to frame and map the region as Southeast Asia.[58]

In the immediate postwar decades, the term *Southeast Asia* rapidly diffused from military and political to academic circles. Scholars of the region were already contending that its history was by no means merely derivative from India and China, and the indigenous peoples rightly objected to such classifications as "farther India" or "Indochina."[59] Yet these were never primary considerations in the evolution of the modern nomenclature. As Donald K. Emmerson has argued, the modern view of the region was essentially a geopolitical one: "Southeast Asia turned out to be an aggregate of nations—individually distinct and collectively a battleground in, first, the Pacific War, then the Cold War, including two Indochina Wars, and finally, in Cambodia, a Sino-Soviet 'proxy war.'"[60] Emmerson also notes that the adoption of a political definition represents a triumph of American over Central European scholarly traditions.

In this way, Southeast Asia as a rigidly delimited macroregion came into being as a Cold War expedient: a geopolitical resolution to what was, in a sense, a postimperial crisis of spatial conceptualization.[61] The term neatly designated an important arena for geopolitical rivalry, and thus a suitable geographical platform for military strategists. It also designated a neatly bounded[62] and relatively compact collection of newly constituted "nation-building" states that did not obviously fit into any other region of the world. Finally, its label had the advantage of echoing the familiar terminology of the continental scheme, portraying the area as an unproblematic subcategory of Asia.

THE COHERENCE OF SOUTHEAST ASIA

For the scholars trained and employed in the new centers of Southeast Asian studies, however, the question has persisted: does Southeast Asia really merit world regional status? Does it possess, in other words, the kind of internal historical and cultural bonds that unite the longer-standing Eurasian regional aggregations? Many have concluded otherwise. Victor Savage, Lily Kong, and Brenda Yeoh castigate geographers for adopting the ten-country political definition of Southeast Asia; in their eyes, "a fundamental principle of regional geography is breached: regions should be defined not by their boundaries but by their inherent unique personalities."[63] Can Southeast Asia be so defined?

By most criteria, the answer is no. Southeast Asia lacks the deeply rooted and widely shared religio-philosophical systems that give coherence to Europe, the "Middle East," South Asia, East Asia, and even "Latin" America. On the contrary, its "high" cultural traditions are both diverse and exogenous to the area, having been imported from several other realms of civilization. The ideological heritages of no less than four distinct world regions—South Asia, East Asia, Southwest Asia, and Europe—suffuse the region. As Victor Lieberman writes, "[B]y comparison with Europe or Latin America or even South Asia, [the category Southeast Asia] appears artificial and residual."[64]

The derivative nature of Southeast Asian civilization is a logical outcome of its history. South Asian influence came early to the peninsula. In the early centuries of the first millennium, Mahayana Buddhism and a socially simplified form of Hinduism (lacking the full complement of caste ideology) swept eastward from India through the more densely populated portions of the region. Eventually Hinduism was to fade away everywhere except the islands of Bali and Lombok, but it left behind a substratum of ideas and practices in many other areas. This was overlaid much later by Theravada Buddhism—a doctrine refined in Sri Lanka and among the Mons of the Irrawaddy Delta—which established a firm position in Burma, Thailand, Laos, and Cambodia.

China was the second major source of influence in the region. Vietnam adopted Mahayana Buddhism, Taoism, and Confucianism from China; indeed, Vietnamese culture was so thoroughly Sinified that some scholars have considered it a part of East Asia.[65] But especially in the more southerly reaches of the Vietnamese zone, local practices were assimilated with Chinese imports to create a more "Southeast Asian" version of Vietnamese culture[66]—giving rise to a cultural north-south bifurcation within the country that is still evident today. Chinese migration later brought characteristically East Asian beliefs into the heart of Southeast Asia, especially in Singapore and other major urban areas.

The third world-civilization to influence the region was that of Southwest Asia. Brought by Muslim traders to the Malay Peninsula and Archipelago, as well as to coastal enclaves on the mainland, Islam largely replaced Hinduism and Buddhism between 1250 and 1650, spreading with special intensity in the latter part of the period.[67] In some areas, such as northern Sumatra, relatively "pure" forms of Islam took root; in others, like central and eastern Java, a more syncretic religion emerged. Islam was spreading quickly through the Philippines when the Spaniards arrived, but its progress there was subsequently arrested. Today, the religion of the prophet retains a small but tenacious foothold in the southwestern corner of that archipelago. Most other parts of the Philippines were subjected, rather, to European influences, being effectively Christianized and partially Hispanicized.[68]

Given this history of successive outside influences in the area of religion (as well as other formal aspects of political and literary culture), and given the complex patchwork that is its legacy, Southeast Asia does not constitute a unified cultural unit. The widespread persistence throughout the area of "hill tribes" (peoples who have historically maintained their own local religious and political traditions) adds another layer to the region's diversity.[69] By contemporary economic criteria as well, Southeast Asia is the most divided of modern world regions, encompassing some of the world's most rapidly growing economies (e.g., Singapore) alongside some of the world's poorest (e.g., Burma [Myanmar]). Politically as well, Southeast Asia has since independence been rent by competing systems of communism, capitalism (of several varieties), and Buddhist socialism. Wherever one looks, differences seem to be more prominent than similarities.

Not a few scholars of the area, troubled by this seeming lack of unifying features,[70] have sought to identify traits that could nonetheless be said to unite Southeast Asia's people (and justify its status as a world region). Harry Benda found such a unifying principle in Japanese colonialism.[71] Although brief, the Japanese interregnum elevated a new group of local elites to power everywhere, inducing major social transformations throughout the area and forging a unifying set of social and political concerns. Recent scholarship, however, questions the depth of Japanese influence, emphasizing instead the continuity of elite groups across this historical watershed.[72] Other attempts at pinpointing a locus of regional unity have proven equally problematic. Lea Williams touted the region's "maritime cohesion,"[73] but a truly "maritime region" in this part of the world would have very different boundaries. Jan Broek noted the "'colonial' character" of the region as its prime distinction, adding to this a few other defining features, such as the export of raw material.[74] Such characteristics are widely shared, however, throughout the so-called Third World. More recently, Pradyumma Karan and Wilford Bladen have argued that "the unity of Southeast Asia as a world region is derived from the process of decolonization and the region's transitional character between China and India."[75] Decolonization, however, provides no grounds for separating the area from South Asia, while the notion of an "Indo-Chinese" interstitial identity was rejected as an inadequate formulation decades ago.

Since specific differentiating attributes of Southeast Asia are difficult to locate, European and American historians have for the past several decades labored to uncover a more diffuse "cultural substratum" of an "echt Southeast Asia."[76] Anthony Reid makes the strongest case for regarding the area as forming a distinct cultural region on this basis.[77] He acknowledges that it is not the same kind of region as Europe or South

Asia, for its commonalities have long been overridden by multiple exogenous influences (contributing to important internal differences).[78] But Reid argues that descent from several common ancestral groups, coupled with a long history of intensive exchange in both the material and ideological spheres, generated a commonality that has been occluded but never extinguished by foreign influences. He locates a distinctively Southeast Asian quality in such diverse phenomena as the relatively high position of women, a penchant for commerce (even among ruling elites),[79] and a social order cemented by debt relationships. Such features were more readily visible prior to colonization, but they may still be seen today as common inflections within Sinitic culture in Vietnam, Hispanic culture in the Philippines, and Islamic culture in Malaysia.[80]

Reid's case is persuasive in the main. But however it is defined, Southeast Asia's boundaries are and will remain arbitrary to a certain extent. The conventional division between Southeast Asia and South Asia runs directly through the territories of numerous small-scale highland societies; India's far northeast represents an amalgam of South and Southeast Asian cultural traditions.[81] Only the historical accident of colonialism and subsequent state formation has placed that boundary where it now lies. Moreover, many of Southeast Asia's connections with other parts of the world are, and will likely remain, as powerful as its internal bonds. Craig Reynolds is perhaps on target in suggesting that Western academics have endeavored "to authenticate Southeast Asia as a region and a field of study" in part because of their vested interests as Southeast Asian specialists.[82] In the region itself, Reynolds notes, Southeast Asia is not considered a significant historical unit.[83]

These qualifications do not obviate the utility of the concept of Southeast Asia. But they do suggest that a certain circumspection is in order. In many ways, Southeast Asia is a residual and artificial category. We continue to use it mainly out of institutional and intellectual inertia—as well as because Southeast Asian political leaders themselves are attempting to forge an ever more integrated Southeast Asian region through the institutions of ASEAN. If they succeed, Southeast Asia may have more coherence in the future than it has had in the past.

CENTRAL ASIA IN THE HISTORICAL
AND GEOGRAPHICAL IMAGINATION

While Southeast Asia as a category may be questionable, it is nonetheless ubiquitous in world regionalization schemes. Central or

Inner Asia (the area once called Tartary), on the other hand, is a historically clear-cut region that is nonetheless almost invisible in contemporary
global cartography. European historians have long acknowledged the
unity and importance of this area, although its importance was traditionally stated in entirely negative terms. In the geographical imagination
of Europe, Tartary represented for centuries a repository of barbarism that
threatened the civilized world. Today, such a jaundiced vision has largely
receded, but a coherent alternative image has yet to take its place. Instead,
until very recently, the area seemed to be simply receding out of view altogether, falling into a sort of geographical black hole "between disciplinary cracks."[84]

The occlusion of Central Asia in the European imagination is a striking development in the annals of global geography. As late as the seventeenth and eighteenth centuries, Tartary was a major presence on the European map of the world. Its two vast components—Independent or
Western Tartary and Chinese or Eastern Tartary—stretched across more
than half of Eurasia.[85] Independent Tartary alone was an extensive geographical unit, usually encompassing not just the former Soviet Central
Asia but all of Tibet (and on one map extending even to Assam).[86] But
Tartary was significant at the time for more than its size. It carried the
dual connotation of primitive power (the land of Gog and Magog) and
"nursery of nations," home of the ancestral Europeans.[87] These connotations faded slowly in tandem with the steady rise of military power along
the Eurasian rim. In the nineteenth century, after the Russian and Chinese armies had conquered and partitioned virtually the entire steppelands,
Tartary ceased to be mapped as a major world region.[88]

The pallid idea of Central Asia that eventually replaced it was originally the vaguest of constructs. In 1855 Charles Anthon identified a Central Asia that ran from the Caucasus on one end to Japan on the other.[89]
By the start of the twentieth century, however, geographers influenced
by the doctrine of environmental determinism started to pay closer attention to the area we now call Central Asia and to its precise definition.
Halford Mackinder and Ellsworth Huntington wrote at length about
Central Asia's pivotal role in world history, the latter seeing in its denizens'
past a climatically driven propensity to burst out of their arid confines at
regular intervals, raiding and plundering the prosperous Eurasian periphery.[90] Such theories fell from favor with the passing of environmental determinism in the 1930s and 1940s, after which Central Asia simply
languished in the geographical imagination, disappearing almost entirely
from American scholars' concerns in the decades after World War II.

The erasure of Tartary from global geography took the form of carto-graphical dismemberment. The western portion of Central Asia was mapped as part of the Soviet Union, and the eastern half was subsumed within China. In postwar maps and textbooks, Xinjiang, Tibet,[91] Inner Mongolia, and neighboring areas are typically portrayed as integral por-tions of the Chinese "nation"—the notion of a culturally autonomous "Chinese [controlled] Tartary" having long since vanished. Historical at-lases especially have obscured the local identity of this region. They typ-ically overstate Chinese influence by selecting for portrayal those periods when the Han Chinese were under a single imperium and when each dy-nasty reached its maximum territorial extent. At times under the Han, Tang, and Qing dynasties, China did indeed control sizable areas of east-ern Central Asia, but at other times Chinese power over the region was minimal or nonexistent; under the Yuan (Mongol) and Qing (Manchu) dynasties, one might even argue that it was Central Asia that controlled China. By neglecting to depict such periods of imperial regression and ambiguity, historical atlases create a mistaken impression that the Tarim Basin is simply a natural part of the Chinese realm.

Central Asia has thus been rendered almost invisible on the contem-porary map of the world. The myth of the nation-state precluded its recog-nition until recently, for the simple reason that it contained only a single sovereign country, Mongolia. With the breakup of the Soviet Union and the emergence of five newly independent Central Asian republics, how-ever, the western half of the region has suddenly come back into view.

In the scholarly imagination, the region as a whole seems to be reap-pearing, with Central Asian studies recently experiencing something of a renaissance. Specialists like Thomas Barfield and Christopher Beckwith are again asserting the region's historical significance, adding this time a new insistence on the sophistication of its societies. Their message has been embraced and amplified by the world-systems theorist Andre Gun-der Frank, who emphasizes the "centrality" of the region in historical as well as geographical terms.[92] The new recognition of Central Asia as a discrete region is thus simultaneously a recognition of its historical im-mersion in the broader fabric of Eurasian life.

Despite such attention, Central Asia's spatial form remains highly un-certain. Most historians have regarded the region simply as the arid zone in the interior of the Eurasian continent. This effectively equates Central Asia with the steppe zone: a vast swath of land extending from south-eastern Europe to the borders of Manchuria, but excluding the Amu Darya and Syr Darya river valleys, the Tarim Basin, and Tibet. Not all analysts concur, however. Mackinder, for one, encompassed all these areas and far more; his "heartland" included European Russia and, in its revised form, all of eastern Europe.[93] Modern political definitions, on the other hand, necessarily exclude the steppe regions of Russia and the Ukraine, or at least treat them separately. Soviet scholars in particular insisted on a distinction between Middle Asia (Turkmenistan, Uzbekistan, Kyrgyzstan, Tajikistan, and sometimes Kazakhstan) and Central Asia proper (Xinjiang, Tibet, and Mongolia).[94] Several American historians and most anthropologists, on the other hand, look to cultural affinity, resulting in yet a third set of boundaries. Robert Canfield, for example, has used the term *greater Central Asia* to refer to those areas that have been, at various times, under Turkish influence or domination: a region extending from "Turkey to Sinkiang [Xinjiang] (or Chinese Turkistan) and, on a more southerly latitude, from the Euphrates to North India."[95] This definition, while highlighting a significant and generally ignored Turko-Persian cultural grouping, fails to include Mongolia—which is elsewhere considered to be Central Asia's core.[96] In short, there is as yet no clear consensus on what the term *Central Asia* denotes.

CENTRAL ASIA AS A WORLD REGION

Unlike Southeast Asia, Central Asia forms a highly ambiguous world region in contemporary geographical discourse, one whose boundaries are contested by almost all parties. Like Southeast Asia, it is also ambiguous in terms of its cultural constitution. In past times, the area shared a Turko-Mongolian cultural heritage, underlain in its southwestern reaches by Iranian (specifically, Sogdian) elements. Common ways of life were shaped by the close integration of pastoralism and oasis agriculture,[97] as well as by the continual traffic along the region's trade routes. For a time, too, Central Asians elaborated their own forms of cultural-ideological integration, notably through Manichaeanism and Nestorian Christianity. Yet an overriding regional heritage never took shape in Central Asia, and the area has been deeply split for centuries between Islam and Lamaist Buddhism, with shamanism[98] persisting until recently along its northern fringe.

The cultural division of the region proceeded in stages. First, Islamic influences began to recast the civilization of its western reaches. The Amu Darya and Syr Darya valleys (the Sogdian cultural hearth whose merchants once spanned much of Eurasia) were thoroughly incorporated into the realm of Islam by the ninth century. The same area was later overtaken by Turkish-speaking peoples from farther east, but rather than rolling back the Islamic frontier, they converted to Islam, furthering "Western Tartary's" integration with the cultural world of Southwest Asia. By the sixteenth century, even the Tarim Basin of modern Xinjiang (in northwest China) had been Islamicized deeply enough to generate a series of small-scale theocratic states.[99] Farther east, however, the Mongols steadfastly resisted Islam. While their leaders had experimented with several universalizing religions, the Mongol peoples as a whole opted in the sixteenth century for the Tibetan form of Mahayana Buddhism (Lamaism). This distinctive and highly centralized religion put the Tibetans and Mongolians into a religious sphere of their own, albeit one that is sparsely populated and spatially disjunct (with Turkish and Chinese populations separating its two main subgroups).[100]

As a result of these developments, Central Asia comprises two quite separate religious culture-zones, Islamic in the west and Lamaist in the east. As one observer puts it, "[T]he links which usually hold together or create a cultural entity—such as script, race [sic], religion, language—played only a very moderate role as factors of cohesion."[101] Twentieth-century political developments have been read by many as portending only a furthering of intraregional fragmentation. The Lamaist area has been subject to great assimilationist pressure from both the Chinese and the Russian regimes. Likewise, during the period of Soviet rule, the Islamic western reaches of Central Asia were all but severed from the greater Islamic cultural core of Southwest Asia. Since 1991, however, the former Soviet republics of Uzbekistan, Kyrgyzstan, Turkmenistan, and Tajikistan have all been busy reestablishing their ties to the Islamic heartland.[102] A signal question now is whether secular Turkey or theocratic Iran will exert greater influence over these new states. A comparable breakup of China hardly seems likely, but if such an event were to occur, the Uighurs (Uygurs) of Xinjiang might be expected to establish close ties to their Turkish-speaking kin in the west, while Lamaist Tibet could well revive its own once-strong relationship with Mongolia (and, perhaps, with Russia's Buryatia, Tuva, and the Kalmyk autonomous region).

What is hard to imagine is a "pan-Tartary" movement taking shape. The western and eastern reaches of Central Asia simply have too little in common to make such a scenario credible. Tajikistan is far more closely connected to Iran and Afghanistan than it is to Mongolia or Tibet (more Tajiks actually live in Afghanistan than in Tajikistan). Central Asia is thus perhaps best understood as a world region with a solid past but a tenuous present and a cloudy future. While its salience for historical studies is unquestioned, its deep internal rifts render it a less satisfactory category for contemporary analysis. Still, the delineation of a Central Asian world region has two signal advantages: it highlights the cultural separation of such areas as Tibet and Mongolia from the Chinese realm to which they are habitually attached, and it underscores the as-yet uncertain position of the five former Soviet republics in the area.

IDEOLOGICAL IMPLICATIONS OF WORLD REGIONALIZATION: THE CASE OF LATIN AMERICA

If a primary virtue of the area studies scheme is its dismantling of the false Europe-Asia comparison, a secondary virtue is that it begins to grapple with the cultural configuration of the Americas. This portion of the world was consistently slighted in most earlier metageographical schemes. On the area studies map, by contrast, Latin America is one of the most clear-cut and conceptually cohesive regions of all. Compared with either Southeast or Central Asia, Latin America appears highly integrated, united by widespread religious, linguistic, historical, and even political commonalities. In some ways, the region may even be said to be more coherent than Europe or South Asia. Its appellation is also an old one; "Latin America" is one of the earliest of world regional designations, dating back to the middle nineteenth century. Closer inspection, however, reveals it, too, to be deeply problematic. Like Southeast Asia and the "Middle East," this region was originally defined by military strategists, and our conceptualization of it still bears the taint of imperial thinking.

To begin with, the term *Latin America* is not logically constituted; the only thing "Latin" about the area is the fact that most of its inhabitants speak languages derived from that of ancient Rome.[103] This trait was in fact singled out for strategic reasons. *Latin America* was deliberately coined by French scholars in the middle nineteenth century as a way to refer simultaneously to the Spanish-, Portuguese-, and French-speaking portions of the Americas. At the time, the French government under Napoleon III was plotting to carve out a new empire in the region, and the notion of a "Latin" essence linking France with Spanish- and Portuguese-speaking American countries had great appeal as a way to naturalize such a project. In the understated words of Fernand Braudel, " 'Latin America' was first used about 1865 by France, partly for her own reasons, and later adopted by Europe as a whole."[104]

In the original formulation, then, Latin America included only those American countries in which Spanish, Portuguese, and French were the official languages. Such a definition was usual in the United States before World War II (although some writers of the period also distinguished a Latin America "in the larger sense" that included British and Dutch colonies), and it was occasionally encountered in later decades.[105] But by disinterested criteria this "Latin" region of the French imperial imagination never made sense; certainly Haitians have less in common with residents of Argentina than with the neighboring Jamaicans, who do not happen to speak a neo-Latin language. Strictly speaking, a linguistic definition would mean that Quebec, too, ought to be considered a part of Latin America, as it lies within America and is inhabited by a people who speak a Romance language.[106]

To minimize these problems, careful scholars once preferred to categorize the Spanish- and Portuguese-speaking countries of the Western Hemisphere separately as Ibero-America or Hispanic America. Yet the term *Hispanic America* proved ambiguous; some writers viewed it as completely equivalent to *Latin America*,[107] while others argued that it should exclude Brazil.[108] Víctor Alba provides the most precise set of definitions. "The term 'Hispano-American,' " he argues, "does not include Brazil or Haiti. 'Ibero-American' includes Brazil but leaves out Haiti."[109] The more inclusive term, *Latin America,* he finds problematic, in part because Anglo-Saxons employ it "to degrade the Spanish influence,"[110] yet it is useful for conveying the extent of pan–Latin American cultural unity.[111]

Such fine distinctions as Alba maintained have in any case languished among North American scholars. Terms such as *Ibero-America* are now seldom encountered, while the Latin America construct itself has outgrown its original linguistic definition, gradually coming to include such non-Romance-speaking countries as Belize and Guyana.[112] By the 1960s, in North American academic parlance, *Latin America* was effectively a shorthand way to refer to everything south of the Rio Grande.

This broad idea of a "Latin" America is probably best abandoned. A stronger case can be made for a world region composed of the Spanish- and Portuguese-speaking countries of the Americas, but even "Ibero-America" remains a loaded construct. For one thing, such a term obscures the strong indigenous influence that remains in many portions of the area. For another, it effectively precludes recognition of the equally important world region of "African America," whose boundaries crosscut those of the former European colonial empires of the region.[113] For these and other reasons, metageographical categories remain problematic even in this relatively cohesive part of the world.

World Regions and Geographical Determinism

The preceding discussions of Southeast Asia, Central Asia, and Latin America highlight some of the empirical problems surrounding the definition and delineation of particular world regions. A more fundamental weakness of the world regional framework—at least in its textbook incarnation—is its residual baggage of geographical determinism. Although the linkages are looser here than in the continental scheme, they are not entirely dissolved.

At the time the world regional framework was created, many anthropologists still assumed that regions constituted on the basis of specifically human criteria would correspond with the distribution of natural features. In fact, the mid–twentieth century saw explicit attempts to theorize such linkages under the rubric of the "culture area."[114] While geographers had been schooled by then to deny environmental determinism in such an overt form, these assumptions shaped the drawing of area studies boundaries in the first place and have continued to pervade world regional textbooks at the college level.[115] Historians, too, have often fallen back on environmental criteria in delineating regional frameworks for historical analysis. As a result, geographical determinism continues to misinform standard treatments of world regions, particularly those within the Eurasian landmass.

Consider the following passages on Southeast Asia, selected from two generally superior overviews of that region's history:

> A traveler who journeyed from one end of Southeast Asia to the other would note a similarity of flora and fauna, climate, and human civilization. While the language and architecture would change, he would be struck by the repeating patterns of wet-rice and slash-and-burn agriculture, found from Burma to Bali. . . . Where water was plentiful he would find wet-rice farming; where it was scarce he would find fewer established communities and less intensive cultivation.[116]

> Yet those who travel to Southeast Asia, from China, India, or anywhere else know at once that they are in a different place. In part this is a matter of environment. Physically marked by its warm climate, high and dependable rainfall, and ubiquitous waterways, Southeast Asia developed lifestyles dominated by the forest, the rice-growing river valleys, and fishing. Its people grew the same crops by the same methods, ate the same food in the same manner, and lived in similar houses elevated on poles against the perils of flood or forest animals.[117]

While these passages evoke a vivid sense of place, that place is not coterminous with Southeast Asia. In some ways, the image is too specific. Neither the semiarid scrublands of central Burma, nor the tropical montane rainforests of central Borneo, nor the permanent-field dryland cropping patterns of Cebu, central Burma, and Madura match this description. In other ways, the landscape invoked by these passages is too general. The agricultural pattern of alternating paddy field and swidden plot is just as prevalent in eastern India and southern China as in Southeast Asia—in fact, similar landscapes can be found in central Madagascar and even in parts of western Africa. Southeast Asia, in other words, cannot be defined on the basis of environmental criteria. If it is a unit at all, it is a social unit alone.

The same is true of the world region now known as Southwest Asia and North Africa. Geography textbooks routinely suggest that this entire area is characterized by aridity.[118] In one noted text from the 1960s, the region was labeled the "Dry World"—even though the authors' explicit intention was to map out "culture worlds" rather than environmental realms.[119] A more recent work claims that "SWANA [Southwest Asia-North Africa] is distinguished by three geographical characteristics": Islam, Arabic, and a precarious water supply;[120] another proposes that climate is the most fundamental characteristic giving unity to the Middle East.[121]

This climatic definition simply does not hold up. The vast area encompassed by the label *Southwest Asia and North Africa* is not uniformly parched. Beirut annually receives some 35 inches of precipitation (12 inches more than London), yet no geographer would see this as grounds for placing it outside the region.[122] Similarly, the Iranian province of Gilan, located along the south shore of the Caspian Sea, receives ample year-round rainfall, supporting rice fields and lush broadleaf forests. Yet humid Gilan is as fully a part of a Southwest Asian cultural sphere as is the arid Dasht-e-Kavir. It is shared bodies of thought, social institutions, and quotidian practices that constitute this region, not supposed regularities of climate or topography.

In discussions of Central Asia, environmental determinism is equally common, and equally problematic. Whereas Halford Mackinder defined the region on the basis of drainage patterns (consigning all areas of internal or arctic drainage to Eurasia's vast "heartland"),[123] a focus on climate is more common today. Gavin Hambly defines Central Asia as isolated from marine influences;[124] Daniel Balland cites hypercontinentality ("high thermal amplitude . . . and more or less permanent aridity") and "high climatic variability."[125] But while the criterion may vary, the notion that this region can be delineated in terms of physical geography is widespread. As Andre Gunder Frank writes: "If this introduces an element of geographical determinism á la Huntington and Mackinder, so be it."[126]

Yet consider for a moment the "environment" of Central Asia (which, in Frank's case, includes most of Siberia).[127] Its landscapes range from the thermally volatile Turfan depression to the permanently cold Tibetan plateau, from the saturated boglands of western Siberia to the parched Takla Makan, and from the fertile pastures of Semirechye to the salt flats of the Kara Kum. Where they do exist, aridity and continentality do not necessarily coincide; central Siberia marks the center of annual temperature extremes, but the area's driest spot is in the thermally milder lowlands of the far southwest. Moreover, extremes of aridity and continentality are found well outside the boundaries of Central Asia (in interior Turkey and Iran, for instance), while within the region they are not so exaggerated as is sometimes supposed. Almaty (Alma-Ata), near the region's center, receives more rain (23.5 inches a year) than parts of southeastern Europe or the loess plateau of China—areas adjacent to Central Asia yet seldom included within it, for good historical reasons.[128] Similarly, the average annual temperature range of Almaty is no greater than that of Beijing, and Krasnovodsk, on the Caspian Sea in Turkmenistan, is significantly milder than either.[129] In a word, Central Asia cannot be defined in terms of climate or any other environmental feature. It is a human region, not a geophysical one.[130]

Unlike the regions considered above, South Asia is almost never defined on the basis of its climatic or vegetational patterns. Its diversity in these terms is evident to all. Yet a different kind of environmental determinism surfaces when South Asia is said to be a distinct world region because it forms a natural subcontinent, separated from the rest of the Eurasian landmass by formidable mountain ranges.

Such a claim is inconsistent with the logic used elsewhere to determine world regional status. The Arabian Peninsula is similarly a natural unit, yet is never counted as a world region. It also overstates the degree of physical separation. While mountainous borderlands have certainly influenced the geographical patterning of South Asian civilization, they have not determined it. Such "barriers" have proved highly porous; Central and Southwest Asian influences have made repeated incursions into India, just as Indian cultural patterns have diffused to Tibet and the Central Asian oases. Moreover, sizable areas of modern South Asia—areas well within the great mountain girdle—were not considered by the indigenous inhabitants to be part of India at all until recent times.[131] The spread of

South Asian cultural patterns through (almost) the entire subcontinent was by no means predetermined. When one considers as well the sea routes that have historically connected South Asia to both Southeast and Southwest Asia, as well as to Ethiopia and other parts of eastern Africa, the notion that the geography of Indian civilization was predetermined by its physiographical frame breaks down completely.

In short, world regions are artifacts of human history, not of natural processes. Of course, environmental regions can be delimited, and they can also be tremendously useful—for the consideration of environmental issues. But they should not be conflated with the macroregions of human geography.

TOWARD A REFINED WORLD REGIONAL SCHEME

Clearly, the world regional system has some serious flaws. In most presentations, it is contaminated both by the myth of the nation-state and by geographical determinism. Similarly, though less Eurocentric than the standard continental scheme, it still bear traces of its origin within a self-centered European geographical tradition. More fundamentally, a world regional framework continues to grossly flatten the complexities of global geography. No less than the continental scheme, it implies that the map of the world is readily divisible into a small number of fundamentally comparable units. Nonetheless, we remain convinced that some form of baseline heuristic scheme is necessary for teaching and thinking about metageography, and that a refined version of the world regional framework is the most serviceable alternative available.

The cartographic outlines of our own preferred regions are depicted in map 10. In its contours, this map does not differ dramatically from the standard depiction of world regions already employed in geography and other disciplines; with a few important qualifications, we endorse the global architecture that has emerged within the North American academic world. Where we part company from textbook geography is less in cartographic depiction than in conceptual procedure. Our methodological

---

Map 10. *A Heuristic World Regionalization Scheme.* This is the map that we employ for teaching global human geography. The regions are as follows: East Asia, Southeast Asia, South Asia, Central Asia (subdivided into Islamic and Lamaist zones), Southwest Asia and North Africa, sub-Saharan Africa (with an Ethiopian subdivision noted), Ibero America, African America, North America, Western and Central Europe, Russia-Southeast Europe and the Caucasus, Australia and New Zealand, Melanesia, and Micronesia and Polynesia.

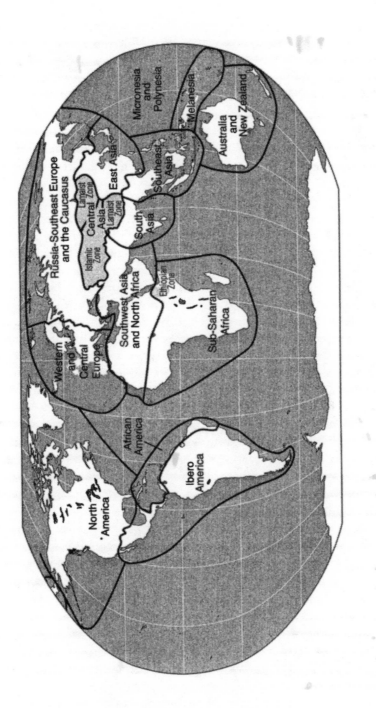

differences with textbook geography can be summed up in three main principles. First, we have avoided defining regions in terms of specific diagnostic traits, focusing instead on historical processes. Second, we have ignored both political and ecological boundaries, giving primacy instead to the spatial contours of assemblages of ideas, practices, and social institutions that give human communities their distinction and coherence. Finally, we have tried to conceptualize world regions not only in terms of their internal characteristics, but also in terms of their relations with one another. For one region's identity has often coalesced only in confrontation with another.

This last point deserves underscoring. As Patrick Manning[132] and others have argued, sub-Saharan Africa emerged as a distinct region of the world in good part through the mechanism of the slave trade, an intrinsically interregional operation implicating Europe, the Americas, and North Africa and Southwest Asia. African unity was later enhanced by the common experiences of colonial rule and anticolonial resistance, and it is maintained today in part by the common yoke of global financial institutions. Southeast Asia, similarly, has coalesced in part through confrontations with the historically expansive civilizations of the Eurasian mainland. And it is emerging as a more coherent region today in part because the majority of its member states are developing a common pattern of economic relationships with East Asia, Europe, and North America. Now more than ever, relational issues are crucial to macroregional identity.

So long as these methodological points are addressed, we believe that the world regional paradigm can be reformed and should be retained. But a number of caveats are still in order. First, as we have insisted throughout this critique, any scheme of global geographical division is only a rough approximation, a convenient but crude device for making sense of particular patterns of human life. World regions are better approximations for most purposes than continents or civilizations, but they are no more naturally given. Second, we would emphasize that this scheme has evolved essentially as a pedagogical tool: a vehicle for talking and teaching about basic global patterns of sociocultural geography at the college level. We claim no authority for it beyond those uses. Third, we would note that while our map by necessity shows seemingly rigid boundaries separating world regions, many of those boundary zones themselves function almost like hybrid regions in their own right.[133] Finally, we would ask the reader to see this scheme, like all similar efforts, as but one contribution to an ongoing dialogue. Our purpose is not to settle the many delicate issues of metageography, but to advance the discussion of those issues.

Conclusion

Toward a Critical Metageography

It is no accident that the global geographical framework in use today is essentially a cartographic celebration of European power. After centuries of imperialism, the presumptuous worldview of a once-dominant metropole has become part of the intellectual furniture of the world. Even postcolonial intellectuals, bent on creating new visions for an alternative global order, find themselves stuck with a collection of parochial geohistorical categories that originated in the Eurasian Far West. Admittedly, those categories have been stretched almost beyond recognition during the past five hundred years. Forced to accommodate a world full of previously unknown lands and peoples, they have also been subjected to increasingly disciplined forms of scientific inquiry and abstract representation. But as much as the metageography of medieval Europe has been bent, broadened, and recast, it has never been completely broken. On the contrary, the very processes of discovery and conquest that forced open its narrow categories served simultaneously to shore up its self-flattering premises.

As recently as fifty years ago, the hegemony of European and European-derived societies seemed absolute. While much of its heartland lay in ruins, the two peripheral giants of the Occidental world dominated the globe. Having just crushed the single major military threat from another quarter, that of Japan, their own rivalry focused on Europe itself, and enormous resources were committed to ensure a rapid recovery for the old core of the West. Not surprisingly, the map of the world that became enshrined at this time — on American campuses and in the U.S. State

Department, and through them eventually in much of the "developing world" as well—both reflected and reinforced the mood of Western triumphalism.

Implicitly, this was a geography of power. Both regional power blocks and individual states were seen as stable elements of a hierarchical order. But cultural and economic space were implicated in the world regions vision as well. For many partisan thinkers of the mid-1900s, the twentieth century belonged to the United States—to the West's own West, where the Occidental spirit of rationality and progress had been distilled to its essence (and from whose vantage point even France might appear as vaguely Oriental). In the reigning American social science paradigm of the time, the rest of the world—if it wanted to enjoy the fruits of modernity and further progress—would have to follow America's lead. On the one hand, this meant perpetuating a capitalist division of labor where the United States and its allies kept a lock on the most advanced industries. (Hence the intent of early U.S. occupation policy that Japan should revert to an agrarian economy, if not to full-fledged Asiatic slumber.)[1] On the other hand, it meant the creation of a stable but progressive cultural landscape, where premodern civilizations retained their distinctiveness— and their geographical place—while bowing to the overarching exigencies of modern life. Select features of the various "great traditions" would surely survive, but only insofar as they proved compatible with nation building, secularization, industrialization, urbanization, universal education, and the rest of the modernization package. Those that proved incompatible with modernity would be shaken off as unnecessary baggage or enshrined as museum pieces and touristic curiosities. Meanwhile, the serious business of modern cultural production would take place in the advanced centers of New York, Paris, and London, from which modernity would radiate outward to a grateful world. The possibility of militant, long-term resistance to such enculturation, much less of forces from the "developing world" rearranging the cultural landscape of the whole, was rarely entertained.

This utopian vision did not, of course, go unchallenged, even in its heyday. But its main rivals, rooted in the same philosophical traditions as modernization theory, inherited a similarly Eurocentric metageography. If anything, classical Marxism's geohistorical vision rested even more explicitly on the thesis of European exceptionality than did that of modernization theory. Although such beliefs were targeted repeatedly by non-European revolutionaries, most Third World nationalisms remained deeply rooted in European ideas and institutional practices. For a while it seemed as if the historical primacy of the West could not but be perpetuated, even by those who wished to dismantle its hegemony.

How different the world appears at century's end. Political upheavals have broken up the stable power blocks of the Cold War era, and nationalist revivals have exposed the vulnerability of existing state borders. Economically, too, the simple global division between a rich dominant core and a poorer dependent periphery no longer seems a sure thing. From some vantage points, the postwar global bifurcation remains very real, as a dominant West asserts financial control more forcefully than ever over various states in the Third World. Such a view is hard to avoid in sub-Saharan Africa, where the failure of development is abundantly clear, and where the International Monetary Fund and the World Bank have assumed unprecedented power over supposedly sovereign states. But the view from Seoul, Guangzhou, Bangkok, or Kuala Lumpur is strikingly different. Recent rates of industrial expansion in South Korea, Taiwan, China, Malaysia, Singapore, and Thailand are almost unprecedented in world history. Even India, with its heritage of sluggish growth, has outperformed western Europe in recent years, and Bangalore has become known as a "silicon plateau." Bombay, while the site of staggering poverty, is also emerging as the financial hub of South Asia—and as one of the major filmmaking centers of the world, along with Hollywood and Hong Kong. Despite the protests of Hindu fundamentalist politicians, India is being steadily integrated into the world economy. While the merits of this multinationalization can be debated, the way it is happening confounds simplistic notions of a stable global core and periphery. When South Korean conglomerates vie with Indian and German firms for Bombay's office space (now the most expensive in the entire world),[2] maintaining the notion of a North versus South or West versus non-West polarity requires some convoluted topological maneuvers.

Indeed, rather than perpetuating straightforward Western hegemony, late-twentieth-century capitalism appears to be characterized by both multicentric growth and fractionated economic differentiation. If some longstanding centers of wealth are indeed growing richer and more powerful, others are stagnating or declining—just as, in the so-called Third World, some traditionally peripheral areas are seeing living standards erode further, while others have turned into centers of economic dynamism. And the closer one looks, the more jumbled the economic landscape becomes. For within the fortunate zones of the new economic order, whether First World or Third, large sectors of the populace are both spatially and socially cut off from the surrounding prosperity. Particularly in the United States, economic differentiation has accelerated greatly since the early 1980s, and "Third World" conditions have begun to spread beyond America's inner cities.

Cultural developments over the last half-century have proven at least as challenging to Western global visions of the early postwar period as have rearrangements of the political-economic landscape. A vibrant international trade in films, music, and "kids' culture"[3] has indeed evolved, but it has not exactly taken the forms envisioned by modernization planners. On the one hand, high-tech cultural production refuses to obey a neat diffusionist paradigm. Such lucrative transnational commodities as action movies, video games, and animated television shows are emanating from studios in Hong Kong and Tokyo as much as from Hollywood. On the other hand, in philosophy, religion, and the performing arts, localism has hardly diminished; booming areas and poor ones alike have seen an efflorescence of indigenous cultural forms, frequently cast in an antimodern or anti-Western idiom. As a result, while cars, cable television, fashion, and fast food constitute the material matrix of an increasingly transnational "world culture," modernity continues to be both inflected and contested in thousands of distinctively local dialects.

Taken together with the crumbling of the Cold War and the shifting geography of capitalist dynamism, the contested character of global culture gives the lie to simple notions of a stable, hierarchical world order. But the challenge to received metageographies goes still deeper. In effect, the geography of social life in the late twentieth century has outgrown not only the particular contours of the postwar world map, but also the very conventions by which we represent spatial patterns in image and text. The cultural territories once confidently mapped by anthropologists are increasingly being crosscut and redefined, by the convulsions of peacetime mobility as well as those of war. Historical border zones in particular have acquired unexpected prominence as sites of cultural innovation. But the accelerating diaspora of merchants, migrants, and refugees around the world has jostled peoples together, not only along traditional boundary zones but deep within every major historical region as well. In response to these dislocations, local conceptions of macro-level identity are mutating rapidly, too. Intensive and innovative work is being done in various disciplines to grapple with fragments of this intriguing puzzle. But devising a new sociospatial lexicon capable of analyzing and representing such developments has yet to receive the sustained scholarly attention it demands.

Professional geographers as a whole have yet to tackle this project. The displacement of the long-cherished continental scheme by that of world regions attracted little geographical notice in the first place. Driven as it was by pedagogical and organizational priorities that were articulated elsewhere, the metageographical revolution of midcentury effectively took the discipline unawares. Nor have many geographers turned their attention to macro-level categories in succeeding decades. While few would claim that the world regional textbooks now in use adequately convey the complex spatial patterns of contemporary social life, global regionalism has yet to become a respectable subject for geographical research. To this day, the ad hoc area studies framework initiated by the wartime Ethnogeographic Board is largely taken for granted, in geography as much as in cognate fields.

This state of affairs may not last long, however, for recent geopolitical developments have begun to undermine the legitimacy of the area studies framework in nonacademic circles. The palpable inadequacy of the existing world regions paradigm in the wake of 1989 has again forced policy makers to seek out new ways of dividing up the world. In particular, the position of Central Asia is in question, as is the eastern boundary of Europe. Meanwhile, major funding agencies have begun to withdraw support from area studies altogether, leaning instead toward thematic studies centers organized under comparative or global rubrics.[4] If this shift of resources represents a potential retreat from rigorous language training and international expertise, as some observers warn, it is disturbing indeed.[5] But to the extent that it serves to call the world area framework into question, the threatened withdrawal of support from area studies centers offers an opportunity to begin talking seriously about global geography. A reorganization of funding will not in itself obviate the utility of such constructs as Latin America or Southeast Asia, but it might provoke a wider debate about the inadequacies of the world regional map.

The present book has been written in hopes of providing both a justification and a historical perspective for such a dialogue. To posit a comprehensive alternative model of the world seems both premature and presumptuous; if a better metageographical paradigm is to emerge, it will come about only through a searching interdisciplinary and international effort. Accordingly, we have pursued a less ambitious agenda here. Convinced that any new paradigm must build on a careful critique of the system it is meant to displace, our goal has been to document and analyze the conventional geographical categories of our day. While we have provided our own alternative maps of both civilizational gradients and world regional boundaries, these have been offered less as solutions than as a way to pose the questions more provocatively. In the same spirit, the remainder of this conclusion sketches our vision of what remains to be done. First, we propose ten criteria for metageographical reform, aiming at the creation of more supple and sophisticated frameworks. Then, with those criteria in mind, we highlight half a dozen areas of particular promise for further metageographical research.

Ten Principles of Critical Metageography

The foregoing chapters have examined a series of specific problems posed by prevailing systems of geographical classification. As we have repeatedly stressed, the plasticity of spatial forms and the plurality of spatial identities mean that some problems of classification are inevitable. Yet others are not. To highlight those areas where change is both possible and desirable, we have distilled a set of ten general principles, which might be viewed as a rudimentary grammar for metageographical modeling. Some of these principles pertain to categorization, others to cartography; some have to do with nomenclature, others with regional definition. Taken together, they effectively denote a set of guidelines for improving on current frameworks, as well as a set of criteria for judging any proposed schema of global geography. While other observers may well come up with additional criteria of their own, we believe that the following ten issues will be central to the collective project of critical metageography.

The first item on our list is a commitment to *combating cartographic ethnocentrism*. Effectively, this means insisting on *logical typing*: assigning the same rank in the spatial hierarchy to roughly comparable units. To be sure, a spatial taxonomy can never be perfect, and scale is forever problematic. Two regions with comparable claims to historical and cultural distinctiveness will not necessarily be commensurable in area or population; both greater Tibet and highland Ethiopia possess many attributes of world regions, but neither can be regarded as fully comparable to South or East Asia. Additionally, some regions (contested frontiers, far-flung island realms, zones of intense cultural miscegenation) defy categorization in conventional terms altogether. For these and related reasons, it is not at all simple to decide how many highest-order zones of cultural affiliation should be inscribed on the globe.

Nonetheless, egregious errors of typing can be both identified and corrected. The prime examples are the spurious comparisons of a historically coherent West with a vast and heterogeneous East, and the related habit of counting the European peninsula as a "continent" on the same order as Asia. These particular violations of taxonomic principle are largely rectified in the world regional scheme, representing in our view its single most important advantage over preceding systems. Any further reform of metageographical frameworks will have to conserve a commitment to comparing like with like.

Second only to fighting residual Eurocentrism is the importance of *combating geographical determinism,* that still-pervasive mode of geographical thought that posits iron links between environmental conditioning and social response. While explicit determinism has been abandoned in mainstream works, the vague notion that cultural regions correspond in some natural, inevitable way to the distribution of physical landscape features persists, both in geography textbooks and in the social sciences at large. Such assumptions are especially hard to escape in our most elementary metageograpical framework, the myth of continents. Since we have examined the problems of geographical determinism at length in the text, we will not belabor them here. We will simply stress that, in our view, this remains a profound challenge for critical metageography.

A third principle might be called *typological honesty:* delineating regions on the basis of consistent criteria, insofar as that is possible, and acknowledging clearly when it is not. Most current textbooks and atlases are compromised by an unacknowledged jumbling of physiographical, cultural, and political categories at the highest level of their spatial taxonomies. The key word here is *unacknowledged.* Because the major spatial groupings of human society are the product of different histories, macroregions inevitably differ to some extent in the logic of their cohesion. But precisely for this reason, the identification of roughly comparable criteria for defining high-order regions deserves more attention than it is customarily given.

To illustrate the cacophony of metageographical discourses that is evident in even the best contemporary efforts to map the world, consider the organization of Rand McNally's *New International Atlas* of 1984.[6] This atlas is one of the few world atlases to diverge from a continental system in organizing its maps, and is certainly one of the best. But aside from cartographic convenience, there is no discernible logic to its highest-order categories. One section is devoted to Europe, a culture region; another to the USSR, a political entity (now defunct); and a third to Asia, a supercontinent. Additional regional headings identify Africa, a passable geological continent; Australia-Oceania, a geological continent plus a dispersed assemblage of islands to which it is minimally related; Anglo-America, a cultural region (albeit one that is "Anglo" only in its majority language); and Latin America, another tolerably coherent cultural region (though one that includes exclaves like Jamaica and Suriname, which have little in common with the "Latin" world). This is but one of many such examples that could be cited.[7] While it would be unwise to impose a single, rigorous logic of division on the world's social space, the multiple logics at work in a given scheme should be both acknowledged and justified.

Our fourth principle for critical metageography is *mastery of the metageographical canon*. While this might seem obvious, it is often ignored. Even among geographers, the highest units of conventional geographical taxonomies are often vaguely understood and imprecisely used; one textbook's "East Asia" may be more expansive than another's entire "Asia." In public discourse, the variation is even wider; it sometimes seems as if world regions can be expanded or contracted at will. For one author, the "Middle East" stretches all the way from Morocco to Afghanistan; for another, it begins only at the Nile Valley; for a third, it might dip southward to encompass sub-Saharan Somalia. Since all human-geographical entities are conventional rather than natural, insistence on a strictly consistent spatial delimitation would be perverse. Yet conventional categories are useful only to the extent that they are clearly understood within particular contexts. The utility of a term like *Southeast Asia* is seriously compromised when for one person it covers little more than Singapore, while for another it extends from Pakistan to Japan. In our experience, it is essential to know with some precision the conventional boundaries of continents, as well as the way world regions are defined by specialists in area studies, if one is to seriously interrogate and improve upon those conventions.

A fifth principle of critical regionalization might be termed *sociospatial precision*. In practice, this means avoiding inaccurate conflations of a given social, economic, or cultural phenomenon with a whole macroregion. The temptation toward broad generalizations is hard to resist when making sound-bite assertions about regional identity. Southeast Asia, for example, is often characterized as a land of remarkable economic success, despite the fact that two of its states, Burma (Myanmar) and Laos, rank among the world's poorest. Equally often, however, one encounters checklist definitions, regional descriptions that attribute a given area's coherence to a finite set of purportedly shared social or material features. This formula is particularly popular in geography textbooks. But checklist definitions present a double problem: discrete cultural traits rarely turn out to have close spatial correspondences, and even when they do, they rarely create significant unity.

Take, for example, a recent textbook that defines the "Middle East" on the basis of "a 'crossroads' location, aridity, oil wealth, Islamic culture, Arabic language, early contributions to civilization, and a recent history of ferocious strife."[8] The authors admit that these attributes are not uniformly distributed in the region and in some cases extend well beyond it, yet argue that these features nonetheless combine to form a "tolerably cohesive area." But consider the list closely. A "crossroads" is a function of history as much as geography; through much of the last millennium, the adjacent zone of Central Asia has arguably been a more significant crossroads than has the "Middle East." The second item on the checklist, "aridity," is by no means prevalent throughout the region's entire extent; the majority of its population in fact lives in areas of reasonably abundant rainfall. "Oil wealth" is a feature of less than half of the countries included as Middle Eastern, and most of the world's oil is produced elsewhere. "Islamic culture" is indeed found through almost all of the area in question, but the region is also home to millions of non-Muslims, and the majority of the world's Muslims live elsewhere. While the "Arabic language" is spoken through a wide swath of the region, well over 100 million "Middle Easterners" speak other languages. Finally, "early contributions to civilization" were made over a vast range of territory, and "ferocious strife" is likewise too widespread to be considered an intrinsic aspect of this or any other region. Indeed, looking at the twentieth century as a whole, Europe has arguably been the most strife-ridden segment of the world. Clearly, a critical metageography should abandon the effort to define regional identity in terms of discrete traits and features. While doing justice to regional complexity takes patience and time, the payoff is high: not only a subtler and more accurate empirical picture of the world, but also a heightened awareness of the variable combinations possible in human social life (and a greater appreciation of how arbitrary macroregional boundaries must be).

A sixth principle we would suggest for metageograpical reform is *definitional integrity*. The most common violation of this principle occurs when what are effectively cultural regions are drawn as if they conformed with current political contingencies. In most representations of the continental scheme, for instance, certain breakpoints between continents are identified with international borders, even when those borders crosscut cultural groupings of long duration. Such a maneuver can result in patent absurdities, as in maps that show the massive supercontinent of Asia terminating along a precise, north-south line through the island of New Guinea. Nor is the continental framework the only one guilty of reifying political boundaries. Geography textbooks that adopt a world regional approach likewise adhere as a rule to the standard of state inviolability, almost never dividing countries into more than one world region.[9] Cleaving to the myth of the nation-state in this way perpetuates some serious cultural distortions: for instance, a trans-Saharan state like the Sudan must be classified in its entirety either with sub-Saharan Africa or as a portion of the Middle East, while the greater Tibetan cultural region ends up being sundered between East Asia (which usually subsumes Tibet proper and Qingdao) and South Asia (typically drawn to include Ladakh, northern Nepal, and Bhutan). To obscure fundamental historical and cultural patterns in this way, simply in order to remain faithful to modern state boundaries, is to naturalize geopolitics to an indefensible extent. In a general-purpose scheme that partitions the world essentially on the basis of shared cultural heritage (language, religion, and intellectual traditions), current political configurations should not be allowed to override those criteria at the boundaries.

A seventh principle for critical metageography is *neutral nomenclature,* or the avoidance of regional designations that carry an unpalatable ideological charge. Traditional European terminology is riddled with implicitly derogatory terms. South Asia (unlike Europe) is still often slighted as a "subcontinent," Southeast Asia has often been reduced to "Indochina," the pre-Columbian history of the Americas is still disparaged in the designation *New World,*[10] and so on. But loaded labeling also besets more recently derived schemes. The division of the world into macroregions may be far less Eurocentric than its division into continents, but it still bears traces of its origin in an imperialistic tradition. The term *Middle East,* as we have seen, is a military designation that makes sense only from the perspective of western Europe, while the name *Latin America* reflects eighteenth-century French territorial ambitions in the Americas. To be sure, geographers have no brief—and no ability—to replace such widespread terms with totally value-free language. Nonetheless, nomenclatural analysis should certainly be on the agenda of a critical metageography.

So, too, should an eighth imperative: *historical specificity,* or the recognition that world regions do not constitute timeless entities (and that therefore a good regionalization scheme will not be applicable across all historical periods). The importance of recognizing historical contingency in the creation of macroregions has been explored at length in the case of

Central Asia, a designation with rich resonance for historical studies but uncertain contemporary relevance. Another way to demonstrate the importance of acknowledging shifts in regional identity over time is to consider the case of Pakistan.

Pakistan is routinely regarded by scholars as part of South Asia, for solid historical and cultural reasons, and many Pakistanis would agree with this designation. The concept of a distinctive South Asian world region is both indigenous and ancient, although its boundaries have varied considerably over the millennia. But with the recent spread of Islamic fundamentalism[11] and with the growing tensions between Hindus and Muslims in India, many Pakistanis are now arguing that their country should be included in the primarily Islamic region centered on Southwest Asia and North Africa. For now, it is perhaps best to regard Pakistan as a Janus-faced country, looking both east to South Asia and west to the "Middle East."[12] In time, local sentiments might resolve this issue clearly; for a future generation, Pakistan's onetime designation as part of South Asia may well seem quaintly anachronistic.[13] Meanwhile, however, scholars must take account of the extent to which regions like South Asia represent contingent historical constructs.

A ninth principle, which might be termed *contextual specificity*, is akin to the eighth. Just as macroregions are habitually generalized across time, so too are they routinely imported into contexts where they simply do not apply. General-purpose regionalization schemes (such as that of area studies) have an important heuristic function, but no geographical scheme can legitimately be deployed as an all-purpose framework. The project of metageographical reform must face the challenge of coming up with multiple crosscutting and overlapping regionalisms for different purposes.

A prime example here can be seen in the patterning of economic activity. As both historical and contemporary research have repeatedly shown, long-distance economic networks often cut across the boundaries of cultural regions. Capital seeks mobility, and the cheapest form of transportation has historically been water-borne. Accordingly, it is essential in some contexts to deploy a regionalization scheme centered on oceans and bays rather than on continents or cultural blocs. The Indian Ocean, the Baltic of the Hanseatic league, the much-traveled Mediterranean Sea, and more recently the Pacific Ocean have all facilitated the creation of complex webs of capital and commodity exchange. Only a sea-centered perspective is capable of revealing these economic regions, which carve up conventional land-centered blocks in unexpected ways.

But as important as it is to recognize alternative ways of drawing regional boundaries, the act of decentering our inherited regional categories must go still deeper. Critical metageography must also tackle a tenth and final challenge: devising a *creative cartographic vision* capable of effectively grasping unconventional regional forms. It is simply no longer tenable to assume that all significant high-order spatial units will take the form of discrete, contiguous blocks; analyzing contemporary human geography requires a different vocabulary. Instead of assuming contiguity, we need a way to visualize discontinuous "regions" that might take the spatial form of lattices, archipelagos, hollow rings, or patchworks. Such patterns have certainly not displaced the older cultural realms, which continue to be important for the multiple legacies they have left behind. But in the late twentieth century the friction of distance is much less than it used to be; capital flows as much as human migrations can rapidly create and re-create profound connections between distant places. As a result, some of the most powerful sociospatial aggregations of our day simply cannot be mapped as single, bounded territories. To grasp these new realities will demand an imaginative approach to regional definition; to map them effectively will demand an innovative approach to cartography.

Like the other items in the foregoing inventory, overcoming the contiguity fetish of prevailing regional schemes will not in itself produce a better map of the world. But taken together, the ten principles we have identified would do much to rectify the most glaring of metageographical mistakes. By vigilantly combating the stubborn remnants of Eurocentrism and environmental determinism, conforming to basic principles of logical typing and typological honesty, mastering the metageographical canon, insisting on sociospatial precision, maintaining definitional integrity, aiming for neutral nomenclature, acknowledging the historical and contextual specificity of world regions, and reaching for creative means to visualize and express unorthodox regional forms, we would at least stand a chance of passing on a less distorted picture of the world than the one we have inherited. This short list cannot claim to be a comprehensive rulebook for generating a perfect spatial taxonomy. Yet it does point the way toward a more accurate and ecumenical vision of the globe.

New Directions

Generating that vision of course requires more than an abstract grammar. To implement these principles effectively, there are many empirical research questions that must be answered as well. We have accordingly decided to conclude this study by sketching out some particularly important directions for future metageographical study.

The first area that cries out for attention is pedagogical research. Efforts to reconceive global geography will have an impact only to the extent that they can reach a broad audience. Here professional geographers have a mandate, since it falls to them to teach students the map of the world. Serious scholarship ought to be brought to bear to enhance the sophistication of basic geographical education, especially at the college level. There is much room for improved textbooks and classroom materials, as well as for more attention to the map as a teaching device. Technical developments in cartography, including geographical information systems (GIS), will be integral to this effort, allowing as they do the depiction of more complex patterns, and more relationships among different patterns, than is possible in conventional cartography. To take full advantage of these possibilities, however, it will be necessary to bring together two of the most estranged subfields in the discipline. To date, most geographers interested in GIS have shown little concern for the kinds of conceptual issues explored in this book, while most of their culturally and historically inclined colleagues have regarded GIS as a forbidding technical specialty, if not a tainted tool of objectification. If our classroom maps are to reflect a truly critical metageographical sensibility, this gap will have to be bridged.

While pedagogy thus constitutes one essential domain for future work, numerous other areas can be identified within what is customarily called the history of geographic thought. One vast and largely unexplored question is how various non-Western peoples have apprehended global human patterns. Herein lies perhaps the most important terrain for future scholarly work in metageography. The present study, like the field as a whole, has concentrated primarily on the history of European-derived regional schemes, touching only briefly on worldviews developed elsewhere. It has also only begun to suggest the manifold ways in which European metageographies have been appropriated and transformed within other intellectual frameworks. Both of these topics are enormously rich and would repay extended study by specialists with the requisite language skills.

A cornerstone for future studies of premodern non-Western global visions will be the massive *History of Cartography* series, edited by the late J. Brian Harley and David Woodward under the auspices of the University of Chicago Press. Harley and Woodward's project, a global survey of cartographic thought, has already resulted in the compilation and analysis of thousands of maps from all parts of the world. Many of the contributors' essays bear directly on metageographical concerns, inasmuch as they examine one or more depictions of the globe as a whole. It may be hoped that other scholars will be inspired by this pioneering work to produce further assessments of metageographical visions from different times and places, including those expressed in discursive as well as pictorial form.

In addition to premodern indigenous traditions, however, the appropriation of European-derived geographies within non-European intellectual frameworks constitutes another essential area for future research. One promising approach here would be to analyze elementary geography textbooks and popular atlases, to see how various modernizing regimes have chosen to represent the world to their school-aged children. Another would be to look at the macroregional categories deployed in journalism (with attention to the different geohistorical visions articulated by competing ideologies). A third avenue would be to look at the global rhetoric employed by such regional organizations as the Association of Southeast Asian Nations (ASEAN), the Organization of African Unity, the Organization of American States, the Arab League, or emerging transregional economic blocs like those set up by the North American Free Trade Agreement (NAFTA) and the Asia Pacific Economic Cooperation forum (APEC). Such organizations' global visions are a complex product of contemporary geopolitics, inherited intellectual schemes, and hoped-for realignments envisaged by certain powerful groups. Among other things, it would be good to know how widely their categories are adopted outside of their own institutional circles, particularly in cases where metageographical rhetoric incorporates a potent political message.

The political charge of macroregional classification is especially evident today in Turkey, Russia, and China. In each case, different political factions have contending visions of how the world is (and should be) ordered, with a fundamental split between those who stress transregional ties and global connections and those who emphasize a more "traditional" picture of civilizational disparities. In Turkey, contention over the nation's fundamental geographical orientation has emerged as a central issue in electoral maneuverings in recent years. In these instances, and in others like them the world over, scholars would do well to attend to the metageographical substratum of contemporary political debate. It will also be important for American researchers to seek a genuine dialogue with scholars from these areas—especially with those who picture the world in terms strikingly different from our own.

In our view, these and other questions about non-Western modes of apprehending the world are the most important topics for future metageographical research. But numerous frontiers remain to be explored within the Euro-American tradition as well. Coverage of that tradition to date has been uneven, as reflected in this work. It is not only because both of us were trained in Asian studies that most of the examples in this book are drawn from the historical geography of Eurasia. Another factor is simply that conceptions of areas outside the "Old World ecumene" have not received the same level of scholarly attention as those within it. There are some signs that this imbalance is being rectified in recent years. One legacy of the 1492 Quincentennial, for instance, was a new interest in perceptions of the Americas among scholars of the Renaissance. The modern history of regional identities and images within Latin America has also begun to attract attention, with highly promising results.[14] Yet the field remains wide open for students of the Americas, sub-Saharan Africa, the Pacific islands, and other regions to explore the shifting metageographical conventions that have helped shape scholarship on those parts of the world.

Another area awaiting research is the entire second tier of the geographical taxonomy: those regional designations that identify areas larger than a country but smaller than a continent. There is much yet to be learned about the intellectual histories, cultural connotations, and strategic deployments of such terms as Middle America, Central America, Mesoamerica, the Andean Realm, the Caribbean borderlands, the Southern Cone, and other designations. These are themselves complex and often loaded concepts, deserving of rigorous study. It would be fascinating to know whether there have historically been close ties between transformations in these second-order regionalizations and shifts in continental or world regional convention.

It is also to be hoped that future work will devote attention to interstitial zones, which have often been slighted precisely because they do not fit easily into established metageographical frameworks. The Caucasus is one case in point. Though home to a singularly dense patchwork of peoples and languages, the Caucasus remains effectively invisible in most schemes, having been designated not a region in its own right but a borderland between Europe, the "Middle East," Inner Asia, and the Russian sphere. Other frequently overlooked areas include highland Ethiopia and greater Tibet, both of which share with the Caucasus the distinction of failing to fit neatly into any one conventional world regional category. Many islands and archipelagos, too, remain for the most part little-known anomalies. Mauritius, for instance, is never included in the standard list of the world's "newly industrializing nations," despite its impressive record in social and economic development.[15] One reason, surely, is the fact that our standard frameworks literally have no place for such a culturally hybrid and oceanically isolated country. All such areas are worthy subjects for extended metageographical analysis.

A final issue we would single out as an important locus for new work is the intellectual history of maritime regions in the geographical imagination. While habitually slighted in our stubbornly continental schemes, communities oriented around the world's major seas have become increasingly visible in recent decades, to historians as well as contemporary analysts.[16] The Black Atlantic is a prime example here. Although this region remains fundamentally obscured on most white Americans' maps of the world, it has recently come into sharp focus in African-American and cultural studies. Similarly, the Indian Ocean world has been rendered visible by a score of exciting new studies in early modern economic history, largely the work of South Asian historians. But in popular discourse, the most prominent maritime region to emerge in recent years is undoubtedly the Pacific Rim.

Neither "Asia-Pacific" nor "Pacific Rim" has yet joined the roster of geographers' standard world regions,[17] but both have gained wide currency in journalism and social science research. Spatially, their territories overlap without being entirely congruent. "Asia-Pacific" centers on East and Southeast Asia, sometimes extending southward to include Australia and New Zealand, and occasionally reaching as far into the Eurasian interior as India and eastern Russia; the "Pacific Rim" is an equally plastic construct, theoretically encompassing all the lands that border on the Pacific Ocean but conceptually anchored in the Japan-U.S. trade nexus. In fact, both Asia-Pacific and the Pacific Rim are implicitly conceived in economic terms. Both are premised on the profound economic linkages uniting Tokyo, Seoul, Hong Kong, Singapore, and Sydney with markets and workers from Vancouver to Tijuana and beyond. What neither adequately conveys, however, is the fragmentary, uneven structure of the Pacific economy: the disjunct distribution of sweatshops, research facilities, and markets that are now linked through transpacific flows of capital, commodities, and labor. Nor does either term convey the extent to which the processes that join them simultaneously differentiate them into rich and poor, dynamic and impoverished segments. The Pacific Rim as a unified, progressive bloc exists primarily in propaganda; as Alexander Woodside has observed, the term is more prophetic than descriptive.[18] For these reasons and many more, sharp attacks have been launched against "Rim-speak" and its ideological connotations.[19]

The debate over how to conceptualize the Pacific community goes to the heart of the challenge before us. First, it conveys a sense of the immense difficulties entailed in rightly grasping—and naming—the elusive spatial structures of contemporary life. All geographical divisions share with these neo-categories the quality of being artificial simplifications, more-or-less convenient devices for advancing analysis rather than reflections of natural, wholly knowable spatial structures. At the same time, this debate exposes the extent to which spatial categories are embedded in a discourse of power. Like "Asia Pacific" and "Pacific Rim," all regional designations need to be subjected to political critique.

Yet we must not stop there. While deconstructing received metageographical categories is essential, it is only the first step toward constructing a more appropriate sociospatial lexicon. Faced with the daunting challenge of that larger project, it might be tempting to succumb to the postmodern mood and declare the end of metageography altogether. But metageography refuses to die. If one is to think seriously about the world, one must have recourse to a spatial vocabulary. In the absence of better alternatives, exhausted metageographical concepts will creep back into use, even in acutely critical postmodern texts. So while the epistemological and political limitations of spatial discourse should humble us, they need not force us to throw all of our maps to the wind. On the contrary, they should persuade us to retain and reform those maps, recognizing that weighty ideological issues are at stake in the way we conceptualize the world. Metageography matters, and the attempt to engage it critically has only begun.

Notes

Preface

1. Murray Hiebert, "Wizard of Oz: Australia's Evans Redraws the Map of Asia," *Far Eastern Economic Review*, 17 August 1995, page 26.

2. Our term *metageography* recalls Hayden White's term *metahistory*, but their meanings are dissimilar. White is concerned with the "deep" *poetic structures* through which history is emplotted and its narratives cast. Our concerns are far more prosaic; we are largely interested in the primary *spatial structures* around which we habitually conceptualize global geography. A different kind of metageography, of course, could be constructed around the poetics of geographical writing (see, for example, Jonathan Smith 1996). Nor are we much concerned with the "meta-narratives" that undergird our understanding of global history for their own sake, but we are very interested in the geographical stage on which those same narratives are played out. Historical metanarratives do imply, at some level, spatial meta-structures, although many authors fail to recognize it. White's own *Metahistory* (1973) is actually constructed around an entirely conventional global geographical vision, one that differentiates absolutely the "Western world"—populated by "Western men" and coincident with both modernity and industrialism—from all of the rest of the planet: "[T]he historical consciousness on which Western man has prided himself since the beginning of the nineteenth century may be little more than a theoretical basis for the ideological position from which Western civiliza-tion views its relationships not only to cultures and civilizations preceding it but also to those contemporary with it and contiguous with it in space. In short, it is possible to view historical consciousness as a specifically Western prejudice by which the presumed superiority of modern, industrial society can be retroactively sub-stantiated" (page 2).

3. On the Cold War origins of the idea of the Three Worlds, see Pletsch 1981, page 569.

4. See Rozman 1991, page 13.

5. See Frank and Gills 1993.

6. See S. Huntington 1993.

7. See Cheney's attack on the new American and world history standards in her article "The End of History," *Wall Street Journal,* 20 October 1994, page A-22. Cheney's position, as Ross Dunn notes in the Internet List Service "World-L" (15 November 1994), is that world history and Western history should be regarded as being essentially the same thing. For the proposed world history standards, see *World History: Exploring Paths to the Present* (Los Angeles: UCLA National Center for History in the Schools, 1994).

8. See Fukuyama 1992.

9. See Barber 1995.

10. Kaplan 1994, page 75.

11. See, for example, Thrift 1993.

12. By the term *construct* we do not wish to imply—as some postmodernists would—that such notions merely reflect the interests of specific social groupings, with no grounding empirical conditions. Rather, we follow I. G. Simmons (1993, page 3) in the belief that "there is indeed a 'real' cosmos out there but we are too limited to comprehend its true nature. . . . In order therefore to reduce the mass of information to something which we can tell ourselves that we understand . . . we make constructions of various kinds. . . . we shall accept from this moment onwards that they are all imperfect and can be only provisional."

13. Geographical illiteracy is unfortunately often evident even among scholars attempting to reconceive global metageography. Benjamin Barber, for example, writes of North Korea as an economic "tiger," and he evidently believes that Indonesia is a predominately Buddhist country (1995, pages 51, 70). Furthermore, the fact that he calls the various struggles of ethnic nationalism now being waged "jihads" betrays a poor understanding of Islam; even in its most violent manifestation, the *jihad* is motivated by religious universalism, not ethnic particularism.

14. This information is derived from student surveys that we conducted while teaching at Duke University, 1992–93.

15. See Grosvenor 1995; Downs 1994. The National Geographical Society sponsors a series of "geographical alliances" in which professional geographers work with primary and secondary teachers in order to improve geographical education. While the alliances have received much praise, they have also been subjected to some harsh criticism (see, for example, Gary Fuller 1989).

Introduction

1. See Rand McNally 1991–92.

2. See Jameson 1986; Ahmad 1987.

3. Criticisms of the notion of a Third World are rapidly mounting and are appearing from all points along the political spectrum. As Linda McDowell (1993, page 313) writes, from a leftist perspective, "It is apparent, for example, that our conventional practice of mapping the first world/third world dichotomy onto a territorial division of the world not only embodies imperialist assumptions but is also an increasingly inadequate portrayal of the world's socioeconomic geog-

raphy." For a recent examination of how Western intellectuals constructed the category of the Third World in the post–World War II era, see Arturo Escobar (1995). The resulting geographical image, Escobar cogently writes, "universalizes and homogenizes Third World cultures in an ahistorical fashion" (page 8). Aijaz Ahmad (1992) also traces the origin of the term, and he provides a particularly cogent examination of the confused thinking upon which the concept of the Third World ultimately rests. See also N. Harris 1986, Arnold 1993, and Pletsch 1981.

4. See, for example, the maps in Barke and O'Hare 1984 (page 1) and Dickenson et al. 1983 (frontispiece). The latter writers define the Third World, in largely continental terms, as the "independent nations of Africa, Asia, and Latin America" (page 1)—although in practice they exclude Japan. A more complete cartographical exposition of the Third World is found in Kurian (1983). Here South Africa, China, and Taiwan are excluded from the category; otherwise the mapping is entirely conventional, with both Singapore and South Korea counted as Third World countries.

Interestingly, many Chinese discussions of the Third World from the 1960s insisted upon China's unproblematic Third World status, yet categorized the Soviet Union, alongside the United States, as a First World power. In this formulation, the "non-superpower" industrialized countries were seen as forming the Second World (see Ahmad 1992, page 306).

5. Hermassi (1980, page 5) thus defines the Third World in terms of "the three continents" of Asia, Africa, and Latin America, while Hadjor (1992, page 2) writes that the term is "a reflection of the reality of the continents which have been excluded from power."

6. See Pletsch 1981, page 565.

7. See "Letters," *New York Review of Books*, 24 March 1994, pages 60–63.

8. Even more confusion is apparent when one considers the so-called Fourth and Fifth Worlds, which many writers take to be those countries (or areas) that are not even wealthy and powerful enough to be counted within the Third World (Ridker 1976). To others, however, the Fourth World refers mainly to poor people living in rich countries (Fourth World Movement 1980). On the other hand, Bernard Nietschmann, the graduate school adviser for one of us (Lewis), insisted that the Fourth World consisted of indigenous peoples struggling for autonomy, while the Fifth World consisted of refugees.

9. The demise of the Third World is also controversial. Nigel Harris (1986) maintains that the Third World has disappeared because of economic growth in the Newly Industrializing Countries, or NICs; Arnold (1993, page viii), to the contrary, maintains that it has disappeared because in the "new world order" there is room only for a "dominant North" and a "dependent South." Hettne (1995, page 265) writes that the "Third World is disappearing," although he still considers it analytically useful for the present. Other contemporary writers, however, vigorously maintain the idea of Third World identity. Kafi Buenor Hadjor (1992, page 11), for example, regards the Third World as a "vast and living social organism." Even the term *Second World* continued to be used well after the demise of Soviet and eastern European communism (see, for example, Merchant 1992, page 26 ff.).

10. On the former party, see Pletsch 1981; on the latter, see N. Harris 1986.

Originally the term *Third World* was used mostly by leftist writers (see Hadjor 1992, page 4).

11. The North-South divide is generally dated back to the 1980 report of the Independent Commission on International Development Issues, chaired by Willy Brandt. Here the authors realized the difficulties involved with making such a monumental global bifurcation, and they were uncertain where to slot China (pages 31–32). They also used the term *Third World* interchangeably with *South* (page 21, for example). See R. J. Johnston, Taylor, and Watts (1995, page 16) for the idea that the Third World has "been replaced, after a decade of severe discipline meted out by the global regulatory agencies, by a map of North-South dependency."

Generalizations about the South are difficult to make, but the gambit often proves irresistible for concerned scholars. Consider, for example, the following passage from Peter J. Taylor (1992, page 12), a highly respected political geographer: "The failure of development is the opportunity for Islam. Does our future hold for us a great battle of *longue durée* dimensions between our 'modern North,' and their 'postmodern South'?" Equating the South with Islam is problematic enough, but the "modern-postmodern" dichotomy employed here is simply bizarre. In an earlier passage, it must be admitted, Taylor expressly limits this challenge to "northern frontier of the poor world," or the area along the "'Islamic Rimland' from North Africa to Southeast Asia" (page 12). Yet India's Hindu "fundamentalists," as well as Singaporean and Thai capitalists, might find this odd classification scheme more than a little objectionable.

12. See, for example, the map in Dickenson et al. 1983 (page 2).

13. As John Ravenhill (1990) argues, the South is now much more differentiated than it was in the 1970s. As a result, he proposes an economic reclassification of the area that would yield five new categories: (1) high-income oil exporters, (2) industrializing countries with strong states and low debt, (3) industrializing countries with debt problems, (4) potentially newly industrializing countries, and (5) primary commodity exporters (pages 745–46). His classification of both the Philippines and Malaysia under category 4 is, however, most curious.

14. According to World Bank statistics (World Bank 1995, pages 18–19), the per capita GNP of Russia in 1993 was $2,350 (when measured in purchasing power parity [PPP], $5,240). These figures are exceeded by those of such "Southern" countries as Argentina (7,290 and $9,130 respectively), Brazil ($3,020 and $5,470), Chile ($3,070 and $8,380), Gabon ($4,050, PPP not available), Malaysia ($3,160 and $8,630), Mexico ($3,750 and $7,100), Panama ($2,580 and $5,940), and several others. Certainly Russia has higher general levels of social welfare than many of these countries, as well as a more sophisticated infrastructure and technological base. Yet even in terms of such measures of social welfare as average longevity and infant mortality, Russia is surpassed by such a quintessential Third World country as Sri Lanka (World Bank 1995, page 9).

As Martin Malia, writing under the pseudonym "Z," argues (1991, page 300): "Even in its most successful decades . . . the Soviet Union was never a great industrial power, and still less a 'modern' society. . . . In reality [it was] never more than a great military-industrial complex and a Party-state superpower." While Malia is no doubt overstating his case, the notion that a wealthy or even a "developed"

North confronts a poor South is as outdated as the division of the globe into First, Second, and Third Worlds.

15. Such failings seem to be particularly common among radical environmentalists. Barry Commoner (1990, page 166), for example, seems to believe that most of the world's population resides in the Southern Hemisphere. In a similar misconception, Carolyn Merchant (1992, page 25) places all of Asia, Latin America, and Africa in the Southern Hemisphere.

16. See, for example, Hermassi 1980. As Pletsch (1981, page 576) shows, the Second World was always the least-used section of the tripartite scheme, and in fact it has often been completely ignored. Note also that Hadjor (1992, page 1) maintains that the terms *East, South,* and *Third World* are essentially interchangeable. Prakash, however, argues that "the Third World is frequently resituated outside such binary oppositions as the Orient and Occident" (1990, page 383); indeed, escaping this polarity is in one sense the precise reason for positing a "third" portion of the world in the first place. But this "resituation" is never complete, and the distinction between East and West remains implicit in virtually all discussions of the Third World.

17. Cohn 1987, page 35. As Cohn explains, under the influence of modernization theory, Western scholars long assumed that the rest of the world would become like the West (see also Slater 1995). Therefore, the gross imbalance in this global division caused them little concern.

18. See Blaut 1993; see also T. Mitchell 1988 (pages 165–68), 1992.

19. For all Blaut's skill in exposing the central features of the "colonizer's model," he remains in many instances mired in a traditional Eurocentric geographical vocabulary. He continues to treat "continents," for example, as nonproblematic units (see pages 51, 94, 115, 152, 161). He also devotes very little attention to the actual spatial form of the model he is criticizing. As a result, he sometimes takes advantage of the locational fungibility of metageographical concepts in a manner reminiscent of the most devoted Eurocentrists. Thus he includes the Near East within Greater Europe when he finds it convenient to do so, yet excludes even Transylvania from the same category on other occasions (see pages 5, 21, 41, 43, 47).

Similarly, Timothy Mitchell (1988) employs the stale and Eurocentric category of the Middle East throughout his work, and at one point seemingly equates it with the "Mediterranean world" (page 173). In its geographical vocabulary, much postcolonial theorizing remains surprisingly colonial.

20. Blaut does discuss inverted Eurocentrism as it is encountered among Afrocentrists (1993, page 208).

21. This term is borrowed from Mikesell 1983.

22. The problems inherent in the "nation-state" formulation are increasingly noted by geographers: see for example, Mikesell 1983 and de Blij 1992 (page 17); on the general problem of nation-state congruence, see Anthony Smith 1986 (pages 221–26). Indeed, in academic geography over the past several years the pendulum has swung so far away from the idea of nation-state identity and permanence that Peter Taylor has felt it necessary to warn readers that it is premature to report the death of the state (1994, page 161).

Not only is the identification of the state with the nation problematic, but so too is the assumption that each state necessarily possesses a clearly demarcated and unambiguous territory—except in relatively recent times. As Thongchai (1994) shows in the case of Thailand, the indigenous state was geographically unbounded, both due to multiple layers of sovereignty and because of shared sovereignty with neighboring countries. The modernization of Thailand, in turn, required not only that its boundaries be regularized, but also that its earlier forms of geographical expression be denied and suppressed. Davies (1989, page 100) perceptively argues that our entire historiographic tradition has a deep "centralist point of departure. . . . [and] bias. . . . in favor of strong government, legal uniformity, and direct and clear lines of command and authority. Such a historiography finds it difficult to come to terms with societies which are institutionally fragmented, fluid in their frontiers, multiple in their loyalties, cultures and laws, and normally well beyond the reach of the practical authority of metropolitan government."

23. Chauncy Harris 1993, page 313.

24. A good example of geographical myopia in a non-Western source is the anonymous *Hudud al-ʿAlam* (1937), originally written in Persia in A.D. 982. Here the author divides the world into fifty-one discrete units—of which some nine cover portions of modern Iran and just one encompasses all of Europe except Spain and Russia.

25. See, for example, *Hammond World Atlas* (Citation Edition, 1992).

26. In the *World Almanac and Book of Facts 1993*, for example, Monaco and Liechtenstein are each allotted approximately one half-column of text, while India rates only some two and a quarter columns of text. Yet India has 30,000 times as many inhabitants as Monaco and is 2 million times as large in area.

27. Some critics of the Enlightenment tradition (for example, Pratt 1992) view the Linnaean taxonomic tradition as an imperial projection of European power from the very beginning and as an endeavor of hegemonic intellectual totalization. We disagree; as long as one accepts the veracity of organic evolution, a taxonomic approach to the diversity of life is essential. We are, rather, concerned here with the misapplication of taxonomy to geography. See also Inden (1990, pages 16, 31, 60) for insightful comments on the role of taxonomic thinking in global historical conceptualizations.

28. See, for example, Masica 1976.

29. There is little novel in this vision of geographical complexity; such is the manner in which the best geographers have always viewed the world. Many of the basic principles employed here were formulated by a small group of conceptually inclined regional geographers in the 1950s. Scholars such as Richard Hartshorne, Derwent Whittlesey, and Robert Minshul recognized that since every mappable phenomenon generally has a unique distributional pattern, the region itself must be a human construct (see M. Lewis 1991). They further argued that any composite region, or area used as a framework for analyzing a variety of natural and social features, is more akin to an artistic creation than to a preexisting given. The need for employing multiple, noncongruent regionalization schemes, paying attention to interstices as well as centers, analyzing gradients and variations in intensity as well as boundaries and other sharp discontinuities, and discerning the hierarchical order of regions defined at different spatial scales—notions

essential to the basic conception of world geography employed in this work—
were all recognized by regional geographers during this period. Perhaps most important, geographers at midcentury insisted on distinguishing between the formal region, based on distributional uniformity, and the functional region, defined
by interconnections (often economic) focused on a single node.

Although the basic principles of regional structure were elucidated decades
ago, they were never fully embraced by the geographic community, let alone by
scholars in other disciplines. Moreover, even Whittlesey and Minshul never systematically applied their ideas to the global scale of analysis. By the 1960s any concern with global geography and large-scale regionalization schemes seemed to
many geographers a quaint remnant of bygone days; geographers in this period
were concerned largely with the mathematical modeling of spatial laws that supposedly determined the locations of all human endeavors.

A much-heralded "new regional" approach appeared in academic geography
in the 1980s, promising to revitalize the field (see Pudup 1988; Gilbert 1988). In
practice, however, this movement seems to have floundered soon after it was an-
nounced, and it may even have further weakened geographical conceptualization
in several key respects. Seeking to establish connections with the mainstream of
the social sciences, the new regional geographers adeptly imported and employed
the latest social theory, but in the process they tended to lose sight of geography
itself. In characteristic works of the genre, conventional and taken-for-granted subnational regions, like Appalachia or the Pacific Northwest of the United States,
are selected as unproblematic spatial containers within which to conduct socio-
logical or economic analysis; thus Alexander Murphy (1991, page 27) accurately
writes of a "general failure of much contemporary geographically informed social theory to confront the ideological and experiential dimensions of regionalism." In several contemporary geographical subdisciplines, the only indication of
actual geographic content is the use of such marker terms as *space, place, location,*
or *region* (this is especially apparent in Gregory 1994).

In geographical studies where a global perspective is adopted, the organizational and theoretical framework most often selected is that of world-systems
theory. While the world-systems paradigm is generally useful for making sense of
economic relationships (but not always, see chapter 5), it sheds little light on issues of cultural affiliation, and is thus of limited use as a general geographical guide.
The most significant global regions, regions such as South Asia or Latin America, are in essence cultural. A "geography of culture[s]," unfortunately, has never
been developed at anything but the most superficial level anywhere in the discipline. In an earlier generation an attempt was made to fashion a global cultural
geography centered around what Donald Meinig (1956, page 204) called the
"structure of culture regions," but with the exception of some notable work conducted on North America (much of it by Meinig himself), such an endeavor remains a "rather vague study." On current trends in cultural geography, see Price
and Lewis 1993.

30. See, for example, the discussion in Stoianovich 1994, page 20.
31. Gregson 1993, page 529.
32. Chaloupka and Cawley 1993, page 5; see also Demeritt 1994.
33. Doel 1993, page 384.

34. For an interesting take on the common roots (and problems) of modernism and postmodernism, from a geographical perspective, see Berg 1993.

35. Cronon 1994, page 42.

36. 1993, page 5.

Chapter 1: The Architecture of Continents

1. Toynbee 1934–61, volume 8 (1954), pages 711–12.

2. See Tozer 1964, page 67; see also Herodotus 1954, page 135. Hecataeus, perhaps owing to his "symmetrical turn of mind," was a strong proponent of the twofold system, subsuming Libya into Asia (see Bunbury, 1959, page 145).

3. It has been suggested that the term *Asia* originally referred only to the plains to the east of Ephesus—the "sunrise" direction from that Greek city of Asia Minor (see Lyde 1926, page 6). This thesis, predicated on the notion that Asia and Europe were derived from words for sunrise and sunset (see note 11, below), is no longer widely held.

4. Toynbee 1934–61, volume 8 (1954), page 718.

5. De Rougemont 1966, pages 36–37.

6. See Tozer 1964, page 69. This position of Europe in Greek thought is tremendously confused. According to one interpretation of the ideas of Euphorus, for example, " . . . Europe means first of all the Greek world or the world occupied by the Greeks" (Van Paassen 1957, page 254). Stoianovich (1994, page 2) argues that originally *Europe* referred only to central Greece.

7. See Aristotle 1932, pages 565–67: "The nations inhabiting the cold places and those of Europe are full of spirit but somewhat deficient in intelligence and skill, so they continue comparatively free, but lacking in political organization and capacity to rule their neighbors. The peoples of Asia on the other hand are intelligent and skillful in temperament, but lack spirit, so they are in continuous subjugation and slavery. But the Greek race participates in both characters, just as it occupies the middle position geographically, for it is both spirited and intelligent."

8. See P. Burke 1980. According to common Greek notions of environmental determinism, however, physical geographical differences could be translated into distinct cultural differences between Europe and Asia. This is notable in the works of Herodotus, Aristotle, and especially Hippocrates (Glacken 1967; Van Paassen 1957).

9. Aujac 1987, page 136.

10. Herodotus, 1954, pages 134–35.

11. Ibid., page 285. The term *Asia* has often been linked to the Phoenician word for "sunrise" (W. H. Parker, 1960, page 278), but modern scholarship more often derives it from the Kingdom of Assuwa in Anatolia or the city of Assos near Troy (Bernal, 1991, page 201); Toynbee (1934–61, volume 8 [1954], page 711) links it ultimately to a local term for "marsh," while others have favored "mainland." Martin Bernal explains how its referent was enlarged: "When the Lydian Kingdom of Western Anatolia was incorporated into the Persian Empire in the 6th century B.C., Ionian geographers extended the meaning of 'Asia' in two ways, to cover

the whole of Anatolia and also to be the name for one of the three continents, with Europe and Libya (Africa)" (1991, page 201; see also Steadman 1969, page 45).

The term *Europe,* which dates back at least to Hesiod circa 800 to 900 B.C., was previously linked to a Semitic term for the setting sun, or to a Greek word meaning "broadfaced" (D. Hay 1957, page 1). Many today trace it to a term meaning simply "mainland" (*Encyclopedia Britannica* 1992, volume 18, page 522), although Traian Stoianovich maintains that it simply means "wide prospect" (1994, page 2). At any rate, the more direct antecedent is Europa, legendary daughter of the king of Tyre, who was abducted and raped by Zeus, and carried off to the island of Crete. Many other origins have also been proposed, however, and one might consider here Denis de Rougemont's (1966, page 25) warning: "Etymology is chiefly a list of errors, mistaken analogies, puns, and boners; practiced as an art, it finds only too many 'significations,' and only its choice among these is significant."

A variety of other controversies surrounded the division of the world throughout the classical period. Some writers, for example, favored an oceanic model that limited the areal extent of the continents and regarded them all as surrounded by water; those favoring the opposing "continental hypothesis" believed that Asia and Africa extended far to the east and south respectively, and that the known oceans were ultimately surrounded by land (see Wright 1925, pages 18–19).

12. Strabo 1854, page 102.

13. De Rougemont 1966, page 41.

14. P. Burke 1980, page 23.

15. De Rougemont 1966, page 19. As Denys Hay (1957, page 9) relates, this idea can be traced back to Josephus in the first century A.D.

16. Woodward 1987, page 334. As Beazley (1949, volume 2, page 577) relates, some medieval maps showed Europe as larger than Asia.

17. See Strabo 1854, page 55.

18. On the importance of Martianus Capella, see Lindberg 1992, page 145.

19. See Capella 1977, pages 231–32.

20. See Kimble 1938, pages 20, 24.

21. See D. Hay 1957, pages 25, 52.

22. Leyser 1992, page 37.

23. See P. Burke 1980; W. H. Parker 1960.

24. See Wright 1925, page 74.

25. As Woodward (1987) explains, "accuracy" of spatial representation was not the aim of the creators of the T-O maps. It should also be noted that medieval cartographers sometimes depicted additional unknown continents (see Beazley 1949, volume 2, pages 571–72; Wright 1925, page 157; J. Friedman 1994, page 67).

26. W. H. Parker 1960, page 281.

27. Glacken 1967, page 365.

28. See P. Burke 1980, page 23.

29. See D. Hay 1957.

30. Toynbee (1934–61, volume 8 [1954], pages 720–21) thus argues that the Renaissance actually saw a deterioration in geographical conception, as continental terms such as *Europe* were imbued with social and cultural rather than mere physiographic qualities. The primary mistake was to assume that boundaries of these

two Europes—physical and civilizational—should coincide, a situation that was only temporarily approached. The wars and population movements of the early twentieth century actually created the closest fit ever seen between these "continental" and cultural areas, at least in the vicinity of the Aegean, and at present we are therefore inclined to view the sea's western shores as Greek, Christian, and (thus) unproblematically European, and its eastern shore as Turkish, Muslim, and (thus) Asian. It is worth recalling, however, that in the seventh and eighth centuries, a time when Asia Minor was largely Greek-speaking and Christian, the Greek peninsula itself was largely non-Greek-speaking and non-Christian (on the non-Greek nature of the Greek peninsula in this period, and its subsequent re-Hellenization, see Threadgold 1988, page 157 ff.; on the Greek nature of Anatolia in this period, see Vryonis 1971, page 42 especially). It is only by virtue of rare historical accidents that such cultural and physical boundaries ever coincide.

31. See O'Gorman 1961.

32. Zerubavel 1992, page 69.

33. See Lach 1977, page 273.

34. Mignolo 1993, page 240. Mignolo contends that the Spaniards preferred the notion of the Indies over that of the Americas in part because it linked its colonial possessions in the Western Hemisphere with the Philippines.

35. See Zerubavel 1992, pages 79–81.

36. See Mignolo 1993. The Americas were often viewed as the distinct realm of nature through the early nineteenth century, and many Enlightenment thinkers considered nature in the Americas to be inferior to that in the Eastern Hemisphere (see Pratt 1992; Glacken 1967).

37. O'Gorman 1961, see especially pages 55, 127–137.

38. Of course, rumors of "monstrous" peoples in the Americas persisted for some time (see Dathorne 1994, page 32).

39. See Ortelius 1570, map 1. His map 50, of Turkey (*Turcicum*), shows both the Asian and the European components of the Ottoman Empire; by the eighteenth century virtually all atlases rigorously separated "Turkey-in-Asia" from "Turkey-in-Europe."

40. See, for example, Sanson 1674.

41. Medieval world maps sometimes had no exact Europe-Asia divide (see Beazley 1949, volume 2, page 565).

42. W. H. Parker 1960, page 282.

43. See Bassin 1991b, page 6.

44. As Forsyth (1992, page 146) elaborates, "It is noticeable that by now [1700s] in the vocabulary of Russian officialdom not only was [Asiatic] 'unreliability' attributed to the Siberian peoples, but 'Asiatic' had come into use as a self-explanatory pejorative term."

45. See Bassin 1991a, page 768. Scholars in western Europe, however, often had different ideas. To Montesquieu (1949, pages 264–65), and most of his contemporaries, for example, Great Tartary was located south of Siberia, constituting, in essence, modern Central Asia. Considering the derivation of the term *Tartary,* this is a more accurate definition. Many western European atlases from the eighteenth and early nineteenth centuries, however, did label Siberia as part of Tartary (see chapter 6).

46. See Bassin 1991b, page 8.

47. See W. H. Parker 1960, page 286. Malte-Brun (1827, volume 1, page 285) contended that Russian geographers in St. Petersburg had recently "proved" that the Urals formed the natural division between the two continents.

48. See, for example, S. Butler 1829, map 2; Cary 1808, map 3; Finley 1826, map 1; D'Anville 1743, map 1; R. Wilkinson 1794, map 1; G. Robert de Vaugondy and D. Robert de Vaugondy 1798, map 1. The Don-Volga-Urals scheme was used as late as 1849 by S. August Mitchell (map 44).

49. Some cartographers, for example, placed Russian Transcaucasia within Europe (W. Johnston 1880, map 2); several others relied on Russian provincial boundaries to include both Transcaucasia and the area immediately east of the southern Urals, while excluding from Europe a small slice of territory to the west of the Ural River (B. Smith 1899, map 73; *Hammond's Modern Atlas of the World* 1909, map 79; Gaebler 1897, map 5; Patten and Homans 1910, page 67). Several atlases of the mid-twentieth century, on the other hand, used political criteria to place the Europe-Asia boundary along the crest of the northern and southern Urals, but in the central portion they pushed the Asian boundary to the west so as to exclude the Perm district from Europe (Rand McNally 1932, page 197; *Encyclopedia Britannica World Atlas* 1949, pages 8–9; *Hammond's Ambassador World Atlas* 1954, page 26).

50. See, for example, S. Hall and Hughes 1856, map 1; Stieler 1865, map 5; Colton 1856, map v.

51. See Dillion 1994, page 19.

52. *Encyclopedia Britannica* 1963, volume 8, page 836.

53. *Encyclopedia Britannica* 1963, volume 21, page 632.

54. Mackinder 1904, pages 428–29. John Kirtland Wright—who argued at a very early date that Europe was but a conventional name for the far west of Asia—similarly included the entire "Africo-Arabian arid region" within the "European area" (1928, pages 4–6). He did so largely for historical rather than racial reasons.

55. Quoted in *Oxford English Dictionary* 1971, volume 1, page 536.

56. Bowen 1752, page 3.

57. Pitt 1680, page 14.

58. Bassin 1991b, pages 9–10.

59. *Oxford English Dictionary* 1971, volume 1, page 536. There is a further problem here, however, in the use of the terms *old* and *new*. While the "New World" is usually taken to encompass North and South America, Australia and New Zealand are not uncommonly placed within it as well. In fact, in the sixteenth century, all areas unknown to Ptolemy (including most of East and Southeast Asia) were classified as part of the New World (see Lach 1977).

In the late nineteenth century, it was not uncommon for mapmakers to use the term *continent* in both the old and the new senses concurrently. Bartholomew (1873, page vii), for example, states in the same paragraph that the world is divided into two continents and into six continents.

60. By the late 1800s, it must be admitted, world maps found in atlases were much less often colored to indicate continental divisions; instead, political divisions were increasingly highlighted (see chapter 6). Still, continents have remained the central *organizational* feature of most atlases up to the present.

61. For North and South America conceived as a single unit, see Sanson 1674; *Geographischer Atlas Bestehen* . . . 1785; Pitt 1680; Bowen 1752; D'Anville 1743; Woodbridge 1824; Butler 1829; for North and South America conceived as two units, see R. Wilkinson 1794; G. Robert de Vaugondy and D. Robert de Vaugondy 1798; Bonne 1771; Kitchen 1773; Palairet 1775; Finley 1826; Cary 1808; for Australia conceived as a part of Asia, see Palairet 1775; Kitchen 1773; Finley 1826; for Australia conceived as a separate division, see R. Wilkinson 1794; G. Robert de Vaugondy and D. Robert de Vaugondy 1798; Bonne 1771; for Australia conceived as an island, see D'Anville 1743; *Geographischer Atlas Bestehen* . . . 1785; Cary 1808; Woodbridge 182.

62. See Montesquieu 1949, volume 1, pages 264–69 especially.

63. Ritter 1864, page 183. Ritter occasionally argued that the distinction between continents and islands was to some extent arbitrary, contending that Australia could be counted either as the world's smallest continent or its largest island, and that even Java or Britain might be considered continents. Elsewhere, however, he virtually reverted to the classical threefold system, holding Europe, Asia, and Africa to be the world's primary landmasses (see Ritter 1863, page 72). In the end he opted for global vision structured around the fourfold continental system, seeing each continent as a divinely planned location for a different part of the story of humanity's rise. On Ritter's essentially religious conviction that continents form natural units, see also R. Dickinson 1969 (page 38).

64. See the discussion in James and Martin 1981, page 129. The linking of continents and racial groups stems from the work of Linnaeus, who differentiated four color-distinguished races, each located on one of the four quarters of the world (see Linnaeus 1735, first table under "*Regnum Animale*"; see also Pratt 1992, page 32). Johann Friedrich Blumenbach (1865) — sometimes called the "Father of Physical Anthropology" — later modified the Linnaean system, employing a five-fold classificatory system with each race linked to, but not exactly identified with, each of the four continents plus Oceania. The Caucasians, Blumenbach maintained, could be found in Europe, North Africa, and Asia west of the Ganges and north of the Amur. (He called this "race" Caucasian, it should be noted, because he considered the Georgians [an "Asian" people] to be its most perfect representatives [page 269].) A belief in the direct correlation between continents and races based on skin color was also popular in China in the late 1800s and early 1900s (see Dikötter 1992, page 78).

65. Guyot 1970, page 28. Guyot embraced a sixfold scheme, according continental status to South America and Australia, a rather rare position at the time. This move was instrumental for his larger theorizing, however, as he often contrasted the three northern continents with the three southern continents, arguing that the former cluster was far superior to the latter. Europe thus stood as the supreme continent in a supreme group of continents.

66. Of course, not all Victorian scholars followed this kind of geographical reasoning to its logical end. Friedrich Ratzel's *History of Mankind*, for example, is of interest for its marked deviation from the standard spatial architecture of continents and civilizations — even though it used standard continental terminology. On a map of "Asiatic and European Civilization," Ratzel distinguished between a zone of "modern (northern) civilizations," which runs from Europe

across southern Siberia to Manchuria, and one of "ancient (southern) civilizations," which encompasses the more densely inhabited areas of Asia and northern Africa (he placed all of insular Southeast Asia, however, into a primitive "zone
of natural races"). On a second map, showing "racial" groups, Ratzel divided the
Eastern Hemisphere into Occidental peoples, East Asiatic peoples (in China,
Japan, Korea, Mongolia, and Vietnam), Persian-Indian peoples (whose territory
stretches from Turkey to the Philippines), and Erythretic peoples (a group encompassing all speakers of Semitic languages as well as most inhabitants of East
and West Africa) (see Ratzel 1896, volume 3, frontispiece maps). These imaginative and rather bizarre divisions do not seem to have had much impact on other
scholars.

    67. Bunbury 1959, page 38.

    68. Ibid., page 163. Most classical geographers held Asia to be the largest con-
tinent; some, like Ptolemy (1932, page 160), also considered Africa to be larger
than Europe, but most regarded Europe as much more extensive than the southern continent.

    69. Maunder 1854, page 30.

    70. See Colton 1856, map v (which includes New Zealand but excludes New
Guinea); Stieler 1865, map 3; Greenleaf 1842, map 1; S. Hall and Hughes 1856,
map 1.

    71. As early as 1827, the noted French geographer M. Malte-Brun adopted a
strict physical definition, based on fixed sea limits, and thus excluded the entire
insular realm of "Southeast Asia" from the continent. Instead, he appended it to
Oceania, a region anchored by Australia, which he regarded as the forming a fifth
portion of the world (see Malte-Brun 1827, volume 1, page 286). Such a view became very common in world atlases; see, for example, S. A. Mitchell 1849 (maps
1 and 71); Greenleaf 1842 (map 1), Rand McNally 1881 (page 211), Bartholomew
1873 (map 1), Cram 1897 (page 216), and *Hammond's Modern Atlas of the World*
1909 (map 101)—all of which included insular Southeast Asia as part of Oceania.
Bartholomew (1873, page vii), moreover, explicitly defined Oceania as a continent.

    Other geographers of the period, however, appended insular Southeast Asia
to New Guinea, Australia, and New Zealand to form the great division of Australasia (for example, W. Johnston 1880, map 1), a maneuver that resubsumes the
area to Asia. But according to *The Columbian Atlas of the World* of 1893, "Australasia" refers only to Australia and Melanesia (page 138), while defining Oceania as a broader region extending from Sumatra to the Galapagos (page 138). The
Rand McNally World Atlas of 1932 actually mapped *mainland* Southeast Asia
within Oceania, although it discussed its various units under the rubric of Asia
(Rand McNally 1932, pages 229, 253). For terminological permutations of this area
found in geography textbooks, see McMurry and Parkins 1921, H. R. Mill 1922,
and Brooks 1926. No other part of the world has endured such incessant metageographical reorientations.

    72. H. R. Mill 1922.

    73. Brooks 1926.

    74. See Van Loon 1937, page 74.

    75. From the mid-1800s onward, however, it was more common to find North
and South America treated as separate landmasses in atlases published in the

United States; in those published in Europe, on the other hand, a united American continent remained the norm. See, for example, Greenleaf 1842, S. A. Mitchell 1849, Colton 1856, Bartholomew 1873 (page vii), and Rand McNally 1881 (page 234) for America views; see Cortambert 1869, Stieler 1865, and S. Hall and Hughes 1856, and W. Johnston 1880 for European views.

Another alternative sometimes employed in the United States was to map both of the Americas as a single continent, but then to emphasize its subcontinental division into North, Central, and South America (see *The Columbian Atlas of the World* 1893, page 150).

76. Charles Beard (1940, page 12) defined this country's traditional, and in his view correct, foreign policy as one of "continental Americanism," a system based on "non-intervention in the controversies and wars of Europe and Asia and resistance to the intrusion of European and Asiatic powers . . . into the western hemisphere." For a contemporary critique of this thesis, see Eugene Staley's (1941) "The Myth of the Continents." Staley advocated maritime rather than continental solidarity, and ridiculed the notion that the United States had any kind of natural relationship with South America based on their common location on the American continent. S. W. Boggs (1945), on the other hand, argued against the notion of a United States–defended Western Hemisphere, contending that the formally delineated Western Hemisphere had to include the Cape Verde Islands, easternmost Siberia, and New Zealand. "Continentalism," as well as "Hemispherism," apparently disappeared soon after the United States entered the war.

77. The old hypothesized *Terra Australis* had sometimes counted as a virtual continent, and in 1680 the author of *The English Atlas* (Pitt 1680, page 4) speculated that land under the South Pole might count as a part of the world on the same taxonomic level as Europe, Asia, and Africa. In S. August Mitchell's atlas of 1849 (map 1) Antarctica is labeled as a continent. Relatively few atlases, however, classified it as such until after World War II. For an early depiction of the now-standard (in the United States) sevenfold scheme, see *Hammond's Ambassador World Atlas* of 1954.

78. When Southeast Asia was conceptualized as a world region during World War II (see chapter 5), Indonesia and the Philippines were perforce added to Asia, which reduced the extent of Oceania, leading to a reconceptualization of Australia as a continent in its own right. This maneuver is apparent in postwar atlases (see, for example, *Encyclopedia Britannica World Atlas* 1949, pages 210–21; Bartholomew 1950, page 104; *Hammond's Ambassador World Atlas* 1954).

79. See Warntz 1968, pages 3, 24. This solution was predetermined once he decided to begin his analysis with the conventionally accepted landmass divisions, assuming that "the land area of any given continent [should] be as close to its center as possible and . . . these centers [should] be as far away from the center of all land as possible"

Coincidentally, the new standard American system of division largely recapitulated an ancient but minor geographical tradition that pictured the world as divided into seven parts, one of which remained covered by water (see Greenblatt 1991, pages 88–89). (Antarctica is, of course, largely covered with frozen water.) Interestingly, several ancient South Asia systems of world geography also divided the world into seven continents (see Tripathi 1969, pages 166, 179 [facing map], 210–11, 214).

80. The most influential Japanese world map of the eighteenth century, known as the *Chikyū bankoku sankai yochi zenzusetsu*, was one of the earliest to depict European-derived continental divisions. This map did, however, categorize Arabia and Turkey as part of Europe. A Russian-influenced map from the same period (*Chikyūzu*), on the other hand, employed the "modern" Ural-Caucasus division—at a time when most European maps still used the older boundaries. (See Kobe City Museum 1989, pages 13, 16.)

81. The most important Japanese work in global geography from the late Edo period, the *Konyo Zushiki Ho*, employed the then-standard fourfold continental scheme (Asia-Europe-Africa-America), appending insular Southeast Asia to Australia and the Pacific islands in a residual fifth category (Mitsukuri Shogo 1845; see also the brief discussion of this text and its accompanying world map in Kobe City Museum 1989). As early as 1875, however, at least one prominent Japanese geography textbook (Shiozu 1875) had adopted the sixfold scheme of Guyot, with its three northern continents paired off against three southern counterparts.

82. See Barthold 1937, page 33.

83. See Yapp 1992, page 139.

84. See Schwartzberg 1992, 1994; Gole 1989, pages 21–22. Some systems of Indic cosmography posited the existence of a series of concentric "continents" and oceans; others a fourfold system of completely discrete island continents. The "known continent," dominated by South Asia, typically covered only a small portion of such maps.

A unique Korean cartographic tradition posited the existence of a central continent (of the "known world"), surrounded by "an enclosing sea ring, which itself is surrounded by an outer land ring" (Ledyard 1994, page 259). Most of the places depicted on these traditional Korean world maps were imaginary.

85. Even such an obviously noncontinental unit as Latin America is still often referred to as a continent (see, for example, Alba 1969, page 4)—showing the persistence of the continental ideal.

86. See, for example, the delightful little juvenile book, *Blast Off to Earth: A Look at Geography* (Leedy 1992), which is structured entirely around the sevenfold continental scheme.

87. See K. Davis 1992, pages 135–36, appendix 3. In another recently published popular work in geography (Grillet 1991), this one constructed in a quiz format, the author similarly takes the continental divisions to be the most basic facts of world geography.

88. In regard to floral realms, one finds most of Africa joined with Southwest, South, and Southeast Asia to form the Paleotropical Kingdom. Northern Africa, on the other hand, is connected to northern Eurasia and northern North America in the expansive Boreal Kingdom, while southwest South Africa stands alone as the diminutive Cape Kingdom. As with fauna, "South American" flora stops not at the isthmus but rather in central Mexico. And while Australia hosts its own assemblage of plant life, New Zealand's floral affinity lies with other such "Antarctic" regions as the southern tip of South America (see Neill 1969, page 99).

89. See, for example, the discussion of "continents" in D. G. Howell (1989,

pages 2–6). Here Howell feels no need to name or number the earth's continents, paying attention instead to underlying structural processes.

90. New Zealand actually sits upon a fairly sizable continental fragment that includes also the (mostly) submerged Campbell Plateau and Lord Howe Rise (see D. G. Howell 1989, page 106).

91. North America and northern Eurasia do, however, have separate geological histories, having grown by accretion around different cores (see D. G. Howell 1989).

92. Interestingly, in *The English Atlas* of 1680 (Pitt 1680, page 1), the authors state that it had not yet been determined whether Greenland should count as a continent or an island.

93. Kurtén 1971, page 214.

94. Wilson 1992, pages 120–21, 250.

95. Lincoln 1994, page 40.

96. See Bartholomew (1950, pages 28–29) for an early example of Eurasia conceptualized as a distinct continent in a world atlas.

97. Halecki 1950, page 65.

98. Dropping Europe and Asia from the continental roster in favor of a single Eurasia only generates new problems in any case. On the one hand, the term *Eurasia* is an ambiguous concept. Since the late nineteenth century, when it was first coined by the Austrian geologist Eduard Suess (see Bassin 1991b, page 10), the term has been used in a number of different contexts to mean very different things. To certain Russian expansionists, *Eurasia* referred not to the whole land mass but rather to a purported middle realm, sandwiched between Europe and Asia proper. Self-proclaimed Russian Eurasianists of the nineteenth century were particularly attracted to this idea and attempted to popularize the term *Eurasia* as a designation for those lands between 30 East (the line of longitude on which St. Petersburg sits) and the Lena River (in east central Siberia). Russian nationalists living farther to the east, not surprisingly, argued for a more expansive Eurasia (see Hauner 1990, pages 158–59), while ardent Pan-Slavists included the entire Slavic zone—and sometimes western Turkey and Syria as well—within this intermediate "third world" (see Bassin 1991b, page 12). As Walicki (1989, page 502) notes, Russian Pan-Slavists viewed Slavdom and Europe as geohistorical antitheses.

Other definitions of the term can also be found. *Eurasia* has often been used to denote the *blending* of Europe and Asia, particularly in human bloodlines. Thus, a Eurasian *person* is one who can claim both European and Asian ancestry.

99. See *Encyclopedia Britannica* 1992, volume 18, page 522.

100. For a notable recent exception, see Rand McNally 1994.

101. Thus Denis Cosgrove (1994, page 281) writes, "Replacing European domination while retaining so many of the old continent's sacred and secular assumptions, the United States inherited the European *mission civilatrice.*"

102. See Reclus 1891, *Asia*, volume 1, page 18.

103. The continental framework has been so firmly embedded in European consciousness that scholars have felt a need to will some sort of uniformity out of this diversity. Even in the 1960s, when the notion of Asian *historical* unity could no longer be seriously entertained, modernization theorists attempted to see a common identity emerging. Thus, in *Approaches to Asian Civilizations,* William

Theodore de Bary argued that the modernization process was already producing "in Asians a sense of identity strangely in contrast to their past disparity" (de Bary and Embree 1961, page viii).

Certainly some scholars in this same period rejected the notion of Asian coherence, for a variety of reasons—some of them highly partisan. Thus in 1954 Dwight Cooke wrote *There Is No Asia*, the ostensible purpose of which was to calm American fears that Asia had already been "lost to communism." If the region did not really exist, he reasoned, then all the countries commonly classified as Asian would not necessarily go the way of China.

104. As Andrew March (1974) shows, Europeans have paradoxically viewed Asia as the land of excess, both in terms of the prodigality of its natural features and the extravagances of its populations ("All is overdone: The Greek mean is lacking"), as well as the land of uniformity ("endless stretches of territory, monotonous climate, masses of unfree people, eons of finished history"). As he goes on to explain, "the two sets of ideas seem to contradict each other, since one aspect of excess is excessive variegation, seemingly the opposite of uniformity" (page 33). Variegation is indeed an "Asian feature"—but only because "Asia" is so large; the notion of Asian uniformity, on the other hand, is simply not supportable. The geographical reasoning behind the "monotony" and "variegation" theses is often strained to the point of absurdity. Thus Ellsworth Huntington (1945, page 385), probably the most famous American geographer of the early twentieth century, claimed that the "backwardness of Asia" could be attributed to its extremes of temperature, its scarcity of midlatitude cyclonic storms, and its seasonal and year-to-year rainfall variability.

105. Cited in de Rougemont 1966, page 150.

106. See March 1974.

107. As is explained in note 71 (this chapter), insular Southeast Asia was often grouped with Oceania instead of with Asia in the middle and late 1800s. In 1896 Keane (pages 1–6), took an intermediate position, defining the Indonesian archipelago as a transitional zone between Asia and Australia. In 1903, however, Herbertson (page xv) insisted that all of this area, up to Timor, was by common convention considered a part of Asia. Yet as late as 1964, *Webster's Geographical Dictionary* (page 74) excluded the Malay Archipelago from its definition of Asia.

A biologically informed definition of Asia's southeastern extent can be far more precisely made, based on land connections during glacial periods. As Peter Bellwood (1992, page 61) writes, "The Sunda Shelf islands west of Huxley's Line form the true eastern limit of Asia, and contain . . . Asian species." Such a delimitation—which slices cleanly through Indonesia (and even the Philippines)—is, however, very rarely encountered outside of zoological and archaeological discussions.

108. This maneuver is rarely encountered before World War II, although one map in *L. L. Poates and Co. Complete Atlas of the World* (1912) does seem to indicate such a division (page 167).

109. The map in question is included on a 1990 advertisement for *Modern Asian Studies*, distributed by Cambridge University Press.

110. See "Asia and Europe: Friends Apart," *Economist*, 9 March 1996, page 33.

111. This view of North America became commonplace relatively recently, but it can be traced back as far as Thomas Kitchen's atlas of 1771 (see map 9). Here

most of Mexico as well as Central America is mapped as part of the West Indies (map 10).

112. See E. Huntington 1907 for a classical expression of environmental determinism.

113. See Bassin 1991b.

114. Toynbee 1934–61, volume 8 (1954),pages 713, 711.

115. Peet 1991b.

116. Murphey 1992; see Spencer and Thomas 1971, pages 1–3, for a similar problem.

117. The excluded southeastern reaches of Siberia are strongly affected by monsoonal airflows, for example, while the included Tarim Basin is not.

118. Murphey (1992, page 5) does indeed offer a list of the region's commonalities: the importance of the extended family, respect for learning, veneration of age, traditional subjugation of women, hierarchical structure of society, awareness of the traditional past, and primacy of group welfare. Such highly general traits, however, are also characteristic of many other parts of the world, and they are by no means universal within the monsoon zone; in fact, there is much less "traditional subjugation of women" in Southeast Asia and in southwestern India than in most other portions of the world.

119. See March 1974, page 44.

120. See, for example, the discussion in Geddes (1982, page 2): "[W]e shall often find that the geomorphic boundary is followed by a human boundary. This is so much more often seen in India's millennial countryside than in a 'new' land, such as North America: more even than Europe in the Middle Ages."

121. In the eighteenth century, America was often considered the land of feeble nature, which was at the time considered evidence of the continent's inferiority (see Glacken 1967, page 680).

122. Buckle 1872, page 87. Carl Ritter similarly argued that Europeans had an advantage over others merely because their continent is the smallest: "Europe was placed between these extremes not to be retarded, but to be favored and urged forward. It received for its portion a much smaller area than the other continents; it was, therefore, much easier known, and was sooner brought into cultivation" (1863, page 348).

123. Quoted in Archer 1993, page 507, emphasis added. There is considerable irony here, as Vidal is often considered to be the founder of the anti-environmental determinist school of geography.

124. Kennedy 1987, pages 17–19. This passage bears strong resemblance to the views of Montesquieu (see 1949, volume 1, page 269).

125. See Kennedy 1987, page xxi.

126. As William Rowe (1984, page 62) writes, "the uniquely efficient water-transport system . . . of preindustrial China allowed it to overcome the barriers of distance and low technology, and to develop a national market by the mid-Ch'ing, even though in Europe and elsewhere such a development may have been conditional upon the advent of steam-powered transportation." (See also McNeill 1995, page 21.)

127. See Gupta and Ferguson 1992, page 17.

Chapter 2: The Spatial Constructs
of Orient and Occident, East and West

1. The debate between Said and his detractors is summarized by Bernard Lewis (1993b) in *Islam and the West*. See Ahmad 1992 for a devastating critique of Said from a Marxist position.

2. See Said's (1978) chapter on "Geographical Imagination."

3. Sylvain Levy, quoted in Schwab 1984, page 1. Many writers have, for some time, noted the absurdity of dividing the world into an East and a West. In an anthology called *Eastern and Western World*, for example, the writer supplying the "Western view" (Vlekke 1953) as well as the writer supplying the "Oriental view" (Som 1953), argued against the notion of any clear-cut global division. But no matter how often the problems associated with longitudinally bifurcating the world are pointed out, the essential idea does not seem to weaken in the public or the academic imagination.

4. See Unno 1994, page 439.

5. Those Europeans who wish to disassociate themselves from the United States, however, have sometimes downplayed the notion of the West in favor of reverting to a strict vocabulary of continents, which allows them to stress the common bonds uniting all of Europe (see Judt 1991, pages 38–39).

6. S. Hay 1970, page 330.

7. Many poststructuralism and postcolonialist writers do attempt to avoid positing such metageographical distinctions. Gyan Prakash (1990, page 384), for example, writes that this new mode of analysis will "unsettle the calm presence that the essentialist categories—east and west, first world and third world—inhabit in our thought." In practice, however, it is much easier to denounce such essentialist categories than it is to avoid them. Prakash himself, in the same essay (pages 383–84), conflates the East with the Third World—just as he is attempting to reject both notions. Yet it is notable that some writers aligned with the "post" movement are aware of such difficulties within their own discourses. As Sara Suleri (1992, pages 12–13) writes, "[C]ontemporary rereadings of colonial alterity too frequently wrest the rhetoric of otherness into a postmodern substitute for the very Orientalism they seek to dismantle." See Rosalind O'Hanlon and David Washbrook (1992), however, for an insightful critique of postmodernism as applied to the Third World, especially India. They argue that it is "foundationalist" historians, not poststructuralists, who are actually breaking down "East-West dichotomies" (page 146).

8. Raymond Schwab (1984, page 1) dates the East-West divide back to the period of the Roman Empire.

9. Even single authors not uncommonly slide from one vision of the West to another in order to maintain the thread of tenuous arguments. While Theodore Von Laue usually limits the West to Britain and France (with the United States joining the club after World War I), at one point he explicitly defines the region as including all of Europe (1987, page 35)—even though he continues to exclude western (European) Russia in practice. Douglas Jerrold (1954, page 57), who excluded Germany from the West, later implicitly defined the West as nothing less

than the historical realm of Christianity—a definition that would have to include Ethiopia. His opponent Arnold Toynbee, who insisted on making an absolute division between Western civilization and the Orthodox realm of Russia and the Balkans, was even less consistent on this point. Toynbee formally defined the West simply as the zone of Western Christianity (that is, Roman Catholicism and its various Protestant offshoots). Following such a criterion, he literally mapped out a maximal West that included not only all of the Americas but also most of the Philippines (1934–61, volume II [1959; Atlas]). Yet Toynbee was just as capable of removing entire populations living in western and central Europe from the zone of Western heritage in spite of their religious traditions. He attempted to explain the restrictions placed on immigration to the United States in the early twentieth century, for example, by reference to the non-Western character of so many of the newcomers: "a majority of the Italian Roman Catholic immigrants in those years were Neapolitans and Sicilians, who still remained crypto-Byzantines even after nine hundred years' experience of an association with Western Christendom into which they had been conscripted originally by a Norman military conquest. The immigrants from the Danubian Habsburg and the Russian Empire included undisguised Byzantines, as well as Jews" (1934–61, volume 8 [1954], page 215).

10. On the positioning and repositioning of the Greeks along the East-West axis, see Anthony Smith 1986, page 203. While Greece is often (but by no means always) considered to be the birthplace of Western culture, the postclassical Greek world is usually classified with the non-West—with the notable exception of geopolitical discourse during the Cold War period. If this ploy creates certain undeniable contradictions, it has nonetheless proven highly convenient. For instance, while a narrative structured around the grand march of *European* civilization falters when the crucial role of medieval Byzantium is considered, such a problem is obviated by consigning this Greek realm to the East. "Senility and vigor, orient and occident, the Greek and Latin churches," one author tells us, are the "contrasts involved in the Fourth Crusade of 1204" (Schevill 1991, page 133). But despite the fondest desires of generations of "Far Western" scholars to relegate Byzantium to the Orient, the fact is undeniable that "their culture [was] an unbroken continuation of ancient Greek culture" (Threadgold 1988, page 384). If one wishes to propound a continuous narrative of Western civilization dating back to the ancient Greeks, it is also difficult to exclude modern Greece from the Western realm; the continuities between Athens of 445 B.C.E. and Athens of 1997 C.E. are orders of magnitude greater than the continuities between Athens of 445 B.C.E. and Los Angeles of 1997 C.E. One can thus exclude Greece from the West only if one dates Western civilization, as did both Otto Spengler (1926) and Toynbee, to the medieval period and views its connection to the ancient Hellenic world as one of affiliation only.

11. See Bartlett 1993.

12. See Szücs 1983, pages 133, 156.

13. See Wolff 1994.

14. During the Protestant Reformation, however, the East-West division was occasionally mapped upon new split within Western Christendom. In Spenser's Protestant imagery, for example, the pope himself was portrayed as a figure of Oriental tyranny (Goldberg 1992, page 217).

15. See Rupnik 1991. A common post–World War II definition of the West was Europe (less Russia and its allies), the Americas, Australia, and South Africa (see, for example, A. Weber 1948, page 2).

16. Only with the rise of the pacifist Green movement in Germany in the 1980s was this vision seriously challenged. All that these new voices did, however, was to revert to a strictly continental framework, contending that the United States, as well as the Soviet Union to a degree, were external, non-European powers and that Europe could reemerge as an integral unit by escaping from superpower rivalry (see Rupnik 1991, page 255; Judt 1991, page 39). The notion that the United States and Russia are comparable units in their relationship with Europe dates back to the early nineteenth century (see Johnson 1991).

17. J. Williams 1960, page 1. Such a complete political reordering of basic metageographical concepts, it should be stressed, failed to gain widespread acceptance.

18. Slater 1995, page 67.

19. A recent example of this maneuver can be seen in the geographical work of Ó Tuathail and Luke (1994, page 391), who describe Lester Thurow's *Head to Head: The Coming Economic Battle Among Japan, Europe, and America* (1992) as a book contending "that the real conflicts will be over who is 'the best in the West.'"

Occasionally, one finds authors expanding the West to include not only Japan but other economically successful Asian regions as well. Thus Har Iqbal Singh Sara (1983) describes Sikhism as a "super-occidental religion," an argument based on parallels between Sikhism and Calvinism—as well as the relative prosperity of the Punjab.

20. "Western Culture Flourishes in a Changed Cambodia," *Raleigh News and Observer*, 26 May 1993.

21. One can, of course, put an opposite moral spin on the same idea. Consider, for example, Benjamin Barber's comments on Japanese electronic games: "The Gameboys are stealth cultural networks reaching into Russian homes and children's minds with a steady diet of Western games, comic characters, and attitudes about competition, violence, consumption, and winning" (1995, page 254).

22. If one contrasts the West with the Third World, yet retains the West's original geographical designation, then one is forced into the incongruous situation of placing Japan within the Third World. Masao Miyoshi thus argues that "Third World narratives"—such as those produced in Japan—are by their very nature radically different from those of the West (1991, page 61; but see also pages 41–43).

23. Yúdice 1992, page 211. See also Pratt 1992 (chapter 8) on the historical relationship between Latin American intellectuals and the metageographical constructs of Europe and Latin America.

24. See Yiengpruksawan 1993, especially page 77.

25. Europeans had almost no knowledge of East Asia until the Mongol period in the thirteenth and fourteenth centuries. Subsequently knowledge of the area again diminished: "[T]o Europeans in the sixteenth century, Cathay remained a mystery" (Philipps 1994, page 42).

26. Discussed in Steadman 1969, pages 42–43.

27. See Chirol 1924, page 5 especially.

28. See Schwab 1984, page 71.

29. While it does make sense to exclude most indigenous Siberian peoples from the mainstream of Asian (or Eurasian) history, the criterion by which this move is made has troubling implications. One of the principal defects of our historical imagination is the fact that peoples without a tradition of literacy have usually been excluded altogether from the flow of history, essentially written off the time line as well as the map of the world. Anthropologists have until recently collaborated in promulgating this notion, insofar as they viewed indigenous cultures as changeless, or at least inaccessible to historical recovery, and thus existing in a timeless "ethnographic present." The resulting picture of the world, at its worst, contrasted an actively historical Occident with a once-historical but now largely static Orient, and deemed the rest of the world as hardly worthy of notice. Such an overt paradigm has largely faded away, and anthropologists, together with historians, are now recovering the histories of peoples once deemed ahistorical (see, for example, Vansina 1990).

30. Guénon 1930, page 179.

31. In a survey of introductory geography students at Duke University, only twelve percent considered Southwest Asia to lie within the Orient. (Nor would these students label the cultural artifacts of the Middle East "Oriental," with the notable exception of carpets.) Even South Asia is being conceptually excised from the region; only a quarter of the Duke students placed India in the Orient. Such surveys are problematic, of course, because they measure not only common geographical conceptions but also sheer geographical illiteracy.

32. The notion that the division of the Roman Empire led to the differentiation of East from West is so deeply ingrained in certain historical circles that no evidence can dislodge it. In Rand McNally's *Atlas of World History* (1993a), for example, we are informed that "The differences between the eastern and western parts of the Roman Empire [are] still clearly visible in the modern world" (page 46). To prove this point we are supposed to compare two completely dissimilar maps, one showing the Empire in A.D. 395, the other the division of Europe between the North Atlantic Treaty Organization and the Warsaw Pact in 1987. Only Bulgaria fits the purported pattern.

33. As Paul Dukes (1990, page 4) writes in a recent history of Russia: "The myth persists that Russia was cut off from Western civilization until the reign of Peter the Great. In fact, ties were often close with the West." Even in the years of Tartar rule, Dukes insists, " Russia's culture . . . was open . . . to influences from both Asia and Europe" (page 40). Most scholars interested in dividing the world into a series of distinctive civilizations of macrocultural regions, however, insist that Russia does not belong in the same category as western and central Europe. This is indeed a legitimate position, provided that one classifies other eastern European countries of Orthodox Christian heritage, such as Serbia and Bulgaria, in the same category as Russia. But it must also be recognized that the Latin West and the Greco-Russian East also have strong common roots (in both Hellenism and Christianity) and have historically maintained a relatively close, if fitful, connection that resulted in periodic rounds of Russian and Balkan "Westernization." As Fernand Braudel writes, "[T]he [former] Soviet Union, now the CIS, . . . despite what is often said has always remained European, even in its ideology" (1994, page 302).

Whether one chooses to classify these two halves of historical Christendom as distinct or merely as variants of a larger whole depends on whether one is a taxonomic lumper or splitter. If one prefers to split, however, one should carry the principle forth to other portions of the world as well. In doing so, one would have to consider distinguishing Persian and Turkish civilizations from Arabic civilization, a south Indian civilization from a north Indian civilization, a Korean civilization from a Chinese civilization—and so on throughout the world. Yet those who insist on separating Russia from the West are usually content to switch into the role of lumpers when examining the rest of the world. Such taxonomic inconsistency is a symptom of Eurocentric geographical myopia.

34. Oswald Spengler went so far as to argue that the very term *Europe* ought to be expunged from our vocabularies, since it implies a false linkage between Russia and the realm of the Latin church. On the contrary, Spengler informs us, not Asia and Europe but East and West "are the notions that contain real history" (1926, volume 1, page 16, note 1).

35. See Kiernan 1980, page 45. Intriguingly, Russia was evidently partially "de-Asianified" in the European imagination when it cooperated with other European countries in the subjugation of China.

36. Such impulses are still current; David Aikman asks in a recent *Time* magazine essay whether Russia will now opt for the West—meaning freedom and democracy, or the East—meaning slavery and subjugation ("Russia Could Go the Asiatic Way," *Time*, 6 July 1992, page 80).

37. See Wolff 1994. Wolff argues that by the end of the Enlightenment the transition from a North-South to an East-West split in European civilization had been completed, but the evidence he presents does not fully warrant this conclusion. Only a few writers of the Enlightenment actually used the formulation "Eastern Europe." Moreover, even Wolff's own evidence suggests that the North-South division remained important throughout the entire period. Robin Okey (1992, page 110) goes so far as to argue that the North-South division of Europe strengthened during the Enlightenment, owing in part to the "Europeanization of Russia."

The Enlightenment's common disparagement of eastern Europe as half-barbarous was also applied to such non-eastern areas as the highlands of Scotland. To a certain extent, Western writers at this time were distinguishing a "core" from a "peripheral" Europe, although such terminology was not used (on medieval antecedents of a core-periphery view of Europe, see Phillips 1994, page 50). In 1929, however, Francis Delaisi explicitly theorized a division of the "continent" into a core "Europe A" and a peripheral "Europe B," the latter being located mainly in the east but also in the far north and south: "S'il cherchait à les délimiter, la ligne frontière passerait approximativement par Stockholm, Dantzig, Cracovie, Budapest, Florence, Barcelone, Bilbao, puis contournant la France, passant entre l'Angleterre et l'Irland, elle s'en irait par Glasgow rejoindre Bergen et Stockholm" (page 20).

38. Admittedly, from 1815 to 1830, a period of French "Russophilia," very few French scholars viewed Russia as Oriental, but with the "Polish Revolution of 1830–1 . . . the European [*sic*] frontier once more retracted west to Poland" (Woolf 1992, page 92).

39. Burton 1973, pages 364–65.

40. See Wolff 1994, page 274 especially.

41. Bassin 1991b, page 13. Many Russian Slavophiles were thoroughgoing Eurocentrists—in the broadest sense of the term—and thus simply expunged Asia (particularly non-Islamic Asia) from their maps of world history: "When he stated that ancient Russia was part of the East, he had in mind the European East; to Pogodin Europe alone was the sole cradle of civilization" (Walicki 1989, page 53). Again we see the dual nature of the East-West divide.

42. Earlier, Catholic Pan-Slavists had argued on similar lines, but many were inclined to stress the Europeanization of Russia (which they viewed as their potential guardian) more than the intrinsic Easternness of Poland, Bohemia, Croatia, and other western Slavic lands (see Bassin 1991b, page 14 on Russian Slavophiles; see Kohn 1960b on the origins of Pan-Slavism among Catholic Slavs). Classical Russian Slavophiles often conflated language and religion, equating Slavdom with Orthodoxy. Such a view led them to denounce the Poles as the great traitors of the Slavic race (but curiously, not so often the Czechs, Slovaks, Slovenians, or Croatians), to seek union with Greeks and Romanians, and to dream of ultimate success in the conquest of Constantinople (see Walicki 1989). Ironically, the entire movement, in both its Pan-Slavist and its Slavophile versions, was deeply rooted in German romanticism.

43. On this "Yellow Russia" movement, see Hauner 1991, page 206. Similar habits of thought remain firmly entrenched in some quarters. Witness historian Raymond Schwab's claim that "proximity to the Asiatic temperament was more evident in the Slavic than in the Germanic nature" (1984, page 449).

Chinese Marxists, it is interesting to note, often saw the Soviet Union as a "new, third civilization rising to mediate between East and West" (Spence 1990, page 306).

44. See Halecki 1950, page 90.

45. Ibid.

46. Ibid., page 11.

47. See T. Mann 1983, page 29 especially. The intellectual lineage of Mann's disassociation from the West can be traced back in German thought to Nietzche and Schopenhauer.

48. See Weigert 1942, page 143. It must be noted that this was only one of many new metageographical ideas propounded by Germany's Nazi-era scholars, several of which were mutually contradictory. A few geopoliticians returned to continents, stressing Germany's role in defending a united Europe against the (non-European) Soviet Union and United States. A more innovative strategy was to argue that the world was in the process of being divided into three or four political-economic "panregions," vast areas dominated respectively by Germany, Japan, the United States, and (in some versions) the Soviet Union. Germany's region, called Eurafrica, was to have encompassed all of Europe and Africa as well as a large chunk of Southwest Asia. The notion of a Eurafrican region has not fared well since the demise of the Nazi state (see O'Loughlin and van der Wusten 1990).

49. Weigert 1942, page 145.

50. See Jerrold 1954, page 51. As Carlo Cipolla (1993, page 278) notes, World War I could appear to Asians rather as "the European Civil War." The Second World War, by this way of thinking, involved two simultaneous "civil wars" (one in Europe and the other between China and Japan) coupled with a "transcivilizational" war.

51. Von Laue 1987, pages 36–39 especially. Von Laue (page 39) explicitly argued that the Germans constitute the "nearest halfway point between East and West, figuratively as well as geographically." One might thus be led to assume that Germany lies at the midpoint between Britain and Japan!

52. Kohn 1960a, pages ix, 5, 10.

53. As political ideology pushed Easternness farther and farther into the heart of Europe, the newly defined non-West made even more dramatic advances on other fronts. Some scholars generalized old categories on a grand scale, arguing that "the Orient includes, in fact, almost everything that is not Western in tradition" (Sinor 1970, page xi). Eventually, however, this meant stretching the East-West framework so far that even the plastic term *Orient* reached the limits of usefulness. At this point, the original metageographical polarity of Occident and Orient yielded to the opposing pair of West and non-West, with the latter increasingly represented by the term *Third World*. The geographical bounding of this non-West, like its predecessors, has shown fantastic variation. Our favorite delineation comes from the frontispiece map of a book optimistically entitled *The Nature of the Non-Western World* (Dean 1957). The "Non-Western World in the Twentieth Century" is taken here to include all of the world except the United States, Canada, and Europe (excluding from the latter category only Russia and Yugoslavia). No explanation is provided as to why Yugoslavia, but not Romania or Bulgaria, should be classified as non-Western.

54. Most Hungarians, however, passionately insist that they are of the West. The historical evidence is clearly in their favor. It is interesting to note that Latin itself was employed as Hungary's administrative language until the 1780s (B. Anderson 1983, page 71)—a time when it had been abandoned almost everywhere else in the region.

55. As early as 1946, certain Western German politicians began "interpreting Nazism as a form of Prussianism" and blaming the Prussians for the enmity existing between France and Germany (see Applegate 1990, page 242).

56. Longworth 1994, page 8. While Longworth's introductory claims are grand, his text actually takes a much more nuanced approach, his ultimate aim apparently being to explain why Eastern Europe never developed the kind of civil society encountered in Britain and France. Longworth never precisely delimits Eastern Europe; an initial map (page x) implies that the term covers only the former communist zone minus Russia and East Germany, but the text itself gives full coverage to Russia, Greece, and even Turkey. Eastern Germany, for its part, is described as merely having an "affinity to eastern Europe" (page 261).

In 1949 Werner Cahnman similarly highlighted the contemporary significance of Charlemagne's eastern frontier. In his view Western Europe was centered on Paris and Eastern Europe on Moscow, with the area between Russia's western boundary and the border between East and West Germany forming a "shatter zone" that could potentially look to either direction. Larry Wolff (1994, page 283), on the other hand, argues that the Iron Curtain merely recapitulated the Enlightenment's division between eastern and western Europe along a line running "from Riga to Trieste." Riga, however, actually lies some 700 kilometers northeast of the traditional northern terminus of the Iron Curtain.

57. As Fry and Raymond (1983, page 1) write, "In college classrooms through-

out the United States, Western Europe has become synonymous with The Big Three." It is interesting to note, however, that geography textbooks written in the 1950s through the 1980s usually included Greece within Western Europe, showing the Cold War geopolitical framework then in ascendance (see, for example, Diem 1979; Clout et al. 1985).

58. See Halecki 1950, page 99.

59. Longworth 1994, page 7; see March 1974 (page 39) for a critical view.

60. It is interesting to note that in certain circles of contemporary popular thought, particularly those associated with radical environmentalism and spirituality, a countervailing tendency expands the West directly into the heartland of the original Orient. Thus one finds a special issue of the periodical *Gnosis*, which bills itself as the journal of the "Western inner traditions," devoted to Sufism, just as one encounters radical environmental theorist Jim Mason discussing the "West's aggressive and rigidly monotheistic Judeo-Christian-Islamic Megareligion" (see *Gnosis: A Journal of the Western Inner Traditions* 30 [Winter 1994]; Mason 1993, page 30). Indeed, as early as 1962 Sidney Gulick (page 19)—a purveyor of some of the most stereotypical East-West dichotomies—defined the West as including the Near East, Europe, and the Americas. It is more than a little ironic to find the West's main historical oppositional ground—the original East—being suddenly culturally annexed by it with virtually no notice.

61. Classical definitions of Central Europe—or, in German, *Mitteleuropa*—were premised on the notion that Germany itself formed the European core (see Meyer 1946, page 180; Ash 1991, page 4), although Sinnhuber (1954, page 20) argued that the core of *Mitteleuropa* itself has actually been Austria, Bohemia, and Moravia. "Central Europe" is, however, one of the more protean of geographical constructs. Various writers have included within it such areas as central France, northern Italy, southern Sweden, and northern Greece (see Sinnhuber 1954, page 19; Mutton 1961, page 3); a French text from 1954 encompassed both Switzerland and Romania (see George and Tricart 1954). In some contexts all of Scandinavia and the Balkans are also included (see Longworth 1994, page 37, note 39), and even Great Britain and the Volga region of Russia have been placed within Central Europe on occasion (Meyer 1946, pages 181 and 191). Sinnhuber maintains that only the Iberian Peninsula is always excluded (1954, page 20), yet Weitzmann (1994, page 38) relates the story of a proposed Central European Federation from the mid-1800s that would have encompassed Portugal!

It has also sometimes been wondered whether all of Germany should be placed within the region. "Is the Rhineland in Central Europe? . . . Certainly not in the way that would be true of Bavaria or Saxony" (Judt 1991, page 42). More recently Central Europe has sometimes been identified with the members of the Central European Free Trade Agreement: the Czech Republic, Slovakia, Hungary, and Poland. Yet this group may itself expand in the near future to include Slovenia, Bulgaria, and the Baltic states (see "Concrete Heads," *Economist*, 16 September 1995, page 60).

On the general disappearance of "Central Europe" after World War II, see Michalak and Gibb 1992 (page 342 especially); on its reappearance in recent years, see Graubard 1991; on the shifting history of geographical ideas pertaining to this region, see Sinnhuber 1954, Meyer 1946, Okey 1992, and Stirk 1994.

62. See Rupnik 1991.

63. See Halecki 1950, pages 120, 125; see also Judt 1991, page 25. This issue is more complex than it appears at first; while many German advocates of the *Mitteleuropa* concept were hardened imperialists, others used the term to advance liberal and federalist ideas (see Beller 1994, pages 73, 82). It should also be noted that many British writers before World War II accepted the idea of a "germanocentric economic Mitteleuropa" (McElligott 1994, page 130).

64. Hobsbawm 1993, page 62. Perhaps the most sophisticated advocate of central Europe's fundamentally Western identity is Hungarian scholar Jenö Szücs (1983, pages 133, 156).

In the interwar years, English-speaking geographers were sometimes inclined to extend the boundaries of western Europe all the way to Poland and the Baltic (see, for example, Cundall 1932). Such a strategy was obviated by subsequent Soviet expansion into this area.

65. This information was obtained in an interview with the members of the Bulgarian Academy of Geography held in Sofia on July 17, 1995. These geographers considered both the categories of East-Central Europe and Southeastern Europe to be appropriate for Bulgaria. One reason why the former seemed to be favored, however, was because it emphasized ties to Germany that many Bulgarians wish to reactivate. The Bulgarian attitude to Germany is markedly unlike that of the western Slavs, as one would expect considering their different historical-political relationships.

66. Okey 1992, page 104.

67. Schöpflin 1991, page 91. Philip Longworth (1994) similarly argues that such countries as the Czech Republic and Slovenia cannot be regarded as part of the West because prior to 1918 they were under the repressive Austro-Hungarian Empire. Yet for some unstated reason he largely exempts that empire's core—Austria—from the same reasoning (see page 7). The only justification for putting Slovenia in the East but Austria (with its sizable Slovene minority) in the West is the now-archaic post–World War II geopolitical settlement—in which Yugoslavia itself never really fit anyway.

68. Joll 1980, page 18.

69. See, for example, Rozman 1991, page 3. Rozman adds Singapore to his list of Eastern states, which is indeed fitting from a cultural perspective.

70. The building in question, adjacent to the former Duke University office of one of the authors (Lewis), was devoted to studies of trade between the United States and Western Europe, on the one side, and Eastern Europe and Russia, on the other. In 1994 it was converted to house the Center for Latin American Studies.

71. A signal text stressing the deviation of Spain from Western history is Americo Castro's *The Structure of Spanish History:* "When examined alongside other European cultures, the history of Spain appears as an utterly aberrant form of life" (1954, pages 651–52). Castro also maintains, however, "In one way or another Spain has never been apart from Europe" (page 5).

In earlier generations, America and British writers often dismissed Spain from the Western world with unveiled contempt: "Thus we must not think of the Spaniards and Portuguese as we would think of the Frenchmen or Englishmen, as being pure Europeans, with purely European traits, but we must think of them

as at least partly Oriental" (Sweet 1919, page 16). An unbiased investigation, however, soon reveals that several centuries of Muslim rule left few social structures
or political institutions in post-Islamic Spain; the Inquisition was, after all, fairly
thorough. Moreover, the Christian kingdoms of northern Iberia were thoroughly
within the original "Western civilization" that emerged in the early Middle Ages.
How else can one explain Santiago de Compostela?

72. Although Patai (1962) referred to the Middle East as a "culture continent,"
no one to our knowledge has defined it as a true continent.

73. Hegel 1956, page 173.

74. Ibid.

75. Ibid., page 113.

76. For instance, one of the principal theorizers of East-West cultural disparity, F. S. C. Northrop (1960, page 8), simply ignored the Middle East, comparing the West exclusively to India, China, and Japan.

77. Lawrence 1989, page 205.

78. B. Lewis and Holt 1962, page 2. The French designation of *Asie occidental*, employed in the *Géographie Universelle* of 1929 (as an alternative to *Moyen Orient*), avoids these conceptual difficulties, but at the cost of remaining faithful to
the continental scheme (see Sion 1929).

79. See Clerk 1944, page 5; see also Martin 1944, page 335.

80. Mahan 1968, page 237: "The middle East, if I may adopt a term which I
have not seen . . . " See also B. Lewis and Holt 1962, page 1.

81. Even at this early date, however, different writers used the term to refer
to different areas. In 1906, for example, Arminius Vambéry included both Central Asia and India in his "Middle East."

82. Chirol was not terribly concerned with delimiting a Middle East, but his
position here is still clear: "[The Middle East encompasses] those regions of Asia
which extend to the borders of India or command the approaches to India, which
are consequently bound up with the problems of Indian political as well as military defense" (1903, page 5).

83. See Davison 1963, page 18.

84. See, for example, Fromkin 1989, page 16.

85. See Beaumont, Blake, and Wagstaff 1976, page 1; see also Davison 1963,
page 20.

86. See W. B. Fisher 1947, page 416.

87. Ibid., page 417.

88. See Patai 1962. Very different boundaries of the Middle East are occasionally encountered. In one text, all of the former British India—including
Burma—is counted as part of the region (see Rivlin and Szyliowicz 1965, page 4).

89. Previously the area was simply a portion of the "Hither Orient" (see B.
Lewis and Holt, 1962, page 1; see also Davison 1963, page 16).

90. See W. B. Fisher 1947, page 414. In a popular text of 1905, D. G. Hogarth delineated a slightly different "Nearer East," which included Greece and the
southern Balkans, Egypt, and Southwest Asia as far east as central Persia, yet he
also slighted his own designation, writing that the Nearer East is "a term of current fashion for a region which our grandfathers were content to call simply The
East" (pages 1–3). Hogarth's scheme was, in essence, little different from that of

Hegel, for he simply divided the Nearer East from the Farther East—conceiving both divisions as still unambiguously Eastern. That he would place eastern Iran (Khorasan) in the Farther East and western Iran in the Nearer East probably indicates a lack of familiarity with cultural geographical patterns.

91. While Greece is often placed within the Near East, it seldom appears within the Middle East—yet even here there are exceptions (see, for example, Morton 1941).

92. Robert Kaplan, however, now tells us that we should revive the old concept of the Near East so that we might see more clearly the historical connections between the Balkans and the Middle East, and so that we might remember the historical (and future?) prominence of Turkey both within this region and within the general orbit of Islamic civilization (see his "Middle East Only a Subplot in the Real Story of the Near East," *Raleigh News and Observer*, 13 March 1994, pages 17A–21A [originally published in the *New York Times*]). Other writers retain the "near" and "middle" distinction to refer to areas centered respectively on the Levant and Iran. Even here, however, locational logic can go far awry; in Anatoly Khazanov's scheme, for example, the Middle East encompasses Turkey while the Near East includes southern Arabia and, in some circumstances, Somalia (1994, pages 53 ff., 102).

93. W. B. Fisher 1947, pages 416, 414.

94. In most other western European languages, terms similar to the English *Middle East* are usually used: *El Medio Oriente, Le Moyen-Orient, and Oriente-Médio*. In German, however, the older term *Vorderasien* ("Anterior Asia") is more commonly employed (see, for example, *Atlas International* 1985, page 184).

95. The catalog in question is labeled "Asian Studies" and was printed in 1993.

96. The Far East, in those days, encompassed the entire swath of land from the Indus Valley to Japan. It might be noted that Winston Churchill, however, regarded the Far East as covering only China and Japan; in his mind, India, Burma, and Malaya—the core of Britain's concern—formed "The East" pure and simple (see Davison 1963, page 21).

97. By strict definitions, Turkey is indeed a partially European country, its largest city lying on the west bank of the Bosporus. Some American scholars now place Turkey in exactly the same conceptual position, as a bridge between Europe and Asia or even between East and West. Ideas pertaining to Turkey's "continental" position, however, are often employed in a rather casual manner. Charles Tilly, for example, discusses the Ottoman Empire as a "peripheral" European power on one page, yet later in the same text reduces it to the position of a "semi-European" country (1990, pages 32, 171). Writers who stressed the Cold War geopolitical divide, on the other hand, usually placed Turkey unambiguously in the West.

A large range of geographical entities have been selected at one time or another to fill the role of the "bridge" between the East and West. The recent president of Georgia, Zviad Gamsakhurdia, for example, "propounded a pseudo-scientific theory of Georgia's historical role as mediator between the civilizations of the Orient and the Occident" (Glenny 1994, page 46).

98. Davison 1963, page 19.

99. Kennan 1993, page 6, emphasis added.

100. See S. Hay 1970, page 1; B. Lewis 1982, pages 59–61. Interestingly, de-

spite their (partial) Greek intellectual heritage, Islamic geographers paid little attention either to continents or to the East-West split (see Yapp 1992, page 139). Instead, they typically downplayed the geographical associations of cultural groupings altogether. But the geographical tradition of the Arabic- and Persian-speaking worlds was rich indeed, and a variety of different global divisional systems were used by different authors associated with different schools of thought (see *Hudud al-'Alam* 1937; Harley and Woodward 1992). Muslim geographers concerned with mercantile practices often devised their own divisions of the world. According to Andre Miquel (1967, page 115), Ibn Rusteh (writing circa A.D. 900), *"se substitue peut-être arbitrairement un classement tout aussi justifié, qui regrouperait les oeuvers par régions traitées, soit: Extrême-Orient (route maritime de l'Inde et de la Chine), Russie du Sud et Asie centrale (Turcs, Hazars, Russes, Bulgares), Empire byzantin (Rum) et Europe, enfin régions méridionales (Nubie et rivages africains de l'océan Indien)."*

It is notable that in the Arabic world, as in Europe, there has long been a notion of an "ancient esoteric wisdom of the east [*ishraq*], (*sharq* [being] the Arabic word for 'east')" (Hourani 1991, page 176).

101. Schwartzberg 1992, pages 389–99. Kautilya's *Arthasastra*, the political classic of the South Asian tradition, at times even seems to limit "the world" to those lands south of the Himalayas (1915, page 372; see also Wink 1986, page 16). Such exaggeration was, no doubt, often intentional. Many Indian monarchs, after all, claimed to rule the entire world (Stein 1977, pages 26–28; Inden 1990, page 246) — even though they were obviously aware of the continuing existence of unsubdued states.

102. See Fitzgerald 1964. Not all Chinese scholars actually placed China at the center of the world; those associated with the Buddhist tradition depicted China as a marginal country on a central continent dominated by India (see Yee 1994, page 173).

103. See S. Hay 1970, page 2. In a similar fashion, the Vietnamese considered themselves "Southern," a position founded in their relationship with China (Reid 1994, page 268).

"Traditional" Japanese metageography is itself a complicated subject. The medieval Japanese view was based on three major components: Honcho (Japan), Shintan (China), and Tenjiku (India). In this view, Japan was a small and insignificant country on the far eastern periphery of the world (Asao 1988, page 2). Japanese global maps constructed by Buddhist scholars continued to depict a world dominated by India, with both China and Japan shown as marginal territories and with Korea sometimes excluded altogether (see Unno 1994, pages 371–76). Confucian influences, however, led many other scholars to organize their worldviews in accordance with the centrality of China—a position that also peripheralized Japan.

According to Asao (1988), what eventually freed the Japanese from this self-denigrating complex, and allowed them to reassess their country's worth, was the notion that while India was the home of Buddhism, and China the home of Confucianism, Japan was the land of the gods. During the early Tokugawa era (1603–1867), Japan began to substitute itself at the center of a Chinese-style "order of civilized and barbarian" (*ka-i chitsujo*), seeing this as the international expression of the assertion of central authority within the state (Kamiya 1989, page

6). Confucian influences continued to be very important, however, throughout the early modern period. One prominent text written in 1708 by Nishikawa Jo-ken, for example, divided the periphery of a China-centered world order into two camps: *gaikoku* (foreign countries, including Korea, Ryukyu, Taiwan, Tonkin) and *gai'i* (barbarian lands, including the Moluccas, Java, Banten, and Holland). The former category was described as "countries which, while outside of Chinese territory, follow the precepts of Chinese civilization, employ Chinese characters, and accept the three teachings" (Kamiya 1991, page 59). For additional information on traditional Japanese conceptions of the global order, see Toby's forthcoming book.

The Chinese traditionally viewed the Japanese as Easterners—Eastern Barbarians, to be precise. In their conception, "West" generally denoted India, which was viewed as far more important than those lands lying farther toward the sunset (see Narla 1981, page 7). India was, to a large extent, the ultimate land in both the Far Eastern and the Far Western traditional geographical imagination.

104. While limitations on time and space prevent us from exploring other Asian metageographical traditions here, we would note that the peoples of maritime Southeast Asia framed a global distinction around the concepts of the "lands below the winds" (Southeast Asia) and those "above the wind" (India and areas farther west). In this case, however, there was evidently no moral coding involved. Moreover, East Asia did not fit at all into this Southeast Asian global divide (see Reid 1994, pages 268–69).

105. Such leaders regarded contemporary Christian Europeans as unworthy successors to their shared classical past (see Kinross 1977, pages 111, 200 especially). Alexander the Great was actually a culture hero as far east as Sumatra; the Hellenic legacy was by no means restricted to Europe or the West by any definition of these terms.

106. See the insightful discussion in Graham Fuller 1993. In the discourse of radical Islam, the guiding figure of Turkish Westernization, Kemal Attaturk, is often cast into the role of Islam's greatest traitor, one who played into the hands of an enemy often portrayed simply as "the West." (According to a recent article in a prominent British magazine, "The propagandists of radical Islam . . . may be the last people in the world who still talk about 'the West' as if it were a clearly identifiable place" ["Islam and the West: Everything the Other Is Not," *Economist,* 1 August 1992, page 34]).

The notion of the unity of all Turkish-speaking peoples, or Pan-Turanianism— dates back to the waning days of the Ottoman Empire. The Turkish-speaking peoples of central Siberia, such as the Yakuts, have seldom been included within such a formulation because they are not Muslims. Yakut nationalism—a growing movement at present—has, however, occasionally appealed to Pan-Turkish sentiments (Forsyth 1992, page 260).

107. See Hashikawa 1980. In contemporary Japan, however, many intellectuals are now arguing that the country must "leave Europe and join Asia," viewing the latter as the dynamic leader of the twenty-first century (see Kogawa 1988, page 59).

108. See Inkster 1988, page 122.

109. Hay 1970, page 315.

110. The geographical coordinates of Asian unity over the last century have been

reoriented in Asian visions just as they have in those of Europe. The peoples of the southwestern portion of the supposed continent are now less often cast as participants in the Oriental drama; in the 1950s the brilliant Indian scholar-diplomat K. M. Panikkar, for example, found profound Eastern commonalities, but only in *"non-Islamic Asia"* (1969, page 322, emphasis added), an area that he viewed as fixed between the poles of China and India. Panikkar isolated the substance of Asian unity in "a community of thought and feeling" (page 322). Yet it is notable that Panikkar also considered Russia to be "permanently in Asia" (page 16), both because of its geographical position and, with overtones of Halford Mackinder, because of its reliance on land rather than sea power.

Despite Panikkar's suggestion, we contend that while there are certainly commonalities linking China with India, such connections are by no means stronger than those linking India to Southwest Asia or even India to Europe. And indeed, many Indian scholars resolutely deny the existence of any fundamental East-West divide and argue against any closely cultural linkages between South and East Asia (see, for example, Anand 1988, page 13).

111. See S. Hay 1970.

112. See F. Drake 1975, page 112; see also Lee 1984, page 108.

113. See F. Drake 1975, pages 68, 112.

114. See Karl 1993, pages 21, 40.

115. On Liang Qichao's global vision, see Karl 1995.

116. Quoted in Panikkar 1969, page 322. Many Japanese thinkers embraced a "psychological" global division similar to that propounded by Tagore. As Nakamura (1960, page 2) argues, "There has long been a tendency in Japan to think in terms of a dichotomy between East and West. The device has been to take two mutually opposed value concepts, labeling one 'Occidental' and the other 'Oriental.' Thus the Oriental way of thinking is represented as 'spiritual,' 'introverted,' 'synthetic' and 'subjective,' while the Occidental is represented as 'materialistic,' 'extroverted,' 'analytic,' and 'subjective.' " Nakamura visualizes the East as centered around China, Japan, India, and Tibet.

117. In practice, most Japanese Orientalists limited their attention to the East Asian cultural sphere, paying particular attention to China. The delineation of the larger region remained the subject of heated debate. One prominent scholar, for example, argued for excluding Southeast Asia from the East while including Southwest Asia; others disputed whether India should be viewed within the same cultural unit as China and Japan. Meanwhile, a vocal minority emphasized an alternative global division along North-South lines. In their view, the central dynamic in Eurasian history was the struggle between the civilized South and the martial North, with Japan—not surprisingly—possessing a uniquely productive balance of these two forces (see Tanaka 1993, especially pages 48, 279).

118. When the Japanese captured Indian soldiers fighting in the British army, they informed them that "the Japanese, Asian like themselves, had been and would be their staunch friends" (Farwell 1989, page 332). Although most Indian soldiers remained loyal to Britain, members of the Japanese-supported "army of liberation" were greeted as heroes when they returned home (many to face military tribunals) in the immediate postwar period.

119. See Tamamoto 1991, page 19.

Chapter 3: The Cultural Constructs
of Orient and Occident, East and West

1. The number of works outlining the essential differences between the East and the West is vast, and we have elected to outline only some of the more prominent ones in the pages that follow. In general, writers on this topic have concurred on the broad topic but have emphasized different distinctions. Western individualism has been stressed by many writers, but probably most strongly by Flewelling (1943, pages 13, 225 especially) and von Beckerath (1942, page 292 especially). Maurice Parmelee (1929) stressed the subjective knowledge of the East, which he contrasted with the objective knowledge of the West, concluding that the West is rooted in action and work and the East in meditation and contemplation. Sidney Gulick (1962) contrarily emphasized the "courtesy" of the "introverted Orientals," which he contrasted with the "moral character" of the "extroverted Occidentals." The simple notion of Western "dynamism" was stressed above all by Alfred Weber (1948, page 2). Paul Cohen-Portheim stressed the "passivity, universality, and intuition" of Orientals; in Asia, he reported, people feel "closely akin to plants and beasts" (1934, pages 46–47). The linkage of the West to revolution is rather more uncommon, and is usually associated with writers of a socialist bent (for example, Braudel 1994, page 356).

2. See Bartlett 1993. Bartlett's survey is most impressive, yet the unique traits of the Occidental cultural entity that he (mis)names "Europe" are in the end surprisingly mundane: "saints, names, coins, charters, and educational practices" (page 291).

3. See, for instance, Kolakowski 1990. Kolakowski is probably the most subtle and convincing exponent of Eurocentrism writing today. While he admits that "some barbaric aspects of Europe are indigenous" (page 25), his rendering of the Western tradition remains highly selective.

4. Charles Krauthhammer, "Flunking Somalia," reprinted in *Raleigh News and Observer*, 10 October 1993.

5. Blaut 1993. Blaut frames his investigation around the question of the West versus the Rest, rather than the West versus the East, but the essential points that he makes are similar to our own. See also Bryan Turner's admirable *Orientalism, Postmodernism, and Globalism* (1994).

6. In much travel lore, the East (like Africa), was also seen—in a tradition dating back to Pliny and earlier—as a land of humanoid monstrosities (see Dathorne 1994).

7. Inden 1990, page 32.
8. Schwab 1984.
9. O'Leary 1989, page 70.
10. Quoted in Kopf 1969, page 238.
11. J. S. Mill 1989, page 70.
12. Townsend 1921, page 29. Townsend's work was important at the time, and not only in the West. Jawaharlal Nehru in particular was influenced by his ideas (see S. Hay 1970, page 292).

The use of religion to differentiate East from West is an old staple that is still commonly encountered. Yet if one had to pick, on historical grounds, the most secular "civilization," it would have to be that of East Asia. Even careful scholars often fall into this trap. Hans Kohn (1934, pages 1–8), for example, while denying the unity of the Orient (holding that the Middle East had to be differentiated from the Far East), nonetheless went on to argue that the entire Orient was, in contrast to the Occident, still in the "religious medieval stage." In fact, Japanese society was in 1934—as it is today—far less religious than U.S. society.

13. Townsend 1921, pages 30–31.

14. Northrop 1960, page 375, emphasis in original.

15. Ibid., page 3.

16. Singer 1973, page 56.

17. Parkinson 1963, page 289.

18. Ibid., page 112. Parkinson may have thought that mirth was uncommon in Asia, but Arminius Vambéry went a step further, arguing: "Humanity in Asia has never known culture and liberty . . . and has therefore never known true happiness" (1906, page 5). Vambéry's book is perhaps the most viciously anti-Asian tract ever written.

19. Rougier 1971, page 176. Rougier's book was published in a far-right-wing book series, entitled "The Principles of Freedom Series."

20. See Henry 1994, page 9.

21. Ibid., page 18. Henry's own knowledge of the encounter between the colonizers and the colonized is minimal, as is evident in his discusssion of disease transmission (page 53).

22. Larson 1988, page 8. Larson and a sizable group of colleagues have undertaken a project to rectify this error, seeking to create a truly cosmopolitan subdiscipline of comparative philosophy. While not all have fully shaken off the notion of a binary split between East and West (see, for example, Nakamura 1988), it is a hopeful sign that comparative philosophers—along with like-minded professionals in other fields (for instance, Krishna 1988, pages 73–74)—are now wrestling with the geographical presuppositions of their disciplines, and are attempting to transcend them.

23. See Baritz 1961, page 618; Tuan 1996, pages 76–79. As Baritz (page 620) notes, the classical world viewed the West as both the site of death and the site of happiness.

24. De Rougemont 1966, page 41.

25. The best discussion is in Lach 1977. A few daring European writers, such as Guillaume Postel (1501–81) actually inverted some of the "traditional" distinctions between East and West. Postel, for example, viewed the East as masculine and ascendant, the West as "feminine, declining, . . . and capricious" (Lach 1977, book 2, page 268).

26. The Tunisian philosopher Hichem Djaït 1985 (page 18) is nonetheless overgenerous when he asks, "How much did the eighteenth century owe its generosity to the fact that it was not infected by a gangrenous will to domination?" Certainly many European intellectuals expressed profound admiration for China and Japan even in the seventeenth century—but such admiration stemmed in part from the mistaken notion that these lands were ripe for Christianization (see Lach 1977, book 3, especially pages 565–66).

27. Schwab 1984, page 15. Voltaire's *Essai sur les moeurs et l'esprit des nations*

was written, in large part, as a rejoinder to Bossuet's (1829) tremendously influential *Discours sur l'histoire universelle*, penned in 1681. Bossuet's concerns were dictated almost entirely by religion, and as a result he had little interest even in classical Greece—let alone in China or India. Donald Lach (1977, book 2, pages 306–23) argues, however, that sixteenth-century Europe did see the initial development of true universal history.

28. Gay 1969, page 392. The amount of space that Voltaire devoted to the non-Western world was greater than that in almost any other attempt at "universal history" up to the present. His lengthy introduction discussed various Asian countries and regions in almost as much depth as those of Europe, and in the first section of his main text he devoted two chapters to China, two to India, and three to Southwest Asia and North Africa. Numerous subsequent chapters return to non-European regions, examining especially their relationships with what would later be styled the "world system." Here Voltaire's focus is precise enough to allow him to write such specific chapters as "*De l'Ethiopie*" and "*Du Japon au XVIIe siècle*" (see Voltaire 1963).

29. B. Lewis 1993b, page 90.

30. Longxi 1988, page 118.

31. See Kopf 1969; see also Schwab 1984, pages 152–53. William Jones, it should be noted, "deliberately chose 'Asiatick' in preference to 'Oriental' because he wanted to study India's civilization on its own terms rather than looking at it from the Western viewpoint implied by the word 'Oriental'" (Kulke and Rothermund 1990, pages 242–43).

32. Voltaire 1937, page 1. There is some doubt as to the authorship of this work.

33. Cited in Durant and Durant 1965, page 656.

34. See, for example, Diderot's entries in his encyclopedia for *Chinois, Arabes*, and *Indiens*. In each article (all of which focus on philosophy), he begins by remarking on the sagacity and erudition of the people in question (1876, volume 14, page 122; volume 13, page 314; volume 15, page 200).

35. See Wolff 1994, page 231. It is hardly coincidental that in the following century many Muslims were to adopt certain key tenets of Enlightenment thought, not least the (somewhat misguided) characterization of Islam as a religion of social equality. On the role of Enlightenment thought in the Islamic world, see Hourani 1962 (page 138 especially).

36. Montesquieu was a virulent racist and general bigot. Consider, for example, his assessment of Africans: "It is hardly to be believed that God, who is a wise being, would place a soul, especially a good soul, in such a black ugly body" (1949, volume 1, page 238). It is essential to realize, however, that "the rising *philosophes* were . . . displeased. They considered *The Spirit of Laws* as almost a manual of conservatism" (Durant and Durant 1965, page 365).

37. Condorcet (1955) thought that the stranglehold of religion prevented progress throughout the entire East, yet he somehow managed (partially) to exempt the West from the same problem (see page 66). The despotism of religion, he argued, doomed the Arabs (and even, to a lesser extent, the Chinese) to "eternal slavery and incurable stupidity" (page 88).

38. Voltaire viewed the Turks with contempt and expressed great enthusiasm for Russian imperial expansion (see Wolff 1994). The Turks were generally hated

both because they had recently threatened Europe and because they were linked to the pastoralist societies of Central Asia ("Tartary"), which Enlightenment geography still saw as a challenge to all Eurasian sedentary civilizations. In the Renaissance, however, positive images of the Turks were sometimes promulgated, in part due to hopes that they might eventually be converted (see Yapp 1992, pages 141–42), and because of widespread contempt for the Orthodox Christians who bore the brunt of the initial Turkish expansion into Europe.

39. See the insightful comments in Woolf 1992, page 84; see also Spence 1990, page 133.

40. Mason 1993, page 30.

41. Nor is it only in environmental studies that one can find, as late as 1989, a description of "the Oriental mind" as one that sees "nature as imbued with divinity" (Nash 1989, page 113). A number of leading anthropologists have forwarded somewhat similar notions, one of them going so far as to quip that "money is to the West what kinship is to the Rest" (M. Sahlins 1976, page 16). Andrew Ross (1992, page 533), a noted critic and "New Age poststructuralist," embraces a similarly strict East-West bifurcation, writing of "Oriental social philosophies" "grounded in naturalistic, holistic, and pantheistic sciences," which have been "systematically displaced and repressed by [a] Western empiricism" bolstered by "the official Western ideology of growth." (See also M. Lewis 1996.)

42. See M. Lewis 1992a. For other examples of the East-West vision in contemporary radical environmentalism, see Plumwood 1993 (page 2) and Merchant 1992 (page 102).

43. Jean 1991, page 62, emphasis added.

44. Escobar 1992, page 27; see also Kang 1992.

45. Yoshimoto 1991, page 247.

46. Ahmad 1992, page 178. As Ahmad further demonstrates, Said seems to conceive of the West along an unproblematic spatial and temporal trajectory stretching from ancient Athens to modern London.

47. Nanda, forthcoming, page 31.

48. Longxi 1988, page 127. Consider, for example, Michel Foucault's (1970, page xix) view of China: "There would appear to be, then, at the other extremity of the earth we inhabit, a culture entirely devoted to the ordering of space, but one that does not distribute the multiplicity of existing things into any of the categories that make it possible for us to name, speak, and think." Foucault derives this bizarre notion from a passage in a Jorge Luis Borges book regarding a mythical Chinese encyclopedia; he evidently did not realize that Borges was writing in an absurdist mode and that no such "encyclopedia" ever existed—or ever could exist.

One might also consider in this regard Roland Barthes's (1982) imaginative work on Japan, *Empire of Signs*, the basic thesis of which seems to be that in Japan, unlike in the West, the center of things is empty. While the work is both vacuous and filled with errors, Barthes provides himself with a pat escape clause; *Empire of Signs*, he proudly contends, is not really about an empirical Japan at all, but rather about an imaginative one that he conjures up to subvert the logical structures that we Westerners take for granted. (One might argue that this applies to Foucault's view of China, although Foucault does not make this explicit.) Moreover, Barthes informs his readers he does not consider Orient and Occident to be real categories. Yet for all of these initial disclaimers, Barthes proceeds to ground all of his arguments in empirical instances taken from a very real Japan, and he litters his text with blanket statements about the "West," "Westerners," and "Western Man." In the end, he merely recapitulates the hackneyed stereotypes of traditional Orientalism, albeit casting them in more rarefied prose. (As James and Nancy Duncan [1992, page 29] perceptively argue, Barthes "succumbs to that long-standing European temptation . . . to appropriate the Other for European purposes.") What is most ironic is that at least traditional Orientalism had the possibility of overcoming its own prejudices through the advancement of knowledge; in the knowledge-disdaining Orientalism of Barthes and his epigones, such a possibility cannot exist.

For a brilliant critique of misplaced East-West dualism in postmodernist-influenced populism in South Asia, see Nanda 1991.

49. See Hodgson 1974, 1993; Radhakrishnan 1956, especially pages 13, 120.

50. Moreover, John Steadman (1969) in *The Myth of Asia* thoroughly discredited most of the supposed differences between Orient and Occident more than twenty-five years ago.

51. Unfortunately, limitations of space preclude us from analyzing at any depth the many other supposed differentiating factors of East and West. We would like to note, however, that Jack Goody (1990) has thoroughly demolished the notion that family, marriage, and inheritance patterns play any role here. He further argues that the erroneous notion that "*[t]heir* marriage [as opposed to ours] has to do with sale, the absence of affection, with immobility, extended families, and so on," has led to an unconscionable "primitivization of the Orient" (pages 467–68). More recently, Goody (1996) has further discredited the notion that rationality is characteristic of the West and not of the East. In an analysis that is complementary to our own, he shows that logical thought, rational bookkeeping, and complex economic organization were all found throughout the literate societies of the Afro-Eurasian ecumene.

Jack Goldstone (1991) has similarly demolished the notion that a history of rebellion and revolution distinguished the progressive West from the static Orient. Rather, as he shows, major revolutions in the early modern period throughout most of Eurasia were rooted in the same kinds of ecological and social structural problems. Goldstone does, however, argue that there was a greater tendency in northwestern Europe for state breakdowns to lead to massive social restructuring (page 415).

Finally, we would also comment on the widespread notion of "freedom of conscience" as a peculiarly Western trait. (Frank Darling, for example, in his massive study *The Westernization of Asia*, lists "religious freedom" as one of seven "distinct Western Historical Inputs" [1979, page 236].) All that one has to do here is to compare the horrific history of religious intolerance in Europe with the generally more humane conditions historically existing in most regions of Asia. Indeed, it is difficult to find a better example of religious freedom in the premodern era than in the Indian state of Kerala (see Woodcock 1967, chapter 6).

52. According to Max Weber, "Experience tends universally to show that the purely bureaucratic type of administrative organization . . . is . . . capable of at-

244   NOTES TO CHAPTER 3

taining the highest degree of efficiency and is in this sense the most rational known means of exercising authority over human beings" (1968, volume 1, page 223).

53. This economic definition of rationality is often closely linked both to capitalism and to organizational efficiency. To quote Max Weber (1968, volume 1, page 224) once again, "capitalism is the most rational economic base for bureaucratic administration, and enables it to develop in the most rational form."

54. See the discussion in Gellner 1988, page 255. The "cunning of reason" stems from Hegel.

55. Alasdair MacIntyre, for instance, insists that "practical reason" can never be defined on purely rational grounds. While one writer may argue that to be practically rational is to "act on the basis of calculations of . . . costs and benefits to one's self of each possible alternative course of action," another might, with equal reason, respond that it is rational only "to act under those constraints which any rational person, capable of an impartiality which accords no particular privileges to one's own interests, would agree should be imposed" (MacIntyre 1988, page 2).

56. M. Weber 1968, volume 2, page 551.

57. Ibid., volume 2, page 555.

58. In fact, pure epistemological rationalism always seemed to its detractors to reveal itself in the end to be a weird kind of irrationalism, based as it is on unquestioned postulates and necessary linkages that have an unfortunate tendency to ignore the facts of nature and arrive at grandiose theories purporting to explain all mysteries of the universe. One might consider here the Cartesian notion that one can reach a certain knowledge of God by a chain of airtight logical connections. Most modern proponents of reason would concur with Ernest Gellner (1992, page 166): "The two seeming opponents [rationalism and empiricism] were in fact complementary. Neither could function without the other."

59. According to historian Peter Gay (1969), the eighteenth century was marked by a dramatic retreat from the wholesale rationalism of Descartes; a reassertion of passion, sensuality, and imagination; and a rejection of all totalizing philosophical "systems." Gay in fact claims that the movement reflected a revolt against dogmatic rationalism as much as against antirationalism (page 189). More surprisingly, he shows that the most prominent Enlightenment thinkers were, as a rule, disinclined to make universal judgments. "The philosophes were relativists to a degree unthinkable before them, but neither their professional situation nor their philosophical convictions permitted them to erect their relativism into an absolute principle. Their absolutes were freedom, tolerance, reason, and humanity" (page 322).

60. See Toulmin 1990.

61. The leaders of the English scientific revolution were for the most part committed Christians who resolutely opposed atheism and other forms of irreligion. But their stance was complicated by political exigencies. As Michael Hunter (1990, page 453) explains: "For even though scientists could claim a particular role in relation to the design argument, they were themselves endangered by the inclusiveness of atheism. . . . It was undoubtedly helpful to those committed to the new science to be able to invoke and refute an atheist menace, thereby asserting their own orthodoxy against more conservative figures who accused them of heresy."

62. Steadman 1969, page 121; see also page 26.

63. Gellner 1992, page 105. Ancient Greek culture, the paradigm of rationality for admirers and detractors alike, was deeply divided, according to Nietzsche at any rate, between calmly rational Apollonian and wildly irrational Dionysian impulses. Nor was Greek philosophy immune from beliefs that would today be commonly regarded as completely irrational, as is thoroughly evident in the mystical doctrines of Neoplatonism, the culmination of Greek philosophy. As Cameron (1993, page 133) writes, "[Neoplatonists] believed in the possibility of divine revelation, especially through the so-called 'Chaldaean Oracles' (second century), which claimed to be revelations obtained by interrogation of Plato's soul."

64. Isaiah Berlin (1991, pages 34–35) sees Western irrationalism as springing from excesses of rationalism: "It seems to me a historical fact that whenever rationalism goes far enough there often tends to occur some kind of emotional resistance, a 'backlash,' which springs from that which is irrational in man." Berlin dates this backlash phenomenon to the fourth century B.C.E.

65. McNeill 1963, pages 411–12.

66. Stephen Toulmin makes a similar point with regard to the French intellectual tradition. "French writers," he concludes, "take 'modern' in a Cartesian sense. For them, formal rationality has no alternative but absurdity; so, for lack of formal grounding, the 20th century situation leaves no room for constructive responses, only for *de*constructive ones" (1990, page 173).

67. The accomplishments of indigenous science and technology in China, before, during, and after the Sung, have been made accessible to an English-speaking audience by Joseph Needham (1954–84) and his colleagues, and can no longer be ignored.

68. Gernet 1982, page 330. While it might be objected that Sung scientific thought was contaminated by mysticism, the same charge could equally be leveled against early scientific exploration in the Far West. Indeed, the mystical ideas of Taoism actually spurred disciplined, empirical inquiry into the secrets of nature in East Asia, just as the hermetical and alchemic traditions helped usher in Europe's scientific age (on the latter issue, see Jacobs 1988; on controversies surrounding such terms as *hermetic, alchemic,* and *occult,* and their role in the formation of modern science, see Copenhaver 1990).

69. Hodgson 1993, pages 103–4, 190.

70. Ibid., pages 103–4.

71. See Robinson 1991, page 107; see also Mazzaoui 1991, page 88, on the "Central Asian Enlightenment."

72. As Jacobs (1988, page 25) shows, the opposition of the Catholic church to the new physics and astronomy associated with Galileo often provoked Protestant support—although many Protestants remained concerned about the ultimate implications of such a strategy. In today's world, however, opposition to scientific doctrines (such as Darwinism) is more often associated with radical (American) Protestantism. While many recent thinkers have continued to link the rise of rationality and science in the early modern West to the Christian heritage (for example, Kolakowski 1990; Wolpert 1991, page 46), such reasoning remains highly strained.

73. See Hodgson 1974, 1993. One of the two basic theological schools of early Islam "was a rationalist-oriented position that emphasized the centrality of rea-

son as an ordering principle in God's being in the human understanding of the universe, and in the governance of human behavior"; moreover, in the nineteenth century, Islamic modernists returned to a similar position in arguing that *true* Islam was by necessity a religion of rationality and science (Lapidus 1988, pages 108, 620–21).

74. See Shaffer 1994, page 7.

75. Wolpert 1991, page 195.

76. Narla 1981, page 84.

77. See P. Gross, Levitt, and Lewis 1996.

78. Leszek Kolakowski (1990), perhaps the most profound defender of Eurocentrism, on the other hand, clearly shows that formal criteria of rationality cannot be applied to political actions. Moreover, he contends, even if one identifies rationality with efficiency one must remain skeptical: "There is no reason to be happy about the rationalization, that is, growth in efficiency, of torture and genocide, and most of us would not like the regimes that practice those measures to become more efficient" (page 197).

79. M. Weber 1968, volume 1, page 223.

80. See John Byrne, "The Virtual Corporation," *Business Week*, 8 February 1993, pages 98–102. See also his article "The Horizontal Corporation," *Business Week*, 20 December 1993, pages 76–81.

81. See, for example, Lieberman 1993b, 1995; G. Parker 1988; Wigen 1995a. For an explicitly Weberian analysis of administrative modernization in eighteenth-century Japan, see Bodart-Bailey 1989. See also L. Darling 1994, page 95 especially.

82. For the strong variant of this argument, see Chisholm 1982, page 161. For a "softer" variant, one arguing that the religious traditions of both the Far East and the Far West lent themselves to capitalist development simply by not getting in its way, see W. Davis 1992. See also Rowe 1984 (page 73) on organizational rationality in preindustrial China.

83. Gellner 1981, page 7.

84. Ibid.

85. See McNeill 1974.

86. On modern democracy as an outcome of political processes rather than as a cultural condition identifiable with the West, see E. Friedman 1994; more generally, see Moore 1966.

For a sophisticated opposing view, see Daniel Bell et al. 1995. Bell and his collaborators argue that while democracy may be slowly and unevenly spreading in East and Southeast Asia, the more fundamental principles of liberalism are not, owing largely to the region's underlying Confucian ideology. We agree that the precise political forms adopted in any one region reflect to some extent the region's cultural background, and that in East Asia a certain "Confucian" inflection can often be expected. But while Bell and company show how Confucian ideas might influence the political structures, their basic thesis remains overstated; liberalism is not nearly so endemic to the Far West, nor foreign to the Far East, as they suppose. Moreover, their argument rests on an act of metageographical legerdemain; their geographical focus is "Pacific Asia" (or East and Southeast Asia), but their ideological focus is Confucianism—which is itself foreign to most of Southeast Asia. One of the authors in this collection (Jones, page 43) attempts to sidestep this problem by arguing that the region's "understanding of authority, obligation, and moral identity emanated from a complex blending of Hindic, Buddhist, Islamic and Confucian traditions," but this formulation involves an unwarranted essentialization of the region. Nowhere in Southeast Asia, it is important to note, have Confucian and Islamic traditions "blended" to any significant extent.

87. Even in China a strong current of liberalism persists—despite the efforts of the country's rulers to crush it (see McCormick and Kelly 1994).

88. Carlo Cipolla (1993, page 211) and others have argued that Europe had clearly emerged as the "most developed area" of the world by 1500, but such a view is no longer easily substantiated across the board. Even in armaments—clearly a European specialty—the Ottomans were fully competitive throughout the sixteenth century. European manufacturing capabilities did not really began to exceed those of East Asia until the 1700s. Only in regard to militarized shipping did Europe have a clear advantage over its Asian rivals in the 1500s and 1600s.

89. See E. Jones 1988.

90. Much recent research has indicated that in the early modern period there were few differences between Eastern and Western trading patterns and general economics (see, for example, Mauro 1990, page 278; Habib 1990, page 398).

91. For more on this issue, see Blaut 1987, 1992; Perlin 1983. See also Washbrook's (1990) impressive work on proto-capitalism in South Asia. As he writes (page 492), "The [erroneous] notion that capitalism is uniquely Western is closely associated with the notions that rationality, achievement, individualism, and 'history' are too; and that, by definition, the 'Orient' was static and enervate, awaiting the coming of the West to 'usher it into history.' "

92. See D. L. Howell 1995.

93. Cipolla 1993, page 120. Certainly there were some instances of premodern urban self-governance in other portions of the world. The Japanese city of Sakai (the port of Osaka), for example, was largely autonomous and run by a merchant oligarchy before national unification in the late 1500s.

94. Ibid., page 70.

95. Cipolla (1993, page 124) claims that celibacy was praised in the West but condemned in the Orient; actually, the Lamaist Buddhist tradition was more devoted to celibacy than was Roman Catholicism. He also argues that Europe, unlike Asia, was not thwarted by "the suffocating pressure of population" (page 136); actually Southeast Asia, Central Asia, and Southwest Asia were all disadvantaged in the early modern period by their relative sparsity of population. Cipolla (page 153) further contends that the cult of saints, which he links to an ideology devoted to the control of nature, was uniquely European; actually the tradition of saints in Sufi Islam, while admittedly informal, was equally significant—and was also connected in certain places to the harnessing of nature (see, for example, Eaton 1993).

96. See M. Porter 1990. Both Italy and Taiwan are noted for their family-led, middle-sized companies.

97. Many aristocratic European thinkers of the seventeenth and eighteenth centuries disdained the Ottoman Empire especially because of its aristocracy's lack of autonomy (see Yapp 1992, page 150).

98. O'Leary 1989, page 66. See also L. Darling 1994.

99. See O'Leary 1989, page 8.

100. On the supposed absence of Asian civil society in Orientalist discourse, see B. Turner 1994, page 23.

101. O'Leary 1989, page 197. It should be noted that a stream of later Marxian scholarship denied the concept altogether; Stalin actually attempted to outlaw it. By the 1970s, as Western Marxists grew increasingly embarrassed over the blatant Eurocentricity of the original Marxian doctrine, the concept of the Asiatic Mode of Production could thrive only if "de-Asianized" and relabeled as something like the "Tributary Mode of Production." In this way, the Asiatic Mode of Production followed the classical trajectory: a spurious "Eastern" trait was first hypothesized, and then generalized to cover virtually the entire non-Western world.

102. Wittfogel 1956, page 154.

103. Wittfogel 1981, page 447. Note that Wittfogel's views on bureaucracy represent an important reversal from those of Max Weber, who viewed the "traditional authority" of Eastern civilizations as fostering the development of patrimonial, but not genuine or fully formalized, bureaucracies (1968, volume 1, pages 228–29 especially).

104. Specifically, Wittfogel argued that in Japan, as in ancient southern Europe, mountainous topography dictated the emergence of multicentric irrigation systems, precluding the concentration of power in a single core. Other mountainous Asiatic areas, however, received little notice in his work.

105. Wittfogel 1981, page 164.

106. Although Wittfogel believed that he could thus account for despotism in the absence of irrigation, he could not resist ferreting out hidden waterworks. He thus asserted, for example, that "Byzantium maintained considerable hydraulic installations, in the main for providing drinking water" (1981, page 173).

107. Dorn 1991.

108. Worster 1985.

109. Peet 1991b, page 98. Peet's adherence to the standard metageographical division of East and West is unambiguous: "Within the context of Eastern absolutism, the Soviet Union and China developed undemocratic Marxist systems now in the final stages of collapse. Within the context of capitalism, characterized by a certain degree of political democracy, Western Marxism has taken a democratic form" (1991a, page 517).

110. Peet 1991b, page 97.

111. Ibid., page 104.

112. Indeed, those Marxists who reject the concept outright face a series of dilemmas in maintaining a consistent theory of historical materialism. For an extended discussion, see O'Leary 1989, especially page 201.

113. Ibid., page 231.

114. Kennedy 1987, page xvi.

115. March 1974, pages 89–90.

116. K. Hall 1985, see page 167 especially.

117. See Butzer 1976. As he writes (page 110): "[T]here is no direct causal relationship between hydraulic agriculture and the development of the Pharaonic political structure and society."

118. See O'Leary 1989. Stein (1989, page 100) also shows that in southern India irrigation did not depend on central administrations.

119. See TeBrake 1985, page 230.

120. O'Leary 1989, page 286.

121. See Bowie 1992.

122. The parcelized sovereignty thesis has been popular with scholars of both Marxian and liberal bents (see, for example, P. Anderson 1974; J. Hall 1985). Japan is often treated as something of a European exclave from this perspective, because it too was characterized by the parcelized sovereignty of feudalism. Such a picture offers a ready explanation for the otherwise incongruous rise of Japan; one of us has in fact invoked this very thesis to help explain local economic development in Tokugawa Japan (Wigen 1995b). Parcelized sovereignty may have been more highly developed in Japan and western Europe than elsewhere, but the general phenomenon seems to have existed throughout Eurasia.

123. See Lapidus 1988, page 333 ff. Consider Hupchick's description of large landholdings in Ottoman Bulgaria: "The new non-military fief-holders viewed their holdings purely as sources of personal enrichment, free of any military obligations to the state. They bent every rule and took advantage of every proffered opportunity to convert their leases into personal, hereditary *mulk* property. . . . Each attempted to invest revenues into purchasing new leases on land or into other profit-generating enterprises . . . with the ultimate goal of casting off completely control of their properties by the central government and transforming them into legally owned sources of personal wealth as capitalistic, market-oriented estates" (1993, pages 32–33). One is reminded here more of England's agricultural revolution than of Oriental absolutism.

124. Richards 1993.

125. Wink 1986, page 157.

126. Ibid., page 189.

127. Jack Goldstone (1991, page 42, note 4) persuasively argues that early modern states in Europe, East Asia, and Southwest Asia were similarly structured, being neither capitalist nor feudal, but rather "agrarian-bureaucratic."

128. M. Mann 1986, pages 521, 526–27.

129. See Wigen 1995b for a related argument.

130. See Cantor 1991, pages 264–65.

131. Kaiser 1994, page 326.

132. "Oriental Renaissance: A Survey of Japan," *Economist,* 9–15 July 1994, pages 3–4.

133. As William McNeill shows, the familiar global metanarrative centered on the rise of freedom and structured around the exceptionality of the West relies on a highly selective reading of European history: "[E]ras of darkness and despotism could properly be skipped over since they made no contribution to the main stream of human achievement" (1995, page 11).

134. It is also essential to note that while certain institutions of modern society have indeed spread from the West to the rest of the world, those very institutions have in the process been reformulated in accordance with local practices and cultural norms. Bryna Goodman, for example, shows how "modernity" in Shanghai was shaped in large part through the "traditional" networks of native-

place associations. She concludes by challenging the standard unilinear and unidirectional view of modernity: "[O]ur standards for 'modernity' . . . too often look like something western. Although we have long been trying to understand Chinese modernity, it has been difficult to see the ways in which 'China' and 'modern' fit together. . . . The notion of modernity as something with fixed standards, as a recognizable threshold, is troublesome, moreover, because it has displaced efforts to understand the process of change with measurements and markers of change, which cannot provide historical explanations" (1995, pages 309–10).

135. It is noteworthy that many Chinese intellectuals have, over the past century, similarly challenged the equating of modernization with Westernization, arguing instead that the process should be seen more in historical than in geographical terms (see Tuan 1996).

136. Said 1978, page 2.

137. Ibid., page 42.

138. Ibid., page 69. Because the Orientalist vision stresses absolute difference, the content of metageographical categories such as Orient and Occident can undergo complete turn-arounds as social and cultural evolution proceed. This is abundantly clear in regard to sexuality. In previous generations, self-proclaimed representatives of the West usually conceived of the East as the land of lavish sensuality, inviting in its lures perhaps, but ultimately repellent in its moral turpitude. Sexual excess was taken to be a pan-Oriental trait, but was considered especially pronounced in the Islamic realm (see Kabbani 1986). Today, the same region is more often regarded in the popular imagination of the West to be repellent by virtue of its puritanism. Non-Western critics of Western culture the world over—and especially those associated with radical Islam—now denounce the West as the land of unbridled sexuality.

Certainly sexual mores in Europe and the United States have loosened considerably over the past decades, while those of many Islamic countries (particularly North African ones) have tightened in the same period. But such changes in customs, and even more important, in characterizations of "the Other," show how perilous it is to associate geographically defined regions of the human community with any particular cultural trait. Actually, it is not uncommon throughout the world to attribute unconstrained sexuality to whatever group of people one frames as exotic. " 'The association between the Orient and sex,'" writes James Millward, "is not unique to western attitudes, but can appear in East Asian attitudes toward the 'Orient'—wherever it is—as well" (1994, page 452; the "Orient" that Millward is discussing is actually Xinjiang, a portion of China's "Occident").

139. See especially B. Lewis 1993b and Lawrence 1993. On the contradictions within Said's own vision, see O'Hanlon and Washbrook 1992 (page 157 especially) and Ahmad 1992.

140. Young 1990, page 19.

141. Said 1978, page 5.

142. Ibid., page 326.

143. MacIntyre 1988, page 395.

144. Said 1985, pages 22–23. Said's disavowal of universalism has met a mixed reception. On the one hand, skeptical postmodernists object to Said's residual universalism, citing his belief that one can invoke a human community rather than

merely a multiplicity of mutually inaccessible universes of meaning (see, for instance, Young 1990, page 131). Other critics regard Said's overtures to global human understanding as disingenuous, arguing that his work "claim[s] the mantle of authentic scholarship only to parade . . . tribal dialect or political ideology as universal discourse" (Lawrence 1993, page 12).

145. The notion of human universals is anathema in certain postmodernist circles; Robert Young (1990, page 121) goes so far as to argue: "Every time a literary critic claims a universal ethical, moral, or emotional instance in a piece of English literature, he or she colludes in the violence of the colonial legacy in which the European value or truth is defined as the universal one." Yet in certain realms the evidence for human universals is impressive. Chomskian linguists and other cognitive scientists have gathered a tremendous array of data in recent years indicating that the basic structure of language is the same for all of humankind. As Steven Pinker (1994, page 43) writes in his brilliant book *The Language Instinct,* "The universal plan underlying languages, with auxiliaries and inversion rules, nouns and verbs, subjects and objects, phrases and clauses, case and agreement, and so on, seems to suggest a commonality in the brains of speakers, because many other plans would have been just as useful." For other possible human universals, see D. E. Brown 1991.

The denial of universality in any sphere of human activity has a long lineage in European thought. Oswald Spengler (1926, page 59), for example, believed that each "culture" (and its decadent successor of a "civilization") possessed its own unique mathematics expressing an essence that was utterly foreign to outsiders. Yet it is mathematical truth, utterly certain within the confines of its own system, that most clearly typifies the universality of reason.

146. Turner 1994, page 7.

147. See, for example, Ahmad's (1992) insightful exploration of this topic. In the targeting of Cartesian epistemology we can find one of postmodernism's most consistent *metanarratives*—even if postmodernists claim that their aim is to annihilate metanarrative. On the continued use of metanarratives in postmodernist discourse, see James F. Harris 1992 (especially page 114).

148. Similar observations have been made by scholars far more sympathetic to poststructuralism than we are. Matthew Sparke (1994, page 119), for example, argues in regard to Robert Young's *White Mythologies*, that "'The West' and an 'anti-West' are seen as the only spaces available and Occidentalism is posed as the only alternative to Orientalism." Sparke, not unfairly, concludes that Young's project is based on an "anemic geography."

Chapter 4: Eurocentrism and Afrocentrism

1. See Amin 1989. While Amin makes some very trenchant points on the general matter of Eurocentrism, his analysis is crucially weakened at several key points by his unbending devotion to neo-Marxian doctrine. For an insightfully critical yet appreciative review of the book, see Kaiwar 1991.

2. Of course, one could remain thoroughly committed to a Eurocentric view while admitting that Europe forms simply a region of social interaction rather

than a geophysically determined continent. The continued prevalence of the continental system in the popular imagination, however, functions to conceal the inordinate degree of attention that Europe receives. Eurocentrism thus takes a covert as well as an overt form. Atlas publishers, for example, may feel that they are remaining reasonably faithful to objective criteria in devoting, on average, 1.3 pages of maps to Europe for every page devoted to Asia (data compiled from *National Geographic Atlas of the World* 1981; *Goode's World Atlas* 1986; Rand McNally's *New International Atlas* 1984; *New York Times Atlas of the World* 1980; *The New Oxford Atlas* 1978; *Reader's Digest World Atlas* 1992; *Hammond World Atlas*, Ambassador Edition, 1991; and *Times Atlas of the World* 1992. For a notable exception, see *Peters Atlas of the World* 1990). If Europe and Asia are both regarded as continents—structurally if not spatially equivalent units—such a ratio does not appear too unbalanced. If, on the other hand, the European and Indian subcontinents are to be considered commensurable units, the highly unequal cartographic coverage of these two regions can only be regarded as distinctly biased.

The close attention given to Europe in atlases is not at all unusual; if anything, world atlases offer a more balanced approach than most other popular geographical media. Most English-language encyclopedias, for example, are heavily focused on Europe and northern North America. The fifteenth edition of the *Encyclopedia Britannica*, for example, devotes 205 pages to its two main articles on Europe, but only 150 to the vastly larger topics of "Asia" and "Asian Peoples and Cultures" (*Encyclopedia Britannica* 1992, volumes 18 and 1). Since Europe and Asia, as distinct continents, are considered commensurate units, "major" Asian countries are allocated roughly the same coverage as "major" European countries—even if they are roughly an order of magnitude larger and more populous. *Britannica* may devote 175 pages to India, yet the United Kingdom weighs in at 139 pages. Similarly biased ratios appear at lower levels of the spatial hierarchy. The important Indian state of Gujarat, geographically comparable to major European countries, receives approximately one-half page in the 1992 *Encyclopedia Britannica*—the same as Italy's Tuscany (see volumes 21, 29, 5, 12).

The situation is little different in the popular geographical media of other Western countries. In the case of France, for example, one might consider the famous *Géographie universelle*. Although, as Hugh Clout claims, the new version "is less Eurocentric" than its predecessor, it retains a certain European focus. While North Africa, the Middle East, and South Asia are all covered in a single volume, Europe is accorded the better part of three volumes (concerned, respectively, with France and Southern Europe, Central and Northern Europe, and Eastern Europe and the Soviet Union) (see Clout 1992, page 426).

3. Following this belief, Carl Ritter (1864, page 64) argued that the population of Asia had necessarily been much greater in the time of Alexander than in his own day.

4. See Baritz 1961; Nordholt 1995.

5. See Baritz 1961; see also Phillips 1994, page 44. As Nordholt (1995) shows, the Germans, the French, and the English all tended to see themselves as history's final destination. The Russians, of course, saw things differently; for many of them the torch of civilization passed eastward from Rome to Constantinople to Moscow.

6. See Hegel 1956, pages 103, 99, 112; Hegel 1942, page 220. Lest one conclude that this thesis has passed on to its deserved oblivion, consider the following passage from an influential news magazine: "The trouble is that, since those days, the West has moved steadily westward, and has left Greece isolated at the bottom right-hand corner of Europe. This does not matter when Greece can present itself as a lonely outpost of western ideas, defying the barbarians; then the West rallies round" ("A Survey of Greece," *Economist*, 22 May 1993, page 3). Or consider this passage from the same magazine, written one year later: "The argument over 'Asian values' is not about whether the tide of history may now be moving east after 500 years of moving west (although that may well be happening)" ("Asian Values," *Economist*, 28 May 1994, page 13). An additional problem with the latter passage is that if one really does believe that the course of history moved west for 500 years, one does not need to reverse its direction to account for the rise of East Asia.

7. Ritter 1864, page 199.

8. Semple 1911, page 390.

9. Fukuyama 1992. Fukuyama, it must be admitted, espouses a rather sanitized, relatively non-Eurocentric form of Hegelian historiography by limiting the "history" of all areas to the period following the discovery of the modern scientific method (see page 135 especially). While he credits Europe with this breakthrough, he correctly argues that, "once having been invented, the scientific method became a universal possession of rational man, potentially accessible to everyone regardless of differences in culture or nationality" (pages 72–73).

For a recent specialized "universal history" structured around an essentially Hegelian metageographical scheme, see Van Doren's (1992) *A History of Knowledge*.

10. Manicas 1987, page 82.

11. See Herder 1968, pages 164–65.

12. G. Weber 1853, page 8.

13. See Amin 1989.

14. De Rougemont, page 247.

15. White 1973, page 164. Note that White also sees Ranke and Burckhardt as archetypical of entirely different schools of historical explanation: Ranke emplotting history as comedy, Burckhardt as satire.

16. Ranke's metageographical views are clearly stated in his *Universal History* (1885, page xi): "From time to time the institutions of one or another of the Oriental nations, inherited from primeval times, have been regarded as the germ from which all civilization sprang. But the nations whose characteristic is eternal repose form a hopeless starting point for one who would understand the internal movement of Universal History."

17. Burckhardt believed that any positive attributes that might once have existed in Asia were extinguished by the Mongol conquests. "In such countries," he informs us, "men will never again believe in right and human kindness" (1943, page 363).

18. To be sure, one can find examples of non-Eurocentric "world histories" composed in the nineteenth century, but most are burdened by similar prejudices. *The History of Mankind*, by the famed German geographer Friedrich Ratzel (1896)

is a prime case in point. Ratzel devotes only a few pages at the end of this massive three-volume work to Europe, but his work as a whole is framed much more as an ethnographic than a historical exploration, and Europeans (or at least those in the region's core areas) have traditionally been considered exempt from ethnographic inquiry (which was until recently concerned primarily with "primitives" and secondarily with the carriers of "stagnant civilizations").

19. Eurocentric history continued its course into the early twentieth century, reaching a high point as late as 1907 with the publication of Henry Smith Williams's massive (25-volume) *Historian's History of the World*. Williams discusses ancient India briefly in volume 2 (entitled *Israel, Persia, Phoenicia, and Asia Minor*), but Indian history after A.D. 664 is relegated to a section on "The British Empire." China and Japan are discussed in the final volume (*Turkey, Minor Eastern States, China and Japan*) — as is Poland! Southeast Asia, bizarrely enough, is treated as one of "The Buffer States of Central Asia." See also J. A. Hammerton's (1927) inordinately ambitious *Universal History of the World*, a work that devoted only a few of its 191 chapters to areas outside of the West.

20. Durant 1954, page ix.

21. Palmer and Colton 1971, page 197.

22. Strayer and Gatzke 1984. Another of the more Eurocentric recent works is Reither's (1973) *World History: A Brief Introduction*. Hugh Thomas, in his *History of the World*, is the most unabashed in his biases: "It will become evident that the 'world' mentioned in the title is not so large as one would expect. There is much more here about Tuscany than Tobago" (1979, page xvii). It is also noteworthy that virtually all world historical atlases are strongly Eurocentric in their cartographic depictions.

23. Published by the Center for History in the Schools, University of California at Los Angeles, 1994. For relatively non-Eurocentric textbooks, see Anglin and Hamblin 1993; also Smith 1988.

24. The university catalogs discussed in this section all pertain to the 1990–91 academic year, when research for the present work began. The figures cited here must be regarded as approximations only, as it is sometimes difficult to ascertain the exact spatial coverage of a specific course based on the information presented in a catalog.

25. We thank Oscars Hamburgs for this observation on the teaching of history in Latvia.

26. It is important to qualify this statement; J. M. Blaut (1987, 1992, 1993) has produced several cogent works demonstrating how the common geographical doctrine of "diffusionism" has unfairly prioritized certain supposed "centers of innovation."

27. Because of this, from a true geographer's point of view, all areas are of intrinsic interest. As Pierce Lewis (1992, page 297) pithily insists, "[T]here is no such thing as a boring landscape."

28. The most literal form of Eurocentrism is actually found in the geographic doctrine that Europe possesses a physically central position in the world's "land hemisphere" — a position that supposedly predisposed it to command the rest of the world. This idea, which originated in the work of Malte-Brun and other French geographers, was strongly advocated by Carl Ritter (1864, page 63 especially) in

the nineteenth century. One still occasionally finds it in geography textbooks (see, for example, de Blij and Muller 1994, page 47).

29. See Cleary and Lian 1991, page 163.

30. The most committed members of the group declared that "spatial laws" were unaffected by the specificities of different places or different historical epochs. As Michael Hill (1981, page 57) argued, "[T]he existence of historical and regional geography cannot be logically defended. Geographical theory must be ahistorical and aspatial."

31. There certainly are exceptions, as some fine work in the positivist school of geography has been carried out in non-Western settings (see, for example, Gould 1963).

32. The main exceptions here have been in the field of Marxist-inspired development geography; see Corbridge 1986 for a critical review.

33. See M. Lewis 1991 and Murphy 1991 for critiques.

34. See Price and Lewis 1993 for a critical review. Again, there have been notable exceptions (see, for example, Duncan 1990).

35. Gregory and Ley 1988, page 116.

36. See, for example, Soja 1989.

37. Bayly 1983, pages 2–4.

38. For one of the best works in this genre, see Timothy Mitchell's *Colonizing Egypt* (1988).

39. The same ideas can, of course, be deployed to show how non-Europeans have interpreted peoples of other cultural traditions (see, for example, Tavakoli-Targhi 1991).

40. Amin 1989, page 147. See also Ahmad 1992 for an insightful review of these trends.

41. As Mary Lefkowitz (1996, page 156) shows, radical Afrocentrism itself is based in part on misguided *European* visions of ancient Egypt.

42. In its more modest guises, Afrocentric scholarship is mainly concerned with ridding the academy of vestigial white racism and with insisting that African peoples be granted due recognition for their contributions to the human endeavor. This effort is wholly admirable. But the most vociferous Afrocentrists promulgate what is essentially a religious dogma. To Molefi Asante (1988, pages 1, 6), one of the movement's academic leaders, Afrocentricity is "the belief in the centrality of Africans in post modern history," and represents nothing less than the "centerpiece of human regeneration."

43. One's "racial position" in ancient Egypt evidently depended in part on what portion of the country one lived in; those residing in Upper Egypt were darker than those living in Lower Egypt. Royal families that came from areas even further to the south (such as the Twenty-fifth, or Nubian, Dynasty), were black by any definition of the term.

44. See Bernal 1987. Mary Lefkowitz (1996), however, maintains that the ancient Greeks were themselves deluded on this score, and that the post-Enlightenment reassessment of Egypt came about in part because of increasing knowledge about the ancient world.

45. See Snowdon 1983.

46. Bernal 1987, page 73.

47. Asante and Mattson 1992, page 9. On continental thinking among radical Afrocentrists, see also Lefkowitz 1996, pages 30–33.

48. See Asante 1988, page 46.

49. Ibid., page 48.

50. See Curtin et al. 1978, page 92.

51. Manning 1990, page 25.

52. See B. Lewis 1990. As Lewis demonstrates, Middle Eastern society was never as racist as United States society prior to the civil rights movement, but it was by no means free from prejudice.

53. Ibn Khaldun 1967, page 59.

54. Ibid., pages 59–64.

55. Ibid., page 59.

56. See Dunn 1986.

57. Agyeman 1985, page 146. Agyeman usually defines himself a pan-Africanist rather than an Afrocentrist; these two terms, however, are sometimes considered interchangeable.

58. Ibid., page 19. Agyeman denies trans-Saharan unity in part on the basis of racial essence, for he views Arabs as a distinct racial group, dismissing the idea that "the term Arab 'has significance in a linguistic and cultural, rather than in a racial sense'" (page 146). While this issue is a complicated one, it is difficult to justify Ageyman's claim that "Arab" is a racial category. Generally speaking, Arabs are simply those people who consider themselves Arabs and who speak Arabic as their primary language. Most are regarded, and regard themselves, as white, but there are also Arabic-speaking peoples living south of the Sahara who are in physical appearance substantially black. Groups like the Baqarra and the Shuwa Arabs, living in modern-day Chad, Nigeria, and the Sudan, have adopted certain sub-Saharan subsistence practices and have intermarried extensively with other sub-Saharan Africans (see Curtin et al. 1978, pages 98–99; Oliver and Atmore 1981, page 38). Moreover, even in southern Arabia some Arabic speakers have a somewhat "African" physical appearance. Nonetheless, complexities remain. As Bernard Lewis (1993a, pages 1–9) demonstrates, in certain contexts the term *Arab* may be restricted to (Arabic-speaking) Bedouins, while in others it may be extended to include even non-Arabic speaking Kurds or Turkomans. Roy (1992, page 75) speaks of "Persian speaking Arabs" in Afghanistan. Moreover, as Hourani notes, the acceptance of Somali and Djibouti into the Arab League is "a sign of the ambiguity of the term Arab" (1991, page 424). The simple linguistic definition we have employed is, however, the one most widely used.

59. Asante 1988, page 96.

60. Diop (1974, page 280) also recognizes a "yellow" race, but he thinks that it formed from ancient "crossbreeding between Blacks and Whites."

61. Ibid., page 111.

62. Ibid., pages 112–13.

63. See Jean 1991, page 96.

64. Acknowledging that this constitutes a revisionist storyline, Jean (1991, page 99) concludes that his "is a contribution to a view of history as it actually happened. This can only be good for everybody."

65. The reversal is so complete that any creativity ascribed to the white race

is ascribed to the hidden influence of black genes and cultural influences (see Diop 1974, pages 117, 152).

66. See Bernal 1991, page 249. The existence of a community of some 30 to 200 persons of mostly African ancestry in Abkhazia in the early twentieth century is undeniable (Blakely 1986, pages 5–7). The most common, and most likely, explanation for their origin is the importation of African slaves by landholders during the period of Ottoman domination, a scenario actually confirmed by several personal histories gathered early in the century (Blakely 1986, pages 12, 75). While Bernal does not deny that the recent movement of African peoples may have bolstered the community, he nonetheless argues strongly that their history goes back to ancient Egyptian invasions. Moreover, he goes on to call them "Black Africans," as if they had maintained a kind of pure racial essence through a 4,000-year sojourn in the Caucasus. Such a view requires more than a little credulity. One of the most striking features of these people was the relatively "pure" African appearance of many individuals (see photography in Blakely 1986, page 6) — even though they lived in an area surrounded by inhabitants described as "narrow faced and fair skinned" (Benet 1974, page 3). We would thus have to assume that a large Egyptian army, composed exclusively of blacks and containing numerous women, settled in the area, adopted the language and customs of the local inhabitants but never intermarried with them, maintained their racial exclusivity for several thousand years, recruited freed African slaves in the nineteenth century (due undoubtedly to the affinity born of racial essence), and then vanished quickly into the general population in the twentieth century.

67. Bernal attributes the Chinese belief in the revocability of the "mandate of heaven," as well as the French art movement of Impressionism, to major eruptions (1987, page 41; 1991, page 38). In a similar vein, he informs us that Japan lacks the "mandate of heaven" conception largely because "there have been no overwhelming events of the order of [the eruptions of] Thera and Hekla III coinciding with an acute political crisis since the foundation of the Japanese empire" (1987, page 316).

68. At one point, for example, Bernal attributes the "weakness" of the "catastrophic tradition" in Greece to "the pleasant consistency of [its] Mediterranean climate," yet later he stresses the devastating effects of that country's "inevitable droughts" (1987, pages 318, 520). His knowledge of global climatic patterns also seems to be woefully limited. He stresses, for example, the "climatic similarities" of Nubia and Colchis [coastal Georgia], evidently unaware of the fact that the coolest month of the year in Nubia is warmer than the warmest month of the year in Colchis, and that any two contiguous months in Colchis see more rainfall on average than Nubia experiences in an entire year (see page 253: according to statistics provided in *World Weather Guide* [Pearce and Smith 1984, pages 86, 435], Sochi [containing the weather station nearest to Colchis] has an average July high temperature of 80° F and an average low of 66° F [August here is equally warm]; Khartoum [of the stations listed, the nearest to Nubia] has an average January high temperature [its coolest month] of 90° F and a low of 59° F. Khartoum receives an average of 6.2 inches of rain a year; Nubia proper is much drier. Sochi, on the other hand, receives 57.3 inches. The driest period in Sochi is midsummer; in July and August the station receives on average 6.4 inches of rain).

In another passage, Bernal implies that there are no seasonal rhythms in the tropics (1987, page 315); a first-year geography student should know that most tropical lands experience a profound alternation between wet and dry times of the year. Even in the equatorial belt many areas show some seasonal discrepancy in precipitation. Manaus, near the heart of the Amazonian Basin, receives 1.5 inches of rain, on average, in August; in March, on the other hand, it usually receives over 10 inches of rain (see Pearce and Smith 1984, page 175).

69. In 1906 Sarda argued not only that "Egypt was originally a colony of the Hindus" (page 149), but also that "[t]he Scandinavians are the descendants of the Hindu Kshatriyas" (page 173) and "[t]he Druids in ancient Britain were Buddhistic Brahmans" (page 176).

70. See Abraham 1962, page 39.

71. Mudimbe 1988, page 185.

72. See Bernal 1991, page 522.

73. As Frank Dikötter (1992, pages 10, 54–56) explains, the Chinese traditionally regarded themselves as white, as did most of the first Europeans to visit their country. They used the color yellow, however, to symbolize their emperor and to some extent their homeland. Europeans began to describe East Asians as yellow in the late 1600s, and the Chinese eventually adopted the usage, in part because of the positive connotations of the color in their iconographic system.

74. Such color categories have been remarkably persistent even in the scholarly literature. In Carleton Coon's posthumously published *Racial Adaptations* (1982, pages 52–53), for example, the author does map northern Europeans as pink and most Amerindians as various shades of brown, but he continues to classify central Africans (along with southern Indians, Melanesians, and indigenous Australians) as black, and East Asians (along with the indigenes of southern Chile) as yellow. It is actually northwestern Europeans who most often exhibit red coloration; in South Asia Europeans are sometimes disparaged as "red monkeys" (Raychaudhuri 1992, page 157), while some early Chinese travelers in the West described European skin color as reddish purple (Dikötter 1992, page 54). East Asian skin shade, for that matter, is quite similar to that found in large areas of Europe (Barnicot 1964, page 188). Indeed, the Japanese of the last century coded color partially in gender terms, viewing Japanese women as distinctly whiter than European men (Alland 1971, page 148).

75. One of the most elaborate—and confused—racial classifications may be found in a 1950 world atlas (Bartholomew 1950, page 30), which divides Eurasia into the following *racial* groups: "Chinese, Other Eastern, Nomadic Mongol [including the inhabitants of Finland and Turkey], Dravidian and Oceanic [including the inhabitants of insular Southeast Asia], Semetic [*sic*], Celto-Teutonic [including the inhabitants of Romania], Romanic, Slavonic, Indo-Iranian, and Greco-Albanian." The same author goes on to subdivide South Asia into no less than eight distinct races (page 71).

76. See Deglar 1991.

77. Marcus 1994, page 100.

78. St. Clair Drake 1987, page xx. Among Nazis and proto-Nazis, on the other hand, the impetus had been just the opposite; namely, to limit the (true) white race to "Nordics" and to regard Europe's "Mediterraneans" and "Alpines" as mem-

bers of distinctly different, and inferior, races. Thus the notorious Madison Grant argued that there were actually three distinct "subspecies" of man in Europe (1918, page 233).

79. See Jackson and Hudman 1990, page 288.

80. See MacFadden, Kendall, and Deasy 1946, page 9.

81. See Coon 1954, page 191.

82. See Coon with Hunt 1966, pages 24–26. Coon (1982, page 13) held to this view to the end of his life.

83. Such patterns probably stem in large part merely from sloppy cartography. Coon was probably trying to map areas of tribal populations in India, which he regarded as exhibiting various mixtures of "Caucasoid" and "Australoid" ancestry.

84. Harper and Schmudde 1978, page 103.

85. Norris 1990, page 38.

86. See Montagu 1964.

87. See *Encyclopedia Britannica* 1963, volume 18, pages 864a–64b. The notion of hair texture as the primary trait for racial classification stems largely from the work of A. C. Haddon (1925, pages 15–16). Haddon devised an elaborate racial taxonomy in which hair texture was used for only the largest-scale division. His finer subdivisions, made on a seemingly ad hoc basis, employed head shape, skin color, stature, and "geographical distribution."

88. Cavalli-Sforza, Menozzi, and Piazza 1994, page 19.

89. Ibid., page 99.

Chapter 5: Global Geography in the Historical Imagination

1. For notable recent works, see P. Sahlins 1989 and Thongchai 1994.

2. Just a decade ago world history was suspect in the discipline—"banished" from the classroom, according to William McNeill (1986, page 82). The field's renaissance in the intervening years has been truly remarkable.

3. McNeill 1990, page 10.

4. The literature on this subject is vast. Both of us have grappled with political and economic geography at length in our respective empirical studies of the Philippine and Japanese highlands (see M. Lewis 1989, 1992b; Wigen 1995b). For a preliminary attempt to posit a "networks" model for economic interaction in the early modern world, see Wigen 1995a.

5. See Toynbee 1934–61, volume 1 (1934), page 149 ff.

6. Ibid., page 157.

7. The same concern drives his present-day followers, members of a group called the International Society for the Comparative Study of Civilizations, who expend considerable energies on the effort to bound discrete civilizations in time and space (see Melko and Scott 1987). The notion of isolation is also still strong in popular literature. Clive Ponting, the author of *A Green History of the World,* for example, tells us: "Before the sixteenth century different areas of the world had evolved to a large extent in isolation" (1991, page 194). Actually, if one defines "civilization" in terms of isolated historical development, then one would

have to agree with S. A. M. Adshead (1993, page 4) that there are really only "four primary civilizations"—those of "Western Eurasia, East Asia, Black Africa, and Meso-America."

8. As Edmund Burke III (1993, page xv) writes, "The textualist position foreshortens history, annihilates change, and levels difference the better to represent the past in dramatic form—either as tragedy, as in the case of Islamic civilization, or as triumph, as in the case of the rise of the West."

9. See Toynbee 1934–61, volume 11 (1959).

10. "In 1958–59 the American Historical Association surveyed department chairmen on the major fields of their graduate students. The total number of graduate students was 1,735; the number reported as concentrating in African history was 1" (Feierman 1993, page 168). Feierman's essay is an eloquent plea for sub-Saharan Africa to be included in stories of world history in its own right—and a warning that doing so will necessarily change the way we perceive that history.

11. See Toynbee 1934–61, volume 12 (1961), pages 560–61.

12. See Hodgson 1993, page 12.

13. See Toynbee 1934–61, volume 8 (1954), page 90.

14. In 1934 Toynbee's taxonomy of civilizations was as follows: "*Wholly Unrelated Societies:* Egyptiac, Andean; *Societies Unrelated to Earlier Societies:* Sinic, Minoan, Sumeric, Mayan; *Infra-affiliated Societies:* Indic (?) [*sic*], Hittite, Syriac, Hellenic (?) [*sic*]; *Affiliated Societies I . . . :* Western, Orthodox Christian, Far Eastern; *Affiliated Societies II . . . :* Iranic, Arabic, Hindu; *Supra-affiliated Societies:* Babylonic, Yucatec, Mexic" (1934–61, volume 1, pages 131–32). The number had increased to 34 by the final volume, but only a handful were posited as existing at any given time. In 1961 (volume 12, pages 558–61), it ran as follows: "*Unrelated to Others:* Middle American, Andean; *Unaffiliated to Others:* Sumero-Akkadian, Egyptiac, Aegean, Indus, Sinic; *Affiliated to Others* [first group]: Syriac, Hellenic, Indic; *Affiliated to Others* [second group]: Orthodox Christian, Western, Islamic; *Satellite Civilizations:* Mississippian, South-Western, North Andean, South Andean, ? [*sic*] Elamite, Hittite, Urartian, Iranian, Korean, Japanese, Vietnamian, ? [*sic*] Italic, South-East Asian, Tibetan, Russian; *Abortive Civilizations:* First Syriac, Nestorian Christian, Monophysite Christian, Far Western Christian, Scandinavian, Medieval Western City-State Cosmos."

15. As K. N. Chaudhuri (1990, page 11) writes, "The specialist historians of Asia, each examining his own narrow chronology and field, are often unable to see the structural totality of economic and social life and they are inclined to treat the experience of their own regions as unique or special." In compensation, Chaudhuri deliberately blurs the boundaries between each civilization in his own work, to good effect.

16. See Stavrianos 1975.

17. Roberts 1993.

18. Southeast Asia fares poorly in many recent college textbooks on world history. F. Roy Willis (1982), for example, attempts to transcend the Eurocentrism once typical of the genre by examining India, China, and the Middle East as well as Europe. In one chapter (9), Southeast Asia is granted a short discussion, yet it is not considered on its own terms; Willis describes premodern Cambodia and Java, for example, as merely exemplars of a "reinvigorated Indian culture" (page 384).

Archibald Lewis (1988), who displays a nearly global sweep in his impressively synthetic *Nomads and Crusaders,* similarly bypasses the region. Lewis opens this book with the assertion: "In the year 1000 five great civilizations existed in the world of Africa and Eurasia: the East Asian, the Indic, the Islamic, the Byzantine-Russian, and the western European" (page 3). While Lewis seems to be adopting the same tired roster of discrete social systems, he does acknowledge the existence of areas, such as Tibet, that do not fit into this scheme, and he is in general highly sensitive to subtle variations in geographical patterning of macrocultural affiliation. Still, the eclipsing of Southeast Asia is noteworthy.

19.   China has managed to retain far more of its early modern empire than have its European counterparts. In the mid-eighteenth century—just as Russia was advancing in the Kazakh steppes, Britain was beginning to build a land-based empire in India, and the Netherlands was starting to carve out an Indonesian imperium—the Manchu rulers of China were expanding vigorously in Central Asia. The actions of the Central Kingdom there were as brutal as those of any European imperialists; it virtually exterminated the Zunghar Mongols, for example (see Bergholtz 1993, page 402). Therefore it is a considerable falsification of history to argue, as Chiang Kai-shek did, that "Xinjiang is Chinese not as a result of Qing [Manchu] conquest . . . but due to 2,000 years of assimilation" (quoted in Millward 1994, page 446). The most ardent Chinese nationalists actually claim that China should rightfully control all territories that ever paid homage to it—including Kazakhstan, Nepal, Russia's far east, and virtually all of mainland Southeast Asia (see Lamb 1968, page 39). For more on China's imperial conquests in Central Asia, see Lattimore 1988, Bergholtz 1993, and Jenner 1992.

20.   In the eighteenth and nineteenth centuries, European atlases often portrayed Korea as a portion of Tartary (see, for example, Palairet 1775; Woodbridge 1824); with the disappearance of this region it was usually simply transferred to China. A few atlases of the late 1800s and early 1900s did, however, portray Korea as an autonomous geographical unit (Bartholomew 1873, map 28; W. Johnston 1880, map 19; *Hammond's Modern Atlas of the World* 1909, map 95; Cram 1897, page 188).

21.   Most of the people of Po-Hai (more properly, in Korean, Parhae) were speakers of Tungusic languages rather than Korean, but the kingdom's rulers were Korean, and they "clearly regarded their state as representing a revival of Koguryo" (Lee 1984, page 72).

22.   This pattern pervades most of the major world historical atlases. In the *Harper Atlas of World History* (Vidal-Naquet 1987, page 3), the initial East Asian maps show a small portion of the Korean peninsula, depicted in outline only. The next treatment (page 55), covering China's Han dynasty, maps most of Korea merely as a province of China while ignoring the rest. A subsequent map, showing the territory of the Tang dynasty, portrays all of Korea as a portion of the Chinese Empire (page 82), while five maps set during the Sung period (page 83) depict the peninsula simply as blank space. On page 85 of the text, Korea appears on a map focused on Japan—but the lengthy text on the facing page makes no mention of it. The next map of Japan (page 115) also includes Korea but leaves it as a featureless outline. Korea again turns up on another map of China during the nineteenth century, but the only information provided on it pertains to foreign aggression.

262   NOTES TO CHAPTER 5

In Rand McNally's *Atlas of World History* (1993a, pages 26, 32, 80–83), Korea appears only in order to show Chinese encroachment in the Han and Japanese invasion, along with tributary status to China, during the late sixteenth and seventeenth centuries. *The Times Atlas of World History* (Barraclough 1984, page 127; see also pages 85, 169, 175, 232, 242, 262, 279, 292) does show Silla on one map of East Asia, but even here emphasis is placed on temporary Chinese occupation during the Tang dynasty; other portrayals of Korea similarly stress Chinese and Japanese aggression. *The Anchor Atlas of World History* (Kinder and Hilgemann 1974, page 176), shows the Korean-ruled kingdom of Po-Hai in Manchuria on one map but little else; as in most other atlases, Korea is never granted depiction in its own right. *The New Cambridge Modern History Atlas* (Darby and Fullard 1970)—which offers the reader seven detailed maps of the Netherlands—shows Korea only as a peripheral portion of the Chinese Empire and as a colony of Japan. Finally, Rand McNally's *Historical Atlas and Guide* (1993b), originally published in Sweden and Norway, does show both Silla and Po-Hai on a map entitled "China around 750" (page 32). Yet this atlas ignores Korea's Three Kingdoms period and never grants the country consideration in its own right.

23. See Tanaka 1993, page 247.
24. See F. Drake 1975, page 73.
25. See Allen 1990, page 799.
26. *The Times Concise Atlas of World History* (Barraclough 1992) does include an admirable set of maps on Southeast Asia (pages 70–71). Another exception is *The Harper Atlas of World History*, which offers a sequence of four maps depicting Southeast Asia's political evolution from 500 to 1400 (Vidal-Naquet 1987, page 86). But even here the territory of Vietnam is left blank, as if there were a void between China and Champa. The long southward drive of the Vietnamese people and their incessant conflicts with the Cham, the Khmer, and the Thai—a vitally important story for world history and geography (as well as for contemporary international relations)—is simply not depicted.
27. Indeed, its Chinese attributes are so marked that Arnold Toynbee categorized Vietnam (like Japan and Korea) as a "satellite civilization" of China, just as he regarded Russia as a satellite of the West (see 1934–61, volume 12 [1961], page 561).
28. See S. Huntington 1993. Huntington's thesis bears a certain resemblance to the arguments put forward by geographer Donald Meinig in a little-known 1956 article. Meinig's piece, which is concerned with the "evolution from nationalism to culturalism," remains in many respects more insightful than that of Huntington.
29. It also seems likely, however, that Huntington—like many other scholars in the field—has exaggerated the religious nature of civilizations, as well as the historical depth of animosity across religious-civilizational boundaries. Barbara Metcalf (1995), for example, convincingly shows that the "civilizational" divide separating Hindus from Muslims in South Asia is largely a modern invention.
30. McNeill 1990, page 9. McNeill has even more recently come to argue that civilizations themselves are actually "pale, inchoate entities" that are "internally confused and contradictory" (1995, page 17). In McNeill's recent works, the dis-

tinction between a civilizational and an integrationist approach to world history has been successfully transcended.

31. In Braudel's third volume of *Civilization and Capitalism* (1984, chapter 5), a global geographical vision is developed around the notion of "world-economies." As it turns out, Braudel's world economies, in the period prior to the establishment of European hegemony at any rate, are essentially the familiar civilizations of world history: the West, Russia, the world of Islam, India, and China. Braudel's pedagogical book, *A History of Civilizations* (1994; first published in 1963) addresses cultural differentiation in more detail than his major works and insists on a thoroughly global scope. Here Braudel outlines the salient characteristics of each of the world's prominent "civilizations," which he delimits in an explicitly geographical fashion: "Civilizations, vast or otherwise, can always be located on a map" (page 9). Although attempting to transcend an overtly Eurocentric approach, Braudel organizes his text so as to emphasize the distinctions between "European Civilizations" (Europe, the Americas, Australia and New Zealand, [white] South Africa, and the Soviet Union) and "Civilizations outside of Europe." Within the latter category, he differentiates "Islam and the Muslim World" from "Africa" and "The Far East." He then further subdivides the Far East into China, India, Japan, and the "Maritime Far East"—a category encompassing "Indo-China, Indonesia, the Philippines, and Korea."

In *A History of Civilizations* Braudel attempts to break down the conventional divisions of the world used in the French educational establishment and to introduce a geographical framework suitable for conveying "the new history" (page xxxv). While the aim is admirable, the project ultimately fails. Braudel's geographical system, usually one of his strong points, appears to be hastily constructed, and it is surely inadequate for the task. He employs the term *Europe* in both its geographical and its cultural sense without differentiating properly between the two. He follows a long-discredited tradition by placing India within a unified "Far Eastern Civilization," considered comparable to those of the Islamic World and Europe. And with the category of the Maritime Far East, he offers a spurious cultural-geographical hodgepodge. Moreover, he is not even consistent within his own categories: at one point he classifies Japan within the maritime grouping, yet he later examines Japan in its own right, as separate subcivilization of the Far East.

32. In particular, Andre Gunder Frank and others in his camp deny Wallerstein's thesis that internal transformations in Europe, associated with a crisis in feudalism, brought about the decisive rupture that led to the creation of the modern world system. Instead, they emphasize systemic continuity—for the span of some 5,000 years—and they downplay any notion of intrinsic European exceptionality (see Frank and Gills 1993).

For a recent critical engagement with the world-systems paradigm as it applies to Africa and Latin America, see Cooper et al. 1993.

33. Prior to the emergence of the European-imposed world order in the early modern period, Wallerstein contends, the large-scale social entities called world economies were "highly unstable structures" that "were always transformed into empires" (1974, page 16; see also page 348). A strict reading of this doctrine would force one to conclude that South Asia never formed a meaningful socioeconomic

system, and that the Islamic heartland in Southwest Asia and North Africa ceased to be a real social entity once the Abbasid caliphate began to decline. See also Voll (1994) for an insightful reading of this central problem in Wallerstein's scheme.

In less capable hands, a Wallersteinian perspective can lead to even more serious geographical misperceptions. This is especially evident in Thomas Richard Shannon's *An Introduction to the World-System Perspective*, a book, lauded by Wallerstein himself, that is designed primarily for undergraduate use. Shannon reveals serious geographical failings in all of his cartography, but nowhere so flagrantly as on a global map intended to illustrate "world-empires and advanced agrarian areas in 1400" (1989, page 41). The initial problem with this map and the text that accompanies it is that the "world-empires" and "advanced agrarian empires" are not adequately distinguished. Mesoamerica and the central Andes—as well as Japan—are left unshaded presumably because at this time they were politically divided rather than united in world empires. Western Europe, however, is shaded despite its lack of imperial unity, in this case apparently because of the "advanced" nature of its agriculture—even though its agriculture was inferior by most measures to that of Japan and the Basin of Mexico. But this is only the most glaring of the map's many geographical misrepresentations. It informs us, for example, that such economically marginal places as the central Sahara, the Nafud, the wild gorge country of western Sichuan and eastern Tibet, and Inner Mongolia were all either portions of world empires or "advanced agrarian areas," while telling us that Java, Poland, and all of Persia must be excluded from the same categories. The failure to include Persia is doubly problematic, since in 1400 C.E. one would have found here both numerous centers of advanced agriculture as well as the extremely powerful empire of Timur. Shannon's map 3.3 ("The world-system in 1763") is almost equally problematic; it portrays the Netherlands as part of the semiperiphery, even though the text (correctly) lists it as part of the core (page 53). Nor are such problems limited to Shannon's interpretation. A popular Dutch exposition of the Wallersteinian view informs the reader that in 1985 Malaysia lay in the periphery of the world system, while Zaire only lay in the semiperiphery—a position it supposedly shared with Norway and Canada (Terlouw 1990)!

Our pointing out of such errors in geographical representation might strike the reader as a petty and mean-spirited act. We consider these mistakes to be highly significant, however, because we believe that they are symptomatic of a far more wide-ranging failure of geographical conceptualization. Errors of a similar magnitude occurring in textual rather than cartographic form would simply not pass in works of similar quality; they would either be viewed as grounds for rejecting a manuscript, or they would be corrected by reviewers or editors. But when it comes to maps, and especially those at the global scale, outrageously sloppy work is not at all exceptional. Because of the massive devaluation of geography—not merely in American public life but also at the highest levels of academia—fundamental geographical ignorance is not regarded as a problem meriting concern. It is significant to note in this context that in one prominent review of "World History as an Advanced Academic Field," no mention whatsoever is made of geography (Vadney 1990).

34. See, for example, Hodgson 1993; McNeill 1963, 1990; Frank 1991; Abu-Lughod 1989. Within this broad school, one of the more empirically rich studies

extending beyond the "Afro-Eurasian ecumene" is that of Eric Wolf (1982). Another attribute of this movement of global history is the rehabilitation of "barbarians" and "semibarbarians" within Eurasia (Turks, Mongols, Tibetans, etc.). In this regard, see especially Christopher Beckwith (1987), who in a fascinating inversion returns to Europe to rehabilitate even the "barbarous" Carolingians.

35. Washbrook 1990, page 492.

36. See Corbridge 1986.

37. Abu-Lughod, 1989, page 361.

38. See Manguin 1993.

39. Indeed, it must be remembered that the Mongol conquests themselves were at first horrifically disruptive for Central Asian trade, which in fact gave impetus to the great upsurge of maritime trade in the same period. The caravan system actually seems to have begun its long decline during the Sung (960–1279) period (see van der Wee 1990, pages 16–17; Rossabi 1990, page 351).

40. The Ming ban on oceanic trade, for example, was lifted in 1567, in a period when Southeast Asian shipping was still competitive with that of Europe. Long-distance trade based in Southeast Asia did indeed begin to wither away at this time, but it was largely because of competition with the Chinese and the Gujaratis, not the Portuguese and the Spanish (see Manguin 1993).

41. For an alternative explanation of the shifting global balance of power between the fifteenth and eighteenth centuries, see G. Parker 1988 and Tilly 1990.

Victor Lieberman (1993a), writing from the perspective of a Southeast Asianist, finds that internal social and economic developments within Europe must at least be considered, instead of being ruled out before analysis even begins, as is the case with Abu-Lughod. While we must take care to avoid Eurocentrism, we should not therefore assume a priori that intra-European developments were themselves insignificant.

42. Sometimes, more narrowly Western biases can be detected as well. In the work of Wallerstein, Europe is given a singular role in the creation of the modern world system, which is directly traced to the crisis of European feudalism. And even in the work of Andre Gunder Frank, for whom debunking Wallerstein's Eurocentrism is a core concern, a related bias can be detected. Frank argues for the "centrality" of three corridors in the world system prior to the sixteenth century: the Nile–Red Sea corridor, the Syria–Mesopotamia–Persian Gulf corridor, and the Aegean–Black Sea–Central Asia corridor (see Gills and Frank 1993, pages 88–90). A glance at a map shows that all of these "central" routes were located far to the west of the demographic and economic pivot of the premodern Afro-Eurasian ecumene.

43. McNeill 1990, page 7.

44. Frank 1991, page 1, emphasis added. "Ecumeno-centrism" is also a common flaw in many non-Eurocentric treatments of world history produced by non-Westerners. Jawaharlal Nehru's impressive *Glimpses of World History* (1942), a 971-page work, for example, barely mentions sub-Saharan Africa and considers pre-Columbian America in but a single short chapter.

45. See McNeill 1976.

46. See Curtin et al. 1978, page 92. The present work may also be justly accused of evincing bias toward this same ecumene of Eurasia and northern Africa.

Our concern, however, is more to examine and criticize widespread geographical notions that are themselves closely focused on this portion of the world, than it is to propose an alternative view.

47. A separate line of criticism, leveled by Marxist analysts, accuses world-systems theorists of downplaying production and giving undue credit to circulation as a cause of historical change. On recent and past debates regarding world-systems theory, see Cooper et al. 1993.

48. See Abu-Lughod 1989, pages 251–53, 259.

49. See Frank 1991, page 12.

50. See Frank and Gills 1993.

51. Frank 1991, page 25.

52. See Meinig 1969 for the use of a "core-domain-sphere-zone of penetration" spatial model of cultural interaction.

53. As Partha Chatterjee (1993, page 75) argues, Indian nationality was constituted from its beginnings by a desire to emulate the West in terms of technology and the institutions of state, but to resist strenuously Western cultural (or spiritual) attributes: "The nationalist response was to constitute a new sphere of the private in a domain marked by cultural difference: the domain of the 'national' was defined as one that was different from the 'Western.' "

54. See Bentley 1993.

55. Braudel 1984, page 65.

56. As William McNeill still maintains, both economic systems and idea clusters are important in grappling with world history, although they do tend to have very different geographical expressions He further argues that the realms of both material life and of ideas may be comprehended under the mantle of "communication nets" (see 1993, pages xi, xii).

57. There were, of course, connections across this line; see Finney (1994), for example, in regard to linkages between Southeast Asia and Oceania.

58. See Crosby 1986.

59. The distinction between sub-Saharan and northern Africa is often overdrawn. As Curtin et al. write: "Even the frontier often drawn between the Sahara and sub-Saharan Africa, one of the sharpest cultural lines anywhere in Africa, is only one step along a continuum. An older view, that the Sahara fringe was a sort of cultural divide, was largely based on racist assumptions" (1978, page 79). In western and central North Africa, it is important to recognize, the Sahara had been a major barrier for several thousand years—until it was surmounted by the Sanhaja Berbers in the eighth century (see McEvedy 1995, page 44).

60. African commonalities seem to be especially prevalent in the domains of religion and statecraft (see Curtin et al. 1978, pages 91, 147, 157). Igor Kopytoff (1987) links certain similarities in sub-Saharan African political systems to ecology and demography through a frontier expansion model. Jack Goody has argued that the Eurasian-North African zone is historically quite distinct from the sub-Saharan zone, particularly in regard to marriage, inheritance, technology, and land tenure patterns (see Goody 1971, pages 25, 76 especially; Goody 1990). Any characteristically "African" traits that one may identify, however, were not necessarily found throughout the sub-Saharan region. Highland Ethiopia, Madagascar, and the Kalahari especially exhibited quite different cultural and political patterns.

61. See Forsyth 1991, page 82.

62. While many traits of material culture, such as metalworking, had diffused throughout almost the entire extent of Eurasia, most ideological and social organizational commonalities stopped short at the central Siberian taiga (see Forsyth 1991, page 75 especially). In Siberia's far west, however, among the Siberian Tartars and to a minor extent even among the taiga-dwelling Mansi and Khanty, Islamic norms had been established before the Russian onslaught (see Forsyth 1992, page 35). The paleo-Asiatic peoples of far northeastern Siberia (such as the Chukchi and Koraks), on the other hand, remained outside of even the metallurgical zone.

63. See Hodgson 1974, 1993.

64. See Turner 1994.

65. The Koreans eventually developed their own alphabet, and the Japanese their own syllabaries, but in both countries (and especially in Japan) Chinese characters continued to play a very important role. Vietnam, of course, adopted the Roman alphabet under French pressure.

66. While cultural historians have traditionally dated modernity back to the Italian Renaissance, and economic historians to the European transoceanic voyages, we reject such chronology as unduly Eurocentric. We follow, rather, Arthur Waldron (1994, page vii), who argues that modern history should ultimately be defined as beginning with the great Mongol expansion. Muslim incursions into South Asia antedate Genghis Khan, but the systematic conquest of (most of) the region was roughly coincident with Eurasia's "Mongol period."

67. As noted above, a major inspiration for this model of the world lies in the work of the late Islamicist Marshall Hodgson. Hodgson never made his geographical scheme explicit, nor did he extend it to a global level. But by devising an implicitly hierarchical order of nesting spatial categories within the ecumene, Hodgson stumbled upon a central principle of regionalization—one that was simultaneously receiving explicit treatment among a handful of theoretically inclined regional geographers. See M. Lewis 1991.

68. Djaït 1985, page 102.

69. See Myers and Peattie 1984; Cumings 1984.

70. See Ahmed 1992, page 200 especially. The distinction between a civilization and a society is that the former refers much more to the learned, elite sphere of life. For a fascinating portrayal of *Dar al-Islam* as an "intercivilizational entity" and as a "special world system," see Voll 1994.

71. See Robinson 1991; Bayly 1983 (page 156 especially). On the division of Islamic India between a Persian-influenced North and Arabic-influenced South, see Wink 1991 (page 360 especially).

72. Recognizing these separate influences, Albert Hourani argues convincingly against the notion of any kind of unified Islamic history. His own scheme (1980, page 184) divides the Islamic core into three parts: a Turko-Iranian east, a Turko-Arab west, and an Arab-Berber far west. It should be noted, however, that even in the far west, only Morocco escaped Turkish influence; in sixteenth-century Algiers, even the official language was Turkish (see Abun-Nasr 1975, page 167).

73. Halperin (1985, page 94) writes that "probably for religious reasons, the Muscovites never adopted any of the models that arose in the Golden Horde

through the influence of the Muslim Near East." Yet if one looks closely enough, many lines of influence between Turkish peoples, on the one hand, and Greeks and Russians, on the other, can be discerned (in the case of Russia, see Spuler 1994, page 87). As Pierce (1993, page 10) notes, "The centuries-long contact of Anatolian Turks with Byzantine civilization . . . had undeniable influences on Ottoman civilization, perhaps particularly in its imperial institutions," while Vryonis (1971, page 496) contends that "Turkish society . . . was strongly Byzantine in its . . . folk culture."

74. See International Centre for Minority Studies and Intercultural Relations Foundation, Sofia, 1994, pages 159–60 especially.

75. See Rorlich 1986, page 179; see also Bacon 1980, pages 113–15.

76. There is also an Iranized group of Mongols, the Hazaras, in what is now Afghanistan (see Barthold 1984, pages 4, 8, 82).

77. See Lattimore 1988, page 124.

78. R. White 1991, page x.

79. Ibid.

80. Ibid., page xi.

81. See, for example, Ferguson and Gupta 1992.

82. On the use of the term *diaspora,* see Clifford 1994. More generally, see the various volumes of the journal *Diaspora: A Journal of Transnational Studies* (Oxford University Press).

83. See Pan 1990.

84. See Curtin 1984.

85. See Bentley 1993.

86. See Pankhurst 1992, pages 54, 113. The Ethiopians also remained in close contact with the Copts of Egypt—from whom they acquired their religious leaders. Ethiopians were also well known in South Asia, where they often played a significant military role (see Richards 1993, pages 32, 112). Europeans in the sixteenth century often classified Ethiopia as a part of India (see Lach 1977, book 3, page 510); intriguingly, the languages of Ethiopia do show certain structural similarities to those of India (see Masica 1976, pages 180, 184).

87. See Toynbee 1934–61, volume 1 (1934), page 90.

88. See International Centre for Minority Studies and Intercultural Relations Foundation, Sofia, 1994; Poulton 1991.

89. On the Hui, see Gladney 1992.

90. See Harding and Myers 1994.

91. On March 2, 1996, city officials in Iida, Japan, expressed to us their frustration with the Internet, noting that few persons in the municipal office have adequate English-language skills to take advantage of it.

92. See Joyce Barnatha, "Passage Back to India," *Business Week,* 17 July 1995, page 44.

93. See Turner 1994, pages 200–201.

94. "Ethnic cleansing," it is important to note, is not a new phenomenon in the Balkans (or elsewhere for that matter). Indeed, the most intensive period of "ethnic cleansing" in this region was the early twentieth century, and Greece was perhaps its most adroit practitioner (see Stoianovich 1994, pages 199–200).

Chapter 6: World Regions: An Alternative Scheme

1. On the geographical influence of Linnaeus, see Wolff 1994 and Pratt 1992. It must be noted, however, that the most thoroughly hierarchical treatment of global geography that we have discovered—that of Sanson (1674)—preceded the work of Linnaeus by several generations.

2. See Palairet 1775, maps 1, 3.

3. See Ortelius 1570, map 48. Ortelius loosely divided Asia into but four regions: Tartary, India, Persia, and Turkey (maps 47–50).

4. See Lach 1977, pages 53, 86, 90, 205. Phillips (1994, page 31) discusses the medieval Europeans' concept of "three Indias": Lesser India in northern South Asia, Greater India in southern South Asia and in Southeast Asia, and Middle India in Ethiopia. (Ethiopia—which to the ancient Greeks denoted modern-day Sudan—was often conceptualized as occupying a vast portion of central Africa even up to the seventeenth century.)

Certain East Asian metageographical conceptions (particularly those of Buddhist inspiration) also pictured India as constituting the bulk of the planet's landmass (see Unno 1994, page 373). As in the Far West, India was often seen as multiple, in this case as "five Indias." Owing to this expanded definition of India, the first European travelers to reach Japan were classified as Indians (Toby 1994, pages 327–28). Even in maritime Southeast Asia, "the Portuguese seemed a variation on the Indian theme" (Reid 1994, page 276). The Portuguese, of course, likewise considered the Southeast Asians to be Indians. Here we have the interesting case of two peoples classifying each other as the same "Other"!

5. Once British control of South Asia was locked up, British atlases begin to limit the term to that area. Cary (1808, map 40), for example, labels the Indian peninsula simply "Hindoostan or India," while referring to Southeast Asia as the "East Indies." S. Hall and Hughes's map of South Asia (1856, map 39) is simply labeled "India."

6. See Finley 1826, map of Asia (unpaginated).

7. See S. A. Mitchell 1849, map 63. By 1861 even such obscure political units as Ladakh began to appear on maps of Asia (see S. Hall and Hughes 1856, map 35).

8. See Stuart-Glennie 1879, page 43; see also Anthon 1855, page 609.

9. Spate and Learmonth (1967, page 6, note 2), for example, argue that "the use of 'India' as a convenient geographical expression implies no disrespect for Pakistan."

10. American atlases did not emphasize European imperial holdings until the very end of the nineteenth century, although European ones sometimes did so earlier (see, for example, Stieler 1865, map 24). British atlases began to emphasize the red-colored expanse of the British Empire in the Victorian period (see, for example, W. Johnston 1880, initial map). For early American atlases that frame the division of the world by colonial territories—downplaying even continents—see B. Smith 1899 (map 1), *Hammond's Modern Atlas of the World* 1909 (map 1), Patten and Homans 1910 (map 1), and *L. L. Poates and Co. Complete Atlas of the World* 1912 (pages 2–3).

11. See Bennett 1947.

12. Ibid., page 30.
13. See Fenton 1947.
14. Fenton 1946, page 697.
15. Bennett 1947, page 3; see also Bennett 1951, page 35.
16. See Fenton 1946, page 700 especially. A plan for the advanced training of a corps of "Asiatic geographers" was considered, but not carried through. (On the role of geographers in the area studies initiative during the war, see also N. Smith 1994, page 498.) The failure of geographers to meet the area studies challenge must be regarded as one of the principal reasons for the discipline's subsequent marginalization in American academia.

One explicit consideration of metageographical concepts by a geographer from this period is that of Stephen Jones (1955). While Jones examined such constructs as hemispheres, the "pan-regions" of the Nazi area, and contemporary economic and geostrategic macroregions, he entirely ignored the emergent system of world areas. His only look at cultural divisions was based on Russell and Kniffen's (1951) idiosyncratic "culture world" system—a scheme that mixes climatological and cultural criteria to isolate such regions as the "Dry World" and the "Oriental World." Russell and Kniffen's text was ultimately based on a world geography course taught at Berkeley by Carl O. Sauer in the 1920s and 1930s. Sauer, however, avoided the climatically based category of the "Dry World" in favor of the culturally construed category of the "Levantine World" (see Hewes 1983, pages 140–41).

17. Several noted geographers claimed area expertise without acquiring linguistic skills; Glenn Trewartha (1952, 1965), for example, wrote authoritatively on Chinese urbanism and Japanese regional geography with no knowledge of Chinese or Japanese. Such amateurism contributed to the marginalization of geography as a discipline. Ironically, Trewartha remained a major figure in the discipline (largely for his climatological works), while genuine area scholars, such as Paul Wheatley, tended to be marginalized.

18. The area concept was shaped above all by anthropologist Clark Wissler (Cahnman 1948, page 236), who was particularly concerned with "trait distribution." A trait preoccupation weakened the world regions concept from the beginning. As an example of the debilitating effect of the emphasis on traits, one may examine the work of Raphael Patai (1952). Patai rigorously attempted to delimit the Middle East as a cultural region, basing his inquiry on a list of fifteen diagnostic traits, traits ranging from the specific attributes of material culture to the prevalence of poetry to the existence of "great poverty." He later expanded these ideas while developing a more sophisticated and nuanced delineation of the region. His revised scheme looked carefully at social structure and cultural background, was attentive to variation in cultural boundary intensity, and ignored the largely artificial dictates of modern state borders. Yet in the end an obsessive attention to trait distribution prevailed. In the northeast portion of the region, for example, Patai argued that "the limits of the Middle East should be drawn where the typical Middle Eastern black hair tent is replaced by the Yurt.. . . . This delimitation excludes a corner of Iran . . . and part of northern Afghanistan from the Middle Eastern culture continent" (1962, page 52).

During the same period, Elizabeth Bacon (1946) attempted to create her own anthropologically informed regionalization system. She tentatively isolated six main and four hybrid culture areas of Asia, a scheme based primarily on technics and objects used in everyday subsistence activities, yet encompassing political and social structure as well as ideology. Her postulates forced her into the unusual position of defining South Asia as merely a hybrid region, since she found that many of its common features of material life were associated primarily with Southeast and Southwest Asia. India's unique cultural attributes thus dissolved away in an object-oriented inquiry more concerned with the distribution of roofing materials than with historical coherence.

In recent decades, cultural anthropologists have moved away from any consideration of world regions. Those devoted to classical ethnography have generally focused on the level of the single village, while those adopting a postmodernist stance have questioned the very existence of such large-scale structures. Richard O'Connor, however, in a powerful article on agricultural and ethnic change in mainland Southeast Asia, argues that "a rigorous anthropology requires regions" (1995, page 989).

19. R. Hall 1947, page 9. Hall—despite his training as a geographer—devoted little attention to the actual geographical delineation of world regions. Yet it is clear from this quote that his conception of global regionalization had by that date come to have more in common with the modern scheme than with the initial categories employed by the Ethnogeographic Board.

20. See Lambert 1984, pages 7–8.

21. See Richards 1995, pages 3–4.

22. Clowse 1981 page 165.

23. See Richards 1995, page 4.

24. See Bennett 1951; Lambert 1984. Oceania does not appear on the list because it has not been the focus of organized area studies in American universities.

25. "South Asia," like similar regional designations, can still occasionally generate substantial ambiguity. In the *World Almanac and Book of Facts* 1993 (page M11), for example, "South Asia" designates a huge zone, encompassing China, Saudi Arabia, and all other portions of Asia outside of the former Soviet Union.

26. See, for example, Stamp and Kimble 1954, a work published in Canada.

27. Some works continued, however, to categorize all of East, South, and Southeast Asia as the "Orient," the "Far East," or "monsoon Asia" (see, for example, James and Davis 1959)—a strategy that has still not been completely abandoned (see Wheeler and Kostbade 1993; J. Fisher 1992). For other textbooks on world regional geography, see de Blij and Muller 1994; de Blij et al. 1989; English 1977 (revised edition, English and Miller 1989); Hepner and McKee 1992; Jackson and Hudman 1982; Norris 1990; Russell, Kniffen, and Pruitt 1969; Helgren, Sager, and Israel 1985; Stansfield and Zimolzak 1990; Morris 1972; James 1964; Fuson 1977; and Benhart and Scull 1985.

The world regional standard was also quickly accepted in anthropology. An ethnographic book series of the 1960s, entitled "Culture Regions of the World," showcased the now-familiar roster of macroregions: India and South Asia, Latin America, Western Europe, Southeast Asia, East Asia ("China, Japan and Korea"), Africa south of the Sahara, Southwest Asia and North Africa, and the Soviet Union and Eastern Europe. The title of the first book of the series, *India and South Asia* (Fersh 1965), involves a major error in logical typing; India is, of course, a part of South Asia. Otherwise, however, the terminology used was quite sophisticated (using "Southwest Asia and North Africa" instead of the "Middle East," for example).

28. See the tables presented in Lambert 1984, pages 326–28.

29. Lambert 1990, page 723.

30. Richard Lambert (1984, page 297) could thus count only four (Western) European studies centers in 1981, compared with some sixteen devoted to Latin America. Moreover, American scholars of Europe have generally resisted the notion of European studies, identifying themselves much more with their disciplines and with the individual countries of their expertise.

31. Canada, by contrast, has remained virtually invisible, dismissed as little more than a cultural extension of the United States and western Europe (except in the handful of institutions, such as Duke University, that support centers for Canadian studies).

32. The concept of a distinctive region constituted by the great peninsula south of the Himalaya, for example, is indigenous to the area. In classical Indian thought, however, it comprised the bulk of the world, rather than being simply one of many world regions. The term *South Asia* is of course Euro-American, and its boundaries have largely been defined within that region's political-geographical discourse. Similarly, although *East Asia* is obviously a term from the European tradition, the notion of a distinctive Confucian region or civilization (or, in Japanese, *ka*) largely coterminous with it has a deep indigenous history (see, for example, Hamashita 1993).

33. Most educated persons from Mexico to Chile consider themselves Latin Americans, for instance, but the term has little resonance for inhabitants of remote villages. Argentina presents a different case, however. According to "Back in the Saddle: A Survey of Argentina": "The adjective 'Latin American' remains a term of racist contempt in Buenos Aires, conjuring up images of shanty towns" (*Economist,* 26 November–3 December 1994, page 18).

34. Some ASEAN officials hope that by the year 2000 the entire "natural cluster" of the ten states of Southeast Asia (including Laos, Myanmar [Burma], and Kampuchea [Cambodia]) will join the organization (see "South-East Asia's Sweet Tooth," *Economist,* 5 August 1995, page 31). On the coherence of the ASEAN group, see Wriggins 1992 (page 4) and H. Hill 1994 (page 833).

35. We are indebted to John Richards for this observation.

36. The usual list is East Asia, Southeast Asia, South Asia, and Southwest Asia (usually joined with North Africa); Central Asia appears less often, and North Asia (Siberia) is almost always (for good reason) appended to European Russia.

Less conventional is "Northeast Asia," a region generally composed of Japan, the two Koreas, northern China, and some unspecified portion of Russia's far east (see the *Journal of Northeast Asian Studies,* a periodical devoted largely to political and strategic interests). Different authors, however, employ different configurations; for the archaeologist Chard (1974), for example, Northeast Asia is essentially a synonym for Japan and Siberia. Allen (1990) offers a view of Northeast Asia centered on Korea. For Tsao and Tsai (1984), on the other hand, the region extends as far south as Hong Kong and even Singapore. The Eleventh Pacific Science Congress, held in 1966, attempted to establish the boundaries of Northeast

Asia by decree. Here it was held that Northeast Asia would extend as far south as the thirtieth parallel, below which one would find Southeast Asia (see Solheim 1985, page 142). Such a convention—which effectively eliminates East Asia—has been largely ignored.

37. The definition of South Asia is relatively straightforward: India, Pakistan, Bangladesh, Nepal, Bhutan, Sri Lanka, and the Maldives. Sometimes Afghanistan is also included (Fersh 1965), and Mauritius has been appended to the region by at least one author (Maloney 1974).

38. In regard to Europe the issue is rather more complicated. In the wake of the breakup of the Soviet Union, Europe should certainly be extended to include former Soviet territories like Latvia that are historically linked to Western Christendom. Russia and other lands of historical Eastern Christendom can, however, either be counted as a part of Europe or as a separate world region. In the former case, Europe would then have to be regarded as a region stretching from the Atlantic to the Pacific. Such a conception, however, is very rare (see, however, Hoffman 1969). If, on the other hand, Russia is removed from Europe, then it is difficult to argue that a country such as Bulgaria (a Slavic, Orthodox country once popularly deemed the sixteenth Soviet republic) should not likewise be excluded. By this way of thinking, southeastern Europe (Serbia, Romania, Bulgaria, and Greece and environs) should be slotted into an "Eastern Europe" that extends through the length of Russia to the Pacific but excludes Poland, Croatia, and other historically Catholic lands (see, for example, S. Huntington 1993). Such an arrangement is, however, is almost equally rare, especially since Greece often receives a "Western" position by virtue of its economic and military ties. At any rate, the general comparison between Europe and South Asia still holds in a historical sense for either definition of Europe.

39. Although the latter now outstrips the former by a fairly wide margin, in 1815 Europe held some 200 million persons to greater India's 190 million, while in 1483 the figures stood at some 73 million and 110 million respectively (see McEvedy 1972, pages 9, 84).

40. See Masica's (1976) *Defining a Linguistic Area: South Asia*, an important work that has unfortunately been largely ignored by geographers. As he shows, the Indo-European, Dravidian, and Austro-Asiatic languages throughout South Asia have come to share a number of basic features.

41. Bhatt 1980, page 43. The comparison between India and Europe, it must be admitted, has often been resisted by Indian nationalists on the grounds that it challenges the unity of the country and thus threatens to divide it into a welter of much less powerful states. Indeed, it is seen by the noted South Asianist Ainslie Embree (1989), among others, almost as a European strategy to maintain hegemony. Embree has also questioned the analogy by arguing that India is quite unlike Europe because "it did not develop a large number of nationalities that became the basis of nation states" (page 21). Such a view is problematic, however, because it risks positing European political evolution as somehow predetermined or paradigmatic. Of course all of the major world regions exhibited divergent patterns of political development, with Europe representing one extreme of national fragmentation and East Asia representing another of unification (albeit never complete) under the auspices of the Chinese Empire. But it is also important to re-

member that Europe's nation-states are recent creations, which were only built with some difficulty, and it is conceivable that the European subcontinent could have taken a very different path of political unification (if, for example, either Charles V or Napoleon I had triumphed in the end). Europe and South Asia are, of course, quite distinctive in their internal social and political structures, but from an ecumenical perspective they are still comparable entities.

42. While Southeast Asia's boundaries are highly formalized in official geographical and political discourses, many writers continue to use the term in a loose, informal sense. Some writers extend the region to cover the western shore of the Pacific as far north as Hong Kong and Taiwan (see Chisholm 1982, pages 162–63; Gellner 1992, page 142; Buell 1962), if not South Korea and Japan, while others shift it eastward to encompass the entire Indian subcontinent (see Mende 1955, page viii; Morgan and Spoelstra 1969, page 3). Nicholas Tarling (1966, page xi) more modestly and appropriately appends the Nicobar and Andaman Islands, conventionally counted as part of South Asia by virtue of their political ties, to Southeast Asia. Lester Thurow (1992, page 217), on the other hand, puts all of Southeast Asia within the category of South Asia.

43. See Savage, Kong, and Yeoh 1993. As they relate, Europeans used a variety of terms related to South Asia to designate the region, including *High India, Greater India, Exterior India,* and *India aquosa. Farther India* was sometimes used to cover all of modern Southeast Asia and sometimes only the mainland portion. In Woodbridge's (1824) atlas, for example, the insular realm is labeled "East India or Asiatic Islands," whereas the mainland is called "Farther India" (map of Asia and facing page, unpaginated). German atlases sometimes distinguished *Vorder-Indien* (anterior India) of modern South Asia from the *Hinter-Indien* of Southeast Asia (see Gaebler 1897, map 24); *Hinter-Indien,* in fact, has continued to be used in Germany up to the present (see *Goldmanns Grosser Weltatlas* 1955, page 134; *Atlas International* 1985, page 208). The 1890 *Scribner-Black Atlas of the World* labels Southeast Asia as "The Indian Archipelago and Further India" (map 35). In some nineteenth-century atlases, the term *East Indies* is extended to cover mainland as well as insular Southeast Asia.

In a few sources, the term *India* seems to have been applied more to Southeast than to South Asia. In Waldseemüller's world map of 1507, for example, the unmodified label "India" is placed on the Malay Peninsula, whereas the area now called India is identified in the old Latin terminology as *"India intra Gangem"* (Fischer and Wieser 1903, map 1); in Greenleaf's atlas of 1842, one map (number 21) labels South Asia as "Hindostan" and mainland Southeast Asia as "India."

44. See, for example, D'Anville 1743, maps 12 and especially 45; in earlier maps the Ganges is often shown running directly south to the delta, which makes it a more reasonable dividing line (*India intra Brahmaputrum* would thus have been a more appropriate designation). Such a division is especially common in historical maps depicting the classical era, where the distinction is between *India intra Gangem* and *India extra Gangem* (see Gole 1983, page 1); see Finley's (1826) map "Orientis Tabula" for a particularly nice division along the Ganges River.

Robert Wilkinson's world atlas of 1794 contains an early depiction (through coloration) of the modern area of Southeast Asia as a distinct region of the world, yet labels it merely "the islands and channels between China and New Holland" (map 42). Wilkinson's map 37, however, which includes both mainland and insular Southeast Asia as a unified region, appears to label it with the inappropriate term *Ava*, which actually applies only to a portion of Burma.

The British began to differentiate Southeast Asia from India once they had effectively conquered the latter area. Yet British cartographers still had an enormously difficult time classifying the former area. J. Cary in 1808, for example, depicted all of Southeast Asia—mainland as well as insular—as the "East India Islands" (map 44).

45. See Wink 1991, page 341. East Asians, on the other hand, generally defined it as a region in its own right, under the label *Nanyang* or *Nampo*, meaning "Southern Seas" (see Emmerson 1984, page 4).

46. *Oxford English Dictionary* 1971, volume 1, page 1420.

47. While *Indochina* has been the common designation, the French geographer M. Malte-Brun argued in the early nineteenth century that the term is misleading, since it implies that the area is basically an extension of China that has been influenced by India. He argued that the region is instead an extension of India that has been influenced by China and should therefore be called "Chin-India" (1827, volume 2, pages 262–63), a term that was never widely accepted. In the Indochina formulation, one can also find Arab precedents, as several Arab geographers viewed the area as an interstitial zone between *al-Hind* and *as-Sin* (China) (see Wink 1991, page 341).

48. See, for example, Rand McNally 1881, page 211; Bartholomew 1873, map 1; W. Johnston 1880 map 1; B. Smith 1899, map 114; Cram 1897, page 216.

49. In cultural and historical terms, Cambodia and especially Laos are far more closely connected to Thailand than to Vietnam, yet in honor of French imperialism they are still commonly slotted into the subregional category of Indochina, from which Siam was removed because it retained independence. (The Lao dialect area of the Thai language group was split in half by this move, as northeast Thailand is Lao-speaking.) As L. Malleret wrote in 1961 (page 301): "It cannot be denied that the territorial concept which has been labeled 'Indo-China' has been no more than a fortuitous construct, created as much in accordance with a political programme . . . as upon a diversity of people grouped together by a simplification of common features."

50. Hauner 1990, pages 170–72.

51. Emmerson 1984, page 5; see also Malcom 1845. The frontispiece map in Malcom's *Travels in South-Eastern Asia* covers the mainland portion of Southeast Asia, all of eastern India and Ceylon, a small area of southern China, but only northern Sumatra and a corner of Borneo within modern-day insular Southeast Asia. In Elton's atlas of 1908, China, Japan, and Korea are mapped as portions of "S-E Asia and the Malay Archipelago" (page 30).

52. See Emmerson 1984, page 12. Some modern scholars do append the Andaman and Nicobar Islands to the region; see note 42 above. In earlier times such seemingly unconventional groupings as Heine-Geldren proposed were not uncommon; in 1918, for example, W. J. Perry (page 1) defined *Indonesia* as an area including "Assam, Burma, the Malay Peninsula, the Philippine islands, and Formosa."

53. Many of the small-scale societies of both northeastern India and southern China are more easily slotted within Southeast than South or East Asia respec-

tively, a situation that was more strongly marked in the past than in the present. Archaeologists in particular have often been inclined to extend the formal boundaries of Southeast Asia as far north as the Yangtze River (see Solheim 1985, page 142), and geographers of human genes inform us that even today the southern Chinese fit into a distinctly Southeast Asia biological grouping (see Cavalli-Sforza, Menozzi, and Piazza 1994, page 234). Historians, on the other hand, are sometimes wont to reduce the area covered under the Southeast Asia rubric. D. G. E. Hall, in his masterful *A History of South-East Asia* (1955), for example, excluded the Philippines in the early editions, owing to its divergent traditions and experiences. Harry Benda tended to agree, noting that the Philippines constitutes a borderline case for inclusion in the region. In the end, however, Benda (1972, page 123) retained the Philippine archipelago and instead subdivided Southeast Asia into three cultural areas: the Indianized realm, the Sinicized realm (Vietnam), and the Philippines.

54. Steinberg 1985, page 5.

55. See Emmerson 1984, pages 7–8.

56. See Freyer 1979, page 2.

57. See Emmerson 1984, page 8. In the postwar period most world atlases began to label the area Southeast Asia and to classify it unambiguously as a part of the Asian continent. As late as 1949, however, the *Encyclopedia Britannica World Atlas* (pages 58–59) labeled the area "East Indian Islands." It is also notable that in 1948 the Social Science Research Council's Committee on World Area Research argued that South and Southeast Asia should be classified together as "Southern Asia" (see Wagley 1948, page 33).

58. See, for example, C. Lewis and Campbell 1951, pages 66–67.

59. See Reynolds 1995. The noted historian D. G. E. Hall (1955, page 4) went so far as to argue that the term *Indonesia* was "open to serious objections" because it rhetorically made the islands a "cultural appendage of India." (The word itself, it should be noted, is based on Greek terms.) The Indonesians, however, have turned the tables by renaming the Indian Ocean *Laut Indonesia*, or the Indonesian Ocean (see Freyer 1979, page 6).

60. Emmerson 1984, page 10. In 1944 Bruno Lasker (page 5) defined Southeast Asia by listing its political territories; he apparently excluded eastern Timor from the region because it was then a Portuguese territory (although occupied by Japan) and not part of the constitutive Netherlands East Indies. After Indonesia conquered eastern Timor, no one to our knowledge has excluded it from Southeast Asia.

61. Fifield (1975) argues that it was precisely the experience of modern war—both World War II and the two Indochina wars—that gave regional coherence to Southeast Asia.

62. The contemporary boundaries of Southeast Asia are most problematic in the region's southeastern corner. The main issue here is whether all of Indonesia should be placed within Southeast Asia or whether that country's eastern periphery (especially western New Guinea [Irian Jaya]) should be excised (this problem was noted almost as soon as a Southeast Asian world region was recognized; see Vandenbosch 1946). Due to the myth of the nation-state, the former option is far more common. A few authors and cartographers, however, go so far as to append

all of New Guinea to the region. If one were to follow the principle of state indivisibility, this gambit would lead into the heart of Melanesia, for both the Bis-
marck Archipelago and the island of Bougainville (physically but not politically
a part of the Solomon Islands) are political appendages of the state of Papua New
Guinea. Such a maneuver is rarely encountered, however, both because the social background of western Melanesia is entirely distinct from that of "Southeast
Asia proper," and because placing an island such as Bougainville in Southeast *Asia*
would insult the myth of continents. The more common expedient is thus cleanly
to slice New Guinea down the middle, disregarding all local cultural affiliations.
But even if one were to disengage from the myth of the nation-state and exclude
western New Guinea from Southeast Asia, the delimitation of the region's southeastern extent would remain tricky. The Maluku (Moluccas) islanders, for example, (and even several of the peoples of the coastal zones of New Guinea itself)
show complex intermixtures of Papuan and Indonesian social and cultural characteristics (see Andaya 1993, page 41).

63. Savage, Kong, and Yeoh 1993, page 233.

64. Lieberman 1993b, page 476. The Indian scholar D. R. Sardesai (1981, page
4) takes such reasoning one step further, arguing that "[t]he Southeast Asian region is not a unit in the religious, historical, geographical, or ethnic senses," and
contending instead that the area contains four separate regions: Indonesia-
Malaysia, the Theravada Buddhist zone, Vietnam, and the Philippines.

65. Chinese cultural patterns were widespread among Vietnam's premodern
elite, who had "inherited a deep admiration for Classical Chinese civilisation and
found many aspects of local culture 'barbaric'"; indeed, "the language of state was
Mandarin Chinese, in comparison to which vernacular Vietnamese was often
thought 'vulgar and inadequate'" (Henley 1995, page 299).

66. See K. Taylor 1993, page 45. Emmerson (1984) notes that the Vietnamese
themselves have been strongly emphasizing their Southeast Asia origins and connections in recent years, while Reid (1993, page 135) shows that Confucianism as
a state ideology was never securely established in Vietnam.

67. See Reid 1993. Hinduism would later return to the Malay Peninsula with
the influx of Indian workers who arrived during the colonial period.

68. See Phelan 1959.

69. See Keyes 1977 on the division between lowland and highland peoples in
mainland Southeast Asia. For an extended look at a modern resolutely "pagan"
community in northern Luzon, see M. Lewis 1992b.

70. See Legge 1992.

71. Benda 1972.

72. See McCoy 1980.

73. L. Williams 1976, page 5.

74. Broek 1944, page 189.

75. Karan and Bladen 1985, page 20. Somewhat earlier, Lucian Pye (1967, page
2) similarly argued: "The contemporary theme that now characterizes the entire
area is the effort to translate diffuse feelings of nationalism into strong loyalties
to the nation." Such attempted "nation building," however, has also characterized most postcolonial regimes of the world.

76. See Reynolds 1995.

77. Scholars for some time speculated about the existence of a buried yet distinctly Southeast Asian cultural strand, but few were able to get a handle on it. In 1944, for example, Broek (1944, page 186) stated: "One can go even further and say that the differences between [Southeast Asian peoples] would be rather small had it not been for the impact of outside forces, first of more advanced Oriental civilizations, later of the western world." Only with the publication of Anthony Reid's *Southeast Asia in the Age of Commerce* in 1988 was the concept of an underlying Southeast Asian way of life convincingly propounded.

78. Note, for instance, Lieberman's (1993b) compelling argument that in the early modern period the historical trajectories of the insular and mainland portions of Southeast Asia were quite distinct, and that in many instances the mainland kingdoms are better compared to mainland states elsewhere in Eurasia than to the states of the archipelagic realm.

79. See Reid 1993, page 270.

80. Several prominent archaeologists have supported the claim to regional identity on the basis of shared material artifacts. Solheim (1985, pages 144–45), for example, argues that particularly Southeast Asian cultural features include "houses built . . . on piles; tatooing . . . ; decoration of the teeth; the chewing of areca nut with betel leaf . . . ; animistic religion with ancestral and nature spirits central to this; bilateral kinship with general equality of the sexes . . . ; land tenure by descent groups; and self-identification of local groups by distinctive elements of material culture."

81. Several groups of Southeast Asian peoples invaded northeast India in historical times, where they came under the influence of Hindu civilization and indeed established Hindu states, yet without discarding all of their earlier ways. See Richards (1993, pages 105–6), for example, on the Ahoms of Assam.

82. See Reynolds 1995, page 437.

83. As Reynolds (1995) notes, Singaporean scholarship is exceptional on this score.

84. J. Gross 1992, page 17.

85. In many accounts, Siberia is mapped instead as Russian Tartary (see Bonne 1771, map 41; Bowen 1752; Sanson 1674; Kitchen 1771; Palairet 1775).

86. See Palairet 1775, map 3.

87. See Marshall and Williams 1982, pages 88, 136, 300.

88. In Samuel Butler's atlas of 1829, Tartary disappears altogether, with the Chinese Empire depicted as extending as far as Herat and the Aral Sea. As late as 1909, small states such as Bukhara were still sometimes depicted as independent (*Hammond's Modern Atlas of the World* 1909, map 95), but it was more common at the time (and more accurate) to show the Russian and Chinese Empires as having engulfed the entire area (for example, B. Smith 1899, map 100).

89. See Anthon 1855, page 608.

90. See Mackinder 1904; E. Huntington 1907.

91. The invasive nature of Chinese rule in Tibet is much more commonly recognized than it is in the case of Turkistan, in part because Americans often regard the Tibetans as more colorfully exotic and as having a noteworthy leader in the Dalai Lama. Yet our basic maps of cultural geography still overstress the connections between China and Tibet. Consider language, for example: virtually all conceptions of global language distribution place Tibetan within the same cate-

gory as Chinese ("Sino-Tibetan"), implying at some level that the current political relationship between the two is rooted in deep cultural connections. But such a view receives little empirical support from students of the two tongues. As Christopher Beckwith concludes, "[I]t is uncertain what language Tibetan *is* related to, but anyone with a knowledge of both comparative-historical linguistics and the Tibetan and Chinese languages (the pillars of the 'Sino-Tibetan' theory) has great difficulty imagining that two such radically different tongues could be genetically related" (1987, page 5, note 3; see also Katzner 1995, page 24).

92. See Barfield 1989; Beckwith 1987; Frank 1992.

93. See Mackinder 1904.

94. See Balland 1992, page 75.

95. Canfield 1991, page xi. Edward Allworth, on the other hand, restricts Central Asia to (the former) Soviet Turkistan, "Chinese" Turkistan, and Afghanistan, delimiting the region primarily on the basis of its "homogeneity at certain levels of being and thinking" (1989, pages xv–xviii); S. A. M. Adshead essentially concurs: "So the Central Asia which has to be considered in relation to world history is not just the three oasis areas of eastern, western, and Afghan Turkestan, but also the associated steppe areas to the north: central Kazakhstan, Semirechhie and Zungharia" (1993, pages 6–7). The journal *Central Asia Survey*, on the other hand, joins religious and linguistic considerations to map a maximal Central Asia centered on all areas of "Muslim Irano-Turkic populations" (including those of the central Volga), to which are appended the historically connected areas of Mongolia and the Caucasus (this coverage is explained on the first page of each issue of the journal).

96. Canfield (1991), it must be noted, has for various other reasons abandoned the term *Greater Central Asia* in preference for the much more accurate term (for his purposes) *Turko-Persia*.

97. See Frank 1992.

98. On the persistence of shamanism in such places as Tuva and Buryatia, see Forsyth 1992.

99. See Lapidus 1988, page 431.

100. The seventeenth-century migration of the Kalmyks created a third Lamaist zone in the southeastern corner of Europe (see Khodarkovsky 1992).

101. Sinor 1990, page 14.

102. Kazakhstan is a somewhat different story, as it is home to almost as many Russians and Ukrainians as Kazakhs.

103. This is occasionally noted in introductory books on the region; thus Schurz (1963, page v) writes, "The Latin Americans are only Latin in so far as the speech of their ruling classes . . . is derived from the tongue of Ancient Rome." In general, however, the "Latin" aspect of this vast cultural region is simply assumed. Some authors (such as Dozer 1962, page 8) further recommend the term because it also credits Italy, which sent many immigrants to certain parts of the region, even if its language did not spread. Yet we might note here the sizable numbers of Germans who settled in southern Brazil—not to mention Africans, Japanese, Chinese, Lebanese, and other "non-Latins"—whose contributions might also be worthy of recognition.

104. Braudel 1994, page 247; see also Collier, Blakemore, and Skidmore 1985, page 9.

105. See, for example, Wilcox and Rines 1917, page 7; Robertson 1922, page 1; Rippy 1958, page 3.

106. In 1922 William Robertson (page 1) did opine that Latin America "might include New France," but few other writers have considered the possibility.

107. See Rippy 1958, page 3; Wilgus 1931.

108. See Kirkpatrick 1939, page ix.

109. Alba 1969, page 47.

110. Ibid.

111. The term *Latin America* has also been the object of some controversy among Latin Americans. Some have argued that the qualifier is unnecessary— that they are simply Americans, a label that has been arrogated by the residents of the United States. Others have embraced the term and have argued that strong linkages connect all of its constituent parts. Luis Sanchez (1945, page 19) writes that "se existe Europe, pese a sus múltiples incompatibilidades, la América Latina, que no las tienne an tal grado, existe con muchísima mayor razón." Sanchez further states (1945, page 10), "En realidid, entre los países que integran la llamada América Latina existen tantas diferencias como antre las provincias que constituyen los Estados Unidos, y menores que entre las naciones de Europa."

112. See Lambert 1984, page 328, for example. One might also note the intrusion of the myth of the nation-state through all of these vicissitudes. Eastern Nicaragua has never been a Spanish-speaking area, yet it is always classified as part of Latin, Ibero-, and Hispanic America merely because it lies within a predominantly Spanish-speaking country.

113. While the term *African America* is not commonly encountered, *Black Atlantic* is, especially in cultural studies. For a provocative look at the cultural hybridization of the "blacks of the West," see Gilroy 1993. For a more comprehensive treatment, see Thornton 1992.

114. One of the most famed collaborative efforts to link the two disciplines, that between geographer Carl Sauer and anthropologist Alfred Kroeber at Berkeley, was cut short in large part because the former objected to the latter's statistical attempt to correlate cultural and natural areas in indigenous North America (see Price and Lewis 1993; see also Patai 1962, page 39). As Canfield (1991, page xiii) writes, the notion of "'culture area' . . . presumed an association between broad cultural systems and their environment. Originally introduced as a means of classifying artifacts of different societies in museum displays, the concept owed major assumptions to zoogeography."

115. Such covert environmental determinism is often encountered in textbooks because, following an old geographical tradition, they seek to provide a synthetic portrayal of all of the varied human and natural attributes of each given region. As a result, natural (and especially, physiographical) factors are often considered just as important as human ones, while economic considerations are often weighed much more heavily than cultural ones. This integrative style of geographical analysis dominated the entire discipline from the 1930s to the 1950s but has since been largely abandoned in scholarly works as intellectually indefensible. Retaining a valuable instructional role, however, such comprehensive description of unproblematized units finds its last refuges in regional geography textbooks.

116. Steinberg 1985, page 5.

117. Reid 1993, page 3. See Purcell (1965, page 3), for a similar view of Southeast Asian environments.

118. See Hepner and McKee 1992, pages 428–29; Wheeler and Kostbade 1993, page 196; Jackson and Hudman 1982, page 404. See also Patai 1952, page 2; 1962, page 16.

119. See Russell, Kniffen, and Pruitt 1969.

120. Nijim 1992, pages 428–29.

121. See Beaumont, Blake, and Wagstaff 1976, page 3.

122. See Pearce and Smith 1984, pages 259, 376.

123. See Mackinder 1904. For a recent critique of Mackinder's thesis, see Drompp 1989. Drompp's interesting article is severely hobbled by an outdated conception of geography, which he essentially takes to mean physical geography. In the end, Drompp seems to be fighting a battle against geographical constraints and influences that was settled in the discipline decades ago.

Owen Lattimore (1953, page 17) also used drainage basins to define "Inner Asia," delimiting it as "a group of regions that have neither frontage on the sea nor navigable rivers leading to the sea"—a definition that included northern but not southern Iran and western but not eastern Manchuria. Recently, several historians have essentially followed Mackinder in delimiting an Inner Asia that includes all of Siberia, as well the area more generally called Central Asia (see Christian 1994). Adshead (1993, page 7) actually takes this definition one step further, defining Inner Asia as a region stretching from "the Carpathians to Korea."

124. See Hambly 1969.

125. Balland 1992, page 75.

126. Frank 1992, page 81.

127. Frank (1992, page 54) diverges from the standard formula to argue: "To the north, there is no identifiable boundary, unless it is where the tundra becomes virtually uninhabitable in the Siberian cold." But there are few solid reasons to group most of Siberia with Central Asia, as this environmental definition leads Frank to do. True, the Mongols once extracted furs from much of the region, and some Siberian groups (notably the Yakut) share common ancestors with the peoples of Turkistan (see Levin and Potapov 1964). But except for its southern fringe (zones like Buryatia and Tuva), Siberia does not easily fit within a Central Asia defined in the usual world regional terms of livelihood, historical interaction, social structures, or religious beliefs.

128. When the peripheral Eurasian steppelands *are* included in Central Asia, it is in recognition of historically specific social developments. The Black Sea coast of Romania, for example, is often discussed under the rubric of Inner Asia by historians of the twelfth century. But if coastal Romania in the twelfth century fits within the Central Asian orbit, it is because the area was then inhabited by Central Asian peoples, the Cumans (Kipchak), not because it was any drier than it is today. As Denis Sinor argues in regard to a nearby area, "The Roman province of Pannonia . . . became 'Inner Asia' when occupied . . . by the Huns" (1990, page 3). When the area was (re)occupied by peoples belonging to an eastern European cultural tradition (the Romanians), it was considered to have rejoined Europe.

129. The disparity between average January low temperatures and average July

high temperatures in both Alma-Ata and Beijing is 74° F. (All climatic data are from Pearce and Smith 1984.)

130. Other environmental definitions of Central (or Inner) Asia are equally problematic. David Christian, for example, emphasizes above all else the unifying nature of the region's "flatness," yet by his own definition, "Inner Eurasia" encompasses such impressive mountain ranges as the Caucasus, Sayan, and Tien Shan (1994, pages 177–79).

131. In particular, the eastern Ganges-Brahmaputra delta (modern Bangladesh) was often considered, prior to the seventeenth century, to lie outside of India's bounds (see Eaton 1993, pages 41, 118).

132. See Manning 1990.

133. See, for example, Alvarez 1995.

Conclusion: Toward a Critical Metageography

1. See Cumings 1993.

2. See *Economist,* 24 February 1996, page 112.

3. See Allison 1995.

4. See Heginbotham 1994; Richards 1995. It is true, however, that a number of schools have recently established new area studies centers, and the pendulum may already be swinging back in this direction.

5. See Richards 1995, page 14.

6. We refer here to the edition we own and use on a near-daily basis, Rand McNally's *New International Atlas* of 1984. The 1994 edition of the same atlas uses a simplified scheme of Eurasia, Africa, Australia/Oceania, Anglo-America, and Latin America.

7. Another prime example would be the mixture of economic and cultural criteria commonly employed for the delineation of world regions in world regional geography textbooks. Most textbooks, for example, isolate an East Asian region centered on China, but many exclude Japan from it, despite the existence of obvious and deep cultural affinities. Japan is instead often treated as a world region of its own, as if its economic prowess had somehow erased all of its East Asian attributes, allying it more closely with western Europe and northern North America. If one were to follow this line of thinking to its logical conclusions, South Korea, Taiwan, and Hong Kong would next have to be removed from East Asia, followed perhaps by Guangdong and Fujian—only here one would have to violate the unacknowledged principle of state indivisibility (see de Blij and Muller 1994 for an attempt to deal with this issue). One world geography textbook forces Japan into a common bed with Australia and New Zealand, evidently by virtue of their wealth and their similar longitudinal ranges. Yet the same work violates its own textual classification scheme by mapping, on one page, Papua New Guinea, the Solomon Islands, Vanuatu, and New Caledonia within this same region, thus forgoing any economic rationale (J. S. Fisher 1992, page xxxii).

8. Wheeler and Kostbade 1983, page 196.

9. There are certainly some minor exceptions (see, for example, de Blij and Muller 1994; Wheeler and Kostbade 1993, page 196).

10. The term *New World* is not necessarily a pejorative term, however, if one takes the long view. The Americas were certainly occupied by human beings much later than Africa and Eurasia.

11. On the position of Islam in Pakistan, see the informative discussion in Kandiyoti 1992 (page 244).

12. Pakistan is already counted as a constitutive part of the Middle East in the *International Journal of Middle East Studies.*

13. In terms of indigenous conceptions of identity, Pakistan's interstitiality varies both spatially and temporally. The Pashto- and Baluchi-speaking portions of the country are, for example, much less South Asian than are the Sindhi- and Punjabi-speaking areas. The latter zones themselves became much less South Asian than they had been following the expulsion of their Hindu populations in 1947 (see, for example, Tandon 1968).

14. See Price 1996.

15. According to statistics from the World Bank (1995, pages 9, 19), Mauritius had a per capita GNP of almost $3,000 in 1993 (measured at an impressive $12,450 by purchasing power parity) and experienced an average annual economic growth rate of 5.8 percent between 1985 and 1993. Its measures of social development are also impressive. Its average life expectancy at birth was 70 years in 1992, and its total fertility rate was only 2.0. In neighboring Réunion, the life expectancy at birth was recorded at 74, while the average infant mortality rate (per 1,000 live births) stood at 7—whereas that of the United States was 9.

16. Japanese historians of Asia have also recently begun to stress regionalization schemes based on maritime connections (see for example, Hamashita 1993).

17. De Blij and Muller (1994) do, however, create a bizarre new world region they call the "Pacific Rim of Austrasia," an area encompassing Japan, South Korea, Taiwan, Hong Kong, China's Shenzen , Singapore, Australia, and New Zealand. The arguments that they use to support this move (page 219) are tenuous at best and cannot be used outside of fairly narrow economic concerns.

18. See Woodside 1993.

19. Dirlik (1993) argues that the Pacific Rim idea serves the interests of Japanese and American capital and disguises the subjugation of the less powerful states in the region (see especially page 10). Finney (1994, page 294) writes that the idea of a Pacific Rim turns the Pacific itself into merely "an immense inconvenience that adds to shipping time and jet lag." Ginsburg (1993) discusses the locational ambiguities of the term. Alwin (1992, page 369) endorses its use, but issues cautionary words about its geographical delimitation. See Brookfield (1973) for a discussion of the related geographical category, "Pacific Realm." See Murphy (1995) for a skeptical examination of "Pacific Asia" as a metageographical and economic entity.

Bibliography

Abraham, W. E.
    1962.   *The Mind of Africa.* Chicago: University of Chicago Press.
Abu-Lughod, Janet L.
    1989.   *Before European Hegemony: The World System, A.D. 1250–1350.*
             Oxford: Oxford University Press.
Abun-Nasr, Jamil.
    1975.   *A History of the Maghrib.* Cambridge: Cambridge University Press.
Adshead, S. A. M.
    1993.   *Central Asia in World History.* New York: St. Martin's Press.
Agnew, John A.
    1989.   "Sameness and Difference: Hartshorne's *The Nature of Geography*
             and Geography as Areal Variation." In J. Nicholas Entrikin and
             Stanley D. Brunn, eds., *Reflections on Richard Hartshorne's "The
             Nature of Geography."* Washington, D.C.: Association of American
             Geographers. Pp. 121–39.
Agyeman, Opoku.
    1985.   *The Panafricanist Worldview.* Independence, Mo.: International
             University Press.
Ahmad, Aijaz.
    1987.   "Jameson's Rhetoric of Otherness and the 'National Allegory.'"
             Social Text 17:3–25.
    1992.   *In Theory: Classes, Nations, Literatures.* London: Verso.
Ahmed, Akbar.
    1992.   *Postmodernism and Islam.* London: Routledge.
Alba, Víctor.
    1969.   *The Latin Americans.* New York: Praeger.
Alland, Alexander, Jr.
    1971.   *Human Diversity.* New York: Columbia University Press.

Allen, Chizuko T.
  1990.   "Northeast Asia Centered around Korea: Ch'oe Namson's View of
            History." *Journal of Asian Studies* 49:787–806.
Allison, Anne.
  1995.   "'Sailor Moon': Japanese Superheroes for Global Girls." Paper
            presented at the Annual Meeting of the American Anthropological
            Association, Washington, D.C., November.
Allworth, Edward, ed.
  1989.   *Central Asia: 120 Years of Russian Rule.* Durham, N.C.: Duke
            University Press.
Alvarez, Robert R.
  1995.   "The Mexican-US Border: The Making of an Anthropology of
            Borderlands." *Annual Review of Anthropology* 24:447–70.
Alwin, J.
  1992.   "North American Geographers and the Pacific Rim: Leaders or
            Laggards?" *Professional Geographers* 44:369–76.
Amin, Samir.
  1989.   *Eurocentrism.* New York: Monthly Review Press.
Anand, Satyapal.
  1988.   *The Asian Identity.* Chandigarh: Sameer Prakashan.
Andaya, Leonard Y.
  1993.   "Cultural State Formation in Eastern Indonesia." In Anthony
            Reid, ed., *Southeast Asia in the Early Modern Era: Trade, Power and
            Belief.* Ithaca, N.Y.: Cornell University Press. Pp. 23–41.
Anderson, Benedict.
  1983.   *Imagined Communities: Reflections on the Origin and Spread of
            Nationalism.* London: Verso.
Anderson, Perry.
  1974.   *Lineages of the Absolutist State.* London: Verso.
Anglin, Jay Pascal, and William J. Hamblin.
  1993.   *World History to 1648.* New York: Harper Perennial.
Anthon, Charles.
  1855.   *A System of Ancient and Mediaeval Geography.* New York: Harper
            and Brothers.
Applegate, Celia.
  1990.   *A Nation of Provincials: The German Idea of Heimat.* Berkeley:
            University of California Press.
Archer, Kevin.
  1993.   "Regions as Social Organisms: The Lamarkian Characteristics of
            Vidal de la Blache's Regional Geography." *Annals of the Association
            of American Geographers* 83:498–514.
Aristotle.
  1932 [fourth century B.C.].
            Politics. Cambridge: Harvard University Press, Loeb Classical
            Library.
Arnold, Guy.
  1993.   *The End of the Third World.* New York: St. Martin's Press.

Asante, Molefi Kete.
    1988.   *Afrocentricity.* Trenton, N.J.: Africa World Press.
Asante, M[olefi Kete], and M. Mattson.
    1992.   *Historical and Cultural Atlas of African Americans.* New York:
             Macmillan.
Asao Naohiro.
    1988.   *Taikei Nihon no Rekishi* [An Outline of Japanese History]. Vol. 8,
             Tenka Itto [Unifying the Realm]. Tokyo: Shogakkan.
Ash, Timothy Garton.
    1991.   "Mitteleuropa?" In Stephen R. Graubard, ed., *Eastern Europe* . . .
             Central Europe . . . *Europe.* Boulder, Col.: Westview Press. Pp. 1–22.
Atlas International.
    1985.   Berlin: Bertelsman Lexicon-Verlag.
Aujac, Germaine.
    1987.   "The Foundations of Theoretical Cartography in Archaic and
             Classical Greece." In J. B. Harley and David Woodward, eds.,
             The History of Cartography, vol. 1, *Cartography in Prehistoric,
             Ancient, and Medieval Europe and the Mediterranean.* Chicago:
             University of Chicago Press. Pp. 130–47.
Bacon, Elizabeth.
    1946.   "A Preliminary Attempt to Determine the Culture Areas of Asia."
             Southwestern Journal of Anthropology 2:117–31.
    1980 [1966].
             Central Asians under Russian Rule: A Study in Cultural Change.
             Ithaca, N.Y.: Cornell University Press.
Balland, Daniel.
    1992.   "Comments on Andre Gunder Frank's 'The Centrality of Central
             Asia.' " *Bulletin of Concerned Asian Scholars* 24:75–76.
Barber, Benjamin R.
    1995.   *Jihad vs. McWorld.* New York: Times Books.
Barfield, Thomas J.
    1989.   *The Perilous Frontier: Nomadic Empires and China from 221 BC to AD
             1757.* Oxford: Blackwell.
Baritz, Loren.
    1961.   "The Idea of the West." *American Historical Review* 66:618–40.
Barke, Michael, and Greg O'Hare.
    1984.   *The Third World: Diversity, Change, and Interdependence.* Edin-
             burgh: Oliver and Boyd.
Barlow, Tani.
    1993.   "Colonialism's Career in Postwar China Studies." *Positions*
             1:224–67.
Barnicot, Nigel A.
    1964.   "Taxonomy and Variation in Modern Man." In Ashley Montagu,
             ed., *The Concept of Race.* New York: Collier Books. Pp. 180–227.
Barraclough, Geoffrey, ed.
    1984.   *The Times Atlas of World History.* Revised Edition. Maplewood,
             N.J.: Hammond.

1992.   *The Times Concise Atlas of World History.* Third Edition. Maple-
          wood, N.J.: Hammond.
Barthes, Roland.
    1982 [1970]. *Empire of Signs.* Translated by Richard Howard. New York: Hill
          and Wang.
Barthold, V. V.
    1937.   Introduction to *Hudud al-'Alam* (The Regions of the World).
             London: Luzak.
    1984 [1971].
             An Historical Geography of Iran. Translated by S. Soucek; edited by
             C. E. Bosworth. Princeton: Princeton University Press.
Bartholomew, John.
    1873.   *A Descriptive Hand Atlas of the World.* Philadelphia: Zell.
Bartholomew, John.
    1950.   *The Advanced Atlas of Modern Geography.* New York: McGraw-Hill.
Bartlett, Robert.
    1993.   *The Making of Europe: Conquest, Colonization, and Cultural Change
             950–1350.* Princeton: Princeton University Press.
Bassin, Mark.
    1991a.   "Inventing Siberia: Visions of the Russian East in the Early
             Nineteenth Century." *American Historical Review* 96:763–94.
    1991b.   "Russia between Europe and Asia: The Ideological Construction
             of Geographical Space." *Slavic Review* 50:1–17.
Bayly, C. A.
    1983.   *Rulers, Townsmen and Bazaars: North Indian Society in the Age of
             British Expansion, 1770–1870.* Cambridge: Cambridge University
             Press.
Beard, Charles A.
    1940.   *A Foreign Policy for America.* New York: Knopf.
Beaumont, Peter; Gerald H. Blake; and J. Malcolm Wagstaff.
    1976.   *The Middle East: A Geographical Survey.* New York: Wiley and Sons.
Beazley, C. Raymond.
    1949.   *The Dawn of Modern Geography,* 2 vols. New York: Smith.
Beckwith, Christopher L.
    1987.   *The Tibetan Empire in Central Asia.* Princeton: Princeton Univer-
             sity Press.
Bell, Daniel A.; David Brown; Kanishka Jayasuriya; and David Martin Jones.
    1995.   *Towards Illiberal Democracy in Pacific Asia.* New York: St. Martin's
             Press.
Beller, Steven.
    1994.   "Germans and Jews as Central European and 'Mitteleuropäisch'
             Elites." In Peter Stirk, ed., *Mitteleuropa: History and Prospects.*
             Edinburgh: Edinburgh University Press. Pp. 61–85.
Bellwood, Peter.
    1992.   "Southeast Asia before History." In Nicholas Tarling, ed., *The
             Cambridge History of Southeast Asia.* Vol. 1, *From Early Times to
             c. 1800.* Cambridge: Cambridge University Press. Pp. 55–136.

Benda, Harry J.
    1972.   *Continuity and Change in Southeast Asia.* New Haven: Yale
             University Southeast Asia Studies Monograph Series, no. 18.
Benet, Sula.
    1974.   *Abkhasians: The Long-Lived People of the Caucasus.* New York: Holt,
             Rinehart, and Winston.
Benhart, John, and C. Robert Scull.
    1985.   *Regions of the World Today.* Dubuque, Iowa: Kendall/Hunt.
Bennett, Wendell Clark.
    1947.   *The Ethnogeographic Board.* Washington, D.C.: Smithsonian
             Miscellaneous Collections (vol. 107, no. 1).
    1951.   *Area Studies in American Universities.* New York: Social Science
             Research Council.
Bentley, Jerry H.
    1993.   *Old World Encounters: Cross-Cultural Contacts and Exchanges in
             Pre-Modern Times.* Oxford: Oxford University Press.
Berg, Lawrence D.
    1993.   "Between Modernism and Postmodernism." *Progress in Human
             Geography* 17:490–507.
Bergholtz, Fred W.
    1993.   *The Partition of the Steppe: The Struggle of the Russians, Manchus,
             and the Zunghar Mongols for Empire in Central Asia.* New York:
             Lang.
Berlin, Isaiah.
    1991.   *The Crooked Timber of Humanity: Chapters in the History
             of Ideas.* New York: Knopf.
Bernal, Martin.
    1987.   *Black Athena: The Afroasiatic Roots of Classical Civilization.* Vol. 1,
             The Fabrication of Ancient Greece 1785–1985. New Brunswick, N.J.:
             Rutgers University Press.
    1991.   *Black Athena: The Afroasiatic Roots of Classical Civilization.* Vol. 2,
             The Archaeological and Documentary Evidence. New Brunswick,
             N.J.: Rutgers University Press.
Bhatt, Bharat L.
    1980.   "India and Indian Regions: A Critical Overview." In David E.
             Sopher, ed., *An Exploration of India: Geographical Perspectives on
             Society and Culture.* Ithaca, N.Y.: Cornell University Press. Pp. 35–61.
Blakely, Allison.
    1986.   *Russia and the Negro: Blacks in Russian History and Thought.*
             Washington, D.C.: Howard University Press.
Blaut, J. M.
    1987.   "Diffusionism: A Uniformitarian Critique." *Annals of the Associa-
             tion of American Geographers* 77:30–47.
    1992.   *1492: The Debate on Colonialism, Eurocentrism, and History.*
             Trenton, N.J.: Africa World Press.
    1993.   *The Colonizer's Model of the World: Geographical Diffusionism and
             Eurocentric History.* New York: Guilford Press.

Blumenbach, Johann F.
    1865 [1775 and 1795].
            The Anthropological Treatises. London: Longman, Green, Long-
            man, Roberts, and Green.
Bodart-Bailey, Beatrice M.
    1989.   "Tokugawa Tsunayoshi (1646–1709): A Weberian Analysis."
            Asiatische Studien / Etudes Asiatiques 43(1):5–27.
Boggs, S. W.
    1945.   "This Hemisphere." *Department of State Bulletin* 12:845–50.
Bonne, –.
    1771.   *Atlas moderne ou collection cartes sur toutes de les parties du globe.* Paris.
Bossuet, Jacques Bènigne.
    1829 [1681].
            Discours sur l'histoire universelle. Paris: Emler Frères.
Bowen, Emanuel.
    1752.   *A Complete Atlas of the Known World.* London.
Bowie, Katherine A.
    1992.   "Unraveling the Myth of the Subsistence Economy: Textile
            Production in Nineteenth-Century Northern Thailand." *Journal
            of Asian Studies* 51:797–823.
Braudel, Fernand.
    1984.   *The Perspective of the World.* Vol. 3, *Civilization and Capitalism
            15th–18th Century.* New York: Harper and Row.
    1994.   [French publication, 1987; originally published, 1963]. *A History
            of Civilizations.* New York: Penguin.
Broek, Jan O. M.
    1944.   "Diversity and Unity in Southeast Asia." *Geographical Review*
            34:175–95.
Brookfield, H. C.
    1973.   "The Pacific Realm." In Marvin Mikesell, ed., *Geographers Abroad:
            Essays on the Problems and Prospects of Research in Foreign Areas.*
            Chicago: University of Chicago Department of Geography,
            Research Paper No. 152. Pp. 70–93.
Brooks, Leonard.
    1926.   *A Regional Geography of the World.* London: University of London
            Press.
Brown, D. E.
    1991.   *Human Universals.* New York: McGraw-Hill.
Brown, Dwight A.
    1993.   "Early Nineteenth Century Grasslands of the Midcontinent
            Plains." *Annals of the Association of American Geographers*
            83:589–612.
Buckle, Henry Thomas.
    1872.   *Introduction to the History of Civilization in England.* New and
            Revised Annotated Edition. London: Routledge and Sons.
Buell, Hal.
    1962.   *Cities of Southeast Asia.* Exeter, England: Wheaton.

Bunbury, E. H.
    1959 [1879].
        *A History of Ancient Geography among the Greeks and Romans from
        the Earliest Ages until the Fall of the Roman Empire.* New York:
        Dover.
Burckhardt, Jacob.
    1943 [1905].
        Force and Freedom: Reflections on History. Translated and edited by
        J. H. Nichols. New York: Pantheon.
Burke, Edmund, III.
    1993.   "Introduction: Marshall G. S. Hodgson and World History."
        In Marshall G. S. Hodgson, *Rethinking World History: Essays on
        Europe, Islam, and World History.* Cambridge: Cambridge Univer-
        sity Press. Pp. vi–xxi.
Burke, Peter.
    1980.   "Did Europe Exist before 1700?" *History of European Ideas*
        1:21–29.
Burton, Richard F.
    1973 [1851].
        Sindh and the Races That Inhabit the Valley of the Indus. Karachi:
        Oxford University Press.
Butler, Samuel.
    1829 (corrected in 1831).
        An Atlas of Modern Geography. London: Longman.
Butzer, Karl W.
    1976.   *Early Hydraulic Civilization in Egypt: A Study in Cultural Ecology.*
        Chicago: University of Chicago Press.
Cahnman, Werner J.
    1948.   "Outline of a Theory of Area Studies." *Annals of the Association of
        American Geographers* 38:233–43.
    1949.   "Frontiers between East and West in Europe." *Geographical Review*
        39:605–24.
Cameron, Averil.
    1993.   *The Mediterranean World in Late Antiquity ad 395–600.* London:
        Routledge.
Canfield, Robert, ed.
    1991.   *Turko-Persia in Historical Perspective.* Cambridge: Cambridge
        University Press.
Cantor, Norman F.
    1991.   *Inventing the Middle Ages: The Lives, Works, and Ideas of the Great
        Medievalists of the Twentieth Century.* New York: Morrow.
Capella, Martianus.
    1977 [circa 420].
        The Marriage of Philosophy and Mercury. In W. H. Stahl, R.
        Johnson, and E. C. Burge, trans. and eds., *Martianus Capella
        and the Seven Liberal Arts.* New York: Columbia University
        Press.

Carter, George F.
　1964.　*Man and the Land: A Cultural Geography*. New York: Holt,
　　　　Rinehart, and Winston.
Cary, J.
　1808.　*Cary's New Universal Atlas*. London: Cary.
Castro, Americo.
　1954.　*The Structure of Spanish History*. Princeton: Princeton University
　　　　Press.
Cavalli-Sforza, L. Luca; Paulo Menozzi; and Alberto Piazza.
　1994.　*The History and Geography of Human Genes*. Princeton: Princeton
　　　　University Press.
Chaloupka, William, and R. McGreggor Cawley.
　1993.　"The Great Wild Hope: Nature, Environmentalism, and the
　　　　Open Secret." In Jane Bennett and William Chaloupka, eds.,
　　　　In the Nature of Things: Language, Politics, and the Environment.
　　　　Minneapolis: University of Minnesota Press. Pp. 3–23.
Chard, Chester S.
　1974.　*Northeast Asia in Prehistory*. Madison: University of Wisconsin
　　　　Press.
Chatterjee, Partha.
　1993.　*The Nation and Its Fragments: Colonial and Postcolonial Histories*.
　　　　Princeton: Princeton University Press.
Chaudhuri, K. N.
　1985.　*Trade and Civilization in the Indian Ocean: An Economic History
　　　　from the Rise of Islam to 1750*. Cambridge: Cambridge University
　　　　Press.
　1990.　*Asia before Europe: Economy and Civilization of the Indian Ocean
　　　　from the Rise of Islam to 1750*. Cambridge: Cambridge University
　　　　Press.
Chirol, Valentine.
　1903.　*The Middle Eastern Question or Some Political Problems of Indian
　　　　Defence*. London: Murray.
　1924.　*The Occident and the Orient*. Chicago: University of Chicago Press.
Chisholm, Michael.
　1982.　*Modern World Development: A Geographical Perspective*. Totowa,
　　　　N.J.: Barnes and Noble.
Christian, David.
　1994.　"Inner Eurasia as a Unit of World History." *Journal of World
　　　　History* 5:173–211.
Cipolla, Carlo M.
　1993.　*Before the Industrial Revolution: European Society and Economy
　　　　1000–1700*. Third Edition. New York: Norton.
Cleary, M. C., and F. J. Lian.
　1991.　"On the Geography of Borneo." *Progress in Human Geography*
　　　　15:163–77.
Clerk, George.
　1944.　"Address at the Annual General Meeting of the Royal Geographi-
　　　　cal Society." *Geographical Journal* 104:1–7.

Clifford, James.
 1992.   "Traveling Cultures." In L. Grossberg, C. Nelson, and P. Treichler,
           eds., *Cultural Studies*. London: Routledge. Pp. 96–116.
 1994.   "Diasporas." *Cultural Anthropology* 9:302–38.
Clout, Hugh.
 1992.   "Vive la géographie! Vive la géographie française." *Progress in
           Human Geography* 16:243–428.
Clout, Hugh; Mark Blacksell; Russell King; and David Pinder.
 1985.   *Western Europe: Geographical Perspectives*. London: Longman.
Clowse, Barbara Barksdale.
 1981.   *Brainpower for the Cold War: The Sputnik Crisis and National
           Defense Education Act of 1958*. Westport, Conn.: Greenwood Press.
Cohen, Saul.
 1991.   "Global Geopolitical Change in the Post-Cold War Era." *Annals of
           the Association of American Geographers* 81:551–80.
Cohen-Portheim, Paul.
 1934.   *The Message of Asia*. Translated by Alan Harris. New York: Dutton.
Cohn, Bernard S.
 1987.   *An Anthropologist among the Historians and Other Essays*. Delhi:
           Oxford University Press.
Collier, Simon; Harold Blakemore; and Thomas E. Skidmore, eds.
 1985.   *The Cambridge Encyclopedia of Latin America and the Caribbean*.
           Cambridge: Cambridge University Press.
Colton, George W.
 1856.   *Colton's Atlas of the World*. New York: Colton.
The Columbian Atlas of the World.
 1893.   New York: Hunt and Eaton.
Commoner, Barry.
 1990.   *Making Peace with the Planet*. New York: Pantheon.
Condorcet, Marie-Jean-Antoine de Caritat, Marquis de.
 1955 [1795].
           Sketch for a Historical Picture of the Progress of the Human Mind.
           Translated by Jane Barraclough. London: Weidenfeld and
           Nicolson.
Cooke, Dwight.
 1954.   *There Is No Asia*. Garden City, N.Y.: Doubleday.
Coon, Carleton S.
 1954.   *The Story of Man: From the First Human to Primitive Culture and
           Beyond*. New York: Knopf.
 1982.   *Racial Adaptations*. Chicago: Nelson-Hall.
Coon, Carleton S., with Edward E. Hunt, Jr.
 1966.   *The Living Races of Man*. London: Cape.
Cooper, Fredrick; Florencia E. Mallon; Steve J. Stern; Allen F. Isaacman; and
           William Roseberry.
 1993.   *Confronting Historical Paradigms: Peasants, Labor, and the Capitalist
           World System in Africa and Latin America*. Madison: University of
           Wisconsin Press.

Copenhaver, Brian P.
1990.   "Natural Magic, Hermetism, and Occultism in Early Modern
         Science." In David C. Lindberg and Robert S. Westman, eds.,
         Reappraisals of the Scientific Revolution. Cambridge: Cambridge
         University Press. Pp. 262–302.
Corbridge, Stuart.
1986.   *Capitalist World Development: A Critique of Radical Development
         Geography*. London: Macmillan.
Cortambert, E.
1869.   *Nouvel atlas de géographie*. Paris: Librairie Hachette.
Cosgrove, Denis.
1994.   "Contested Global Visions: *One-World, Whole Earth,* and the
         Apollo Space Photographs." *Annals of the Association of American
         Geographers* 84:270–94.
Cram, George.
1897.   *Cram's Universal Atlas*. New York: Cram.
Cronon, William.
1994.   "Comment: Cutting Loose or Running Aground?" *Journal of
         Historical Geography* 20:38–43.
Crosby, Alfred.
1986.   *Ecological Imperialism: The Biological Expansion of Europe 900–1900*.
         Cambridge: Cambridge University Press.
Cumings, Bruce.
1984.   "The Legacy of Japanese Colonialism in Korea." In R. H. Myers
         and M. R. Peattie, eds., *The Japanese Colonial Empire 1895–1945*.
         Princeton: Princeton University Press. Pp. 478–96.
1993.   "Japan's Position in the World System." In Andrew Gordon, ed.,
         Postwar Japan as History. Berkeley: University of California Press.
         Pp. 34–63.
Cundall, Leonard B.
1932.   *Western Europe*. London: Cundall.
Curtin, Philip.
1984.   *Cross-Cultural Trade in World History*. Cambridge: Cambridge
         University Press.
Curtin, Philip; Steven Feierman; Leonard Thompson; and Jan Vansina.
1978.   *African History*. London: Longman.
D'Anville, Jean Baptiste Bourguignon.
1743.   *Atlas de D'Anville*. Paris: Cartes Geographiques.
Darby, H. C., and Harold Fullard, eds.
1970.   *The New Cambridge Modern History Atlas*. Cambridge: Cambridge
         University Press.
Darling, Frank C.
1979.   *The Westernization of Asia: A Comparative Political Analysis*.
         Boston: Hall.
Darling, Lynda.
1994.   "Ottoman Politics through British Eyes: Paul Rycaut's *The Present
         State of the Ottoman Empire*." *Journal of World History* 5:71–98.

Dathorne, O. R.
   1994.   *Imagining the World: Mythical Beliefs versus Reality in Global Encounters.* Westport, Conn.: Bergin and Garvey.
Davies, Rees.
   1989.   "Frontier Arrangements in Fragmented Societies: Ireland and Wales." In Robert Bartlett and Angus MacKay, eds., *Medieval Frontier Societies.* Oxford: Oxford University Press. Pp. 77–100.
Davis, Kenneth.
   1992.   *Don't Know Much about Geography.* New York: Morrow.
Davis, Winston.
   1992.   *Japanese Religion and Society: Paradigms of Structure and Change.* Albany: State University of New York Press.
Davison, Roderic H.
   1963.   "Where Is the Middle East?" In Richard H. Nolte, ed., *The Modern Middle East.* New York: Atherton Press. Pp. 13–29.
Dean, Vera Micheles.
   1957.   *The Nature of the Non-Western World.* New York: Mentor.
Dear, Michael, and Gregg Wassmansdorf.
   1993.   "Postmodern Consequences." *Geographical Review* 83:321–25.
de Bary, William Theodore, and Ainslie T. Embree, eds.
   1961.   *Approaches to Asian Civilizations.* New York: Columbia University Press.
de Blij, Harm.
   1992.   "Political Geography in the Post Cold-War World." *Professional Geographer* 44:16–19.
de Blij, Harm; G. Danzer; R. Hart; and D. Drummond.
   1989.   *World Geography: A Physical and Cultural Study.* Glenview, Ill.: Scott, Foresman.
de Blij, Harm J., and Peter O. Muller.
   1994.   *Geography: Realms, Regions, and Concepts.* Revised Seventh Edition. New York: Wiley and Sons.
Deglar, Carl N.
   1991.   *In Search of Human Nature: The Decline and Revival of Darwinism in American Social Thought.* Oxford: Oxford University Press.
Delaisi, Francis.
   1929.   *Les Deux Europes.* Paris: Payot.
Demeritt, David.
   1994.   "Ecology, Objectivity and Critique in Writings on Nature and Human Societies." *Journal of Historical Geography* 20:22–37.
de Rougemont, Denis.
   1966.   *The Idea of Europe.* New York: Macmillan.
De Vorsey, Louis, Jr.
   1992.   "Western Europe: Landscapes of Development." In James S. Fisher, ed., *Geography and Development: A World Regional Approach.* New York: Macmillan. Pp. 206–23.

Dickenson, J. P.; C. G. Clarke; W. T. S. Gould; A. G. Hodgkiss; R. M.
    Prothero; D. S. Siddle; C. T. Smith; and E. M. Thomas-Hope.
    1983.   *A Geography of the Third World*. London: Methuen.
Dickinson, Robert E.
    1969.   *The Makers of Modern Geography*. New York: Praeger.
Diderot, Denis.
    1876 [circa 1750].
                Oeuvres complètes. Paris: Garnier Frères.
Diem, Aubrey.
    1979.   *Western Europe: A Geographical Analysis*. New York: Wiley and Sons.
Dikötter, Frank.
    1992.   *The Discourse of Race in Modern China*. Stanford: Stanford
                University Press.
Dillon, Leo.
    1994.   "Map: The Center of Europe." *Geographic and Global Issues
                Quarterly* 4:19–21.
Diop, C. A.
    1974.   *The African Origin of Civilization: Myth or Reality*. Chicago: Hill.
Dirlik, Arif, ed.
    1993.   *What Is in a Rim? Critical Perspectives on the Pacific Region Idea*.
                Boulder, Col.: Westview Press.
Dixon, Chris.
    1991.   *South East Asia in the World Economy*. Cambridge: Cambridge
                University Press.
Djaït, Hichem.
    1985.   *Europe and Islam: Cultures and Modernity*. Berkeley: University
                of California Press.
Dmytryshyn, Basil.
    1991.   "The Administrative Apparatus of the Russian Colony in Siberia
                and Northern Asia 1581–1700." In Alan Wood, ed., *The History of
                Siberia: From Russian Conquest to Revolution*. London: Routledge.
                Pp. 17–36.
Doel, Marcus A.
    1993.   "Proverbs for Paranoids: Writing Geography on Hollowed
                Ground." *Transactions of the Institute of British Geographers*, New
                Series 18:377–94.
Dorn, Harold.
    1991.   *The Geography of Science*. Baltimore: Johns Hopkins University
                Press.
Downs, Roger M.
    1994.   "Being and Becoming a Geographer: An Agenda for Geography
                Education." *Annals of the Association of American Geographers*
                84:175–91.
Dozer, Donald M.
    1962.   *Latin America: An Interpretive History*. New York: McGraw-Hill.
Drake, Fred.
    1975.   *China Charts the World: Hsu Chi-Yu and His Geography of 1848*.
                Cambridge: Harvard University Press.

Drake, St. Clair.
   1987.   *Black Folk Here and There.* Los Angeles: Center for Afro-American
           Studies, University of California, Los Angeles.
Driver, Harold E.
   1961.   *Indians of North America.* Chicago: University of Chicago Press.
Drompp, Michael.
   1989.   "Centrifugal Forces in the Inner Asian 'Heartland': History
           versus Geography." *Journal of Asian History* 23:135–55.
Dukes, Paul.
   1990.   *A History of Russia: Medieval, Modern, Contemporary.* Durham,
           N.C.: Duke University Press.
Duncan, James S.
   1990.   *The City as Text: The Politics of Landscape Interpretation in the
           Kandyan Kingdom.* Cambridge: Cambridge University Press.
Duncan, James S., and Nancy G. Duncan.
   1992.   "Ideology and Bliss: Roland Barthes and the Secret Histories
           of Landscape." In Trevor J. Barnes and James S. Duncan, eds.,
           *Writing Worlds: Discourse, Text and Metaphor in the Representation
           of Landscape.* London: Routledge. Pp. 18–37.
Dunn, Ross E.
   1986.   *The Adventures of Ibn Battuta: A Muslim Traveler of the Fourteenth
           Century.* Berkeley: University of California Press.
Durant, Will.
   1954 [1935].
           Our Oriental Heritage. New York: Simon and Schuster.
Durant, Will, and Ariel Durant.
   1965.   *The Age of Voltaire.* New York: Simon and Schuster.
Eaton, Richard M.
   1993.   *The Rise of Islam and the Bengal Frontier, 1204–1760.* Berkeley:
           University of California Press.
Edwardes, Michael.
   1971.   *East-West Passages.* London: Cassell.
Elton, Edward F., ed.
   1908.   *The Class-Room Atlas.* Edinburgh: Johnston.
Embree, Ainslie T.
   1989.   *Imagining India: Essays on Indian History.* Delhi: Oxford Univer-
           sity Press.
Emmerson, Donald K.
   1984.   "'Southeast Asia': What's in a Name." *Journal of Southeast Asian
           Studies* 15:1–21.
Encyclopedia Britannica World Atlas.
   1949.   New York: Hammond.
English, Paul W.
   1977.   *World Regional Geography: A Question of Place.* New York: Harper
           and Row.
English, Paul W., and James A. Miller.
   1989.   *World Regional Geography: A Question of Place.* New York: Wiley
           and Sons.

Escobar, Arturo.
  1992.   "Imagining a Post-Development Era? Critical Thought, Develop-
            ment and Social Movements." *Social Text* 10:20–56.
  1995.   *Encountering Development: The Making and Unmaking of the Third
            World*. Princeton: Princeton University Press.
Farwell, Byron.
  1989.   *Armies of the Raj: From Mutiny to Independence, 1858–1947*. New
            York: Norton.
Feierman, Steven
  1993.   "African History and the Dissolution of World History." In R.
            Bates, V. Y. Mudimbe, and J. O'Barr, eds., *Africa and the Disciplines:
            The Contributions of Research in Africa to the Social Sciences and
            Humanities*. Chicago: University of Chicago Press. Pp. 167–212.
Fenton, William Nelson.
  1946.   "Integration of Geography and Anthropology in Army Area
            Studies Curricula." *Bulletin of the American Association of University
            Professors* 32:697–706.
  1947.   *Area Studies in American Universities*. Washington, D.C.: American
            Council on Education.
Ferguson, James, and Akhil Gupta, eds.
  1992.   "Space, Identity and the Politics of Difference" (theme issue).
            Cultural Anthropology 7, no. 1 (February).
Fersh, Seymour.
  1965.   *India and South Asia*. New York: Macmillan.
Fifield, Russell.
  1975.   "The Concept of Southeast Asia: Origins, Development and
            Education." *Southeast Asian Spectrum* 4:42–51.
Finley, Anthony.
  1826.   *A New General Atlas*. Philadelphia: Finley.
Finney, Ben.
  1994.   "The Other One-Third of the Globe." *Journal of World History*
            5:273–97.
Fischer, J., and F. W. Wieser.
  1903.   *The World Maps of Waldseemüller 1507 and 1516*. Innsbruck: Verlag
            der Wagner'schen Universitats Buchhandlung.
Fisher, James S., ed.
  1992.   *Geography and Development: A World Regional Approach*. New
            York: Macmillan.
Fisher, W. B.
  1947.   "Unity and Diversity in the Middle East." *Geographical Review*,
            37:414–35.
Fitzgerald, Charles P.
  1964.   *The Chinese View of Their Place in the World*. London: Oxford
            University Press.
Flewelling, Ralph Tyler.
  1943.   *The Survival of Western Culture: An Inquiry into the Problem of Its
            Decline and Resurgence*. New York: Harper and Brothers.

Forsyth, James.
　1991.　"The Siberian Native Peoples before and after the Russian Conquest." In Alan Wood, ed., *The History of Siberia: From Russian Conquest to Revolution*. London: Routledge. Pp. 69–89.
　1992.　*A History of the Peoples of Siberia: Russia's North Asian Colony 1581–1990*. Cambridge: Cambridge University Press.
Foucault, Michel.
　1970 [1966].
　　　　The Order of Things: An Archaeology of the Human Sciences. New York: Pantheon.
Fourth World Movement.
　1980.　*Dialogue with the Fourth World*. London: Fourth World Movement.
Frank, Andre Gunder.
　1991.　"A Plea for World System History." *Journal of World History* 2:1–28.
　1992.　"The Centrality of Central Asia." *Bulletin of Concerned Asian Scholars* 24, 50–74, 80–82.
Frank, Andre Gunder, and Barry K. Gills, ed.
　1993.　*The World System: Five Hundred or Five Thousand Years?* London: Routledge.
Freyer, Donald W.
　1979.　*Emerging Southeast Asia: A Study in Growth and Stagnation*. New York: Wiley and Sons.
Friedman, Edward, ed.
　1994.　*The Politics of Democratization: Generalizing East Asian Experiences*. Boulder, Col.: Westview Press.
Friedman, John.
　1994.　"Cultural Conflicts in Medieval World Maps." In Stuart B. Schwartz, ed., *Implicit Understandings: Observing, Reporting, and Reflecting on the Encounters between Europeans and Other Peoples in the Early Modern Era*. Cambridge: Cambridge University Press. Pp. 64–95.
Fromkin, David.
　1989.　*A Peace to End All Peace: The Fall of the Ottoman Empire and the Creation of the Modern Middle East*. New York: Avon.
Fry, Earl H., and Gregory A. Raymond.
　1983.　*The Other Western Europe: A Political Analysis of the Smaller Democracies*. Santa Barbara and Oxford: ABC-Clio Information Services.
Fukuyama, Francis.
　1992.　*The End of History and the Last Man*. New York: Avon.
Fuller, Gary.
　1989.　"Why Geographical Alliances Won't Work." *Professional Geographer* 41:484–86.
Fuller, Graham E.
　1993.　"Turkey's New Eastern Orientation." In Graham E. Fuller and Ian

O. Lesser, eds., *Turkey's New Geopolitics: From the Balkans to Western China*. Boulder, Col.: Westview Press. Pp. 37–98.

Fuson, Robert H.
  1977.   *Introduction to World Geography: Regions and Cultures*. Dubuque, Iowa: Kendall/Hunt.

Gaebler, Eduard.
  1897.   *Neuster Hand-Atlas über alle Teile der Erde*. Leipzig: Druck und Verlag.

Gay, Peter.
  1969.   *The Enlightenment: An Interpretation. The Science of Freedom*. New York: Norton.

Geddes, Arthur.
  1982.   *Man and Land in South Asia*. New Delhi: Concept.

Gellner, Ernest.
  1981.   *Muslim Society*. Cambridge: Cambridge University Press.
  1988.   *Plough, Sword, and Book: The Structure of Human History*. Chicago: University of Chicago Press.
  1992.   *Reason and Culture: The Historical Role of Rationality and Rationalism*. Oxford: Blackwell.

Geographischer Atlas Bestehen in 44 Land-Charten Worauf alle Theile.
  1785.   Berlin.

George, Pierre, and Jean Tricart.
  1954.   *L'Europe centrale*, 2 vols. Paris: Presses Universitaires de France.

Gernet, Jacques.
  1982.   *A History of Chinese Civilization*. Cambridge: Cambridge University Press.

Gilbert, A.
  1988.   "The New Regional Geography in English and French-Speaking Countries." *Progress in Human Geography* 12:208–28.

Gills, Barry K., and Andre Gunder Frank.
  1993.   "The Cumulation of Accumulation." In Andre Gunder Frank and Barry K. Gills, eds., *The World System: Five Hundred or Five Thousand Years?* London: Routledge. Pp. 81–114.

Gilroy, Paul.
  1993.   *The Black Atlantic: Modernity and Double Consciousness*. Cambridge: Harvard University Press.

Ginsburg, Norton.
  1985.   Foreword to *Southeast Asia: Realm of Contrasts*. Edited by Ashok K. Dutt. Boulder, Col.: Westview Press. Pp. ix–x.
  1993.   "Commentary on Alwin's 'North American Geographers and the Pacific Rim.' " *Professional Geographer* 45:355–57.

Glacken, Clarence.
  1967.   *Traces on the Rhodian Shore: Nature and Culture in Western Thought from Ancient Times to the End of the Eighteenth Century*. Berkeley: University of California Press.

Gladney, Dru C.
  1992.   "The Hui, Islam, and the State: A Sufi Community in China's Northwest Corner." In Jo-Ann Gross, ed., *Muslims in Central Asia*. Durham, N.C.: Duke University Press. Pp. 89–111.

Glenny, Misha.
   1994.   "The Bear in the Caucasus: From Georgian Chaos, Russian
            Order." *Harper's* 288 (March): 45–53.
Goldberg, Ellis.
   1992.   "Smashing Idols and the State: The Protestant Ethic and Egyptian
            Sunni Radicalism." In Juan R. I. Cole, ed., *Comparing Muslim
            Societies*. Ann Arbor: University of Michigan Press. Pp. 195–236.
Goldmanns Grosser Weltatlas.
   1955.   Munich: Goldmann Verlag.
Goldstone, Jack A.
   1991.   *Revolution and Rebellion in the Early Modern World*. Berkeley:
            University of California Press.
Gole, Susan.
   1983.   *India within the Ganges*. New Delhi: Jayaprints.
   1989.   *Indian Maps and Plans*. New Delhi: Manohar.
Goode's World Atlas, Seventeenth Edition.
   1986.   New York: Rand McNally.
Goodman, Bryna.
   1995.   *Native Place, City, and Nation: Regional Networks and Identities in
            Shanghai, 1853–1937*. Berkeley: University of California Press.
Goody, Jack.
   1971.   *Technology, Tradition, and the State in Africa*. Cambridge: Cam-
            bridge University Press.
   1990.   *The Oriental, the Ancient, and the Primitive: Systems of Marriage
            and the Family in the Pre-industrial Societies of Eurasia*. Cambridge:
            Cambridge University Press.
   1996.   *The East in the West*. Cambridge: Cambridge University Press.
Gould, Peter R.
   1963.   "Man against His Environment: A Game Theoretic Framework."
            Annals of the Association of American Geographers 53:290–97.
Grant, Madison.
   1918.   *The Passing of the Great Race*. New York: Scribner's Sons.
Graubard, Stephen R., ed.
   1991.   *Eastern Europe . . . Central Europe . . . Europe*. Boulder, Col.:
            Westview Press.
Greenblatt, Stephen.
   1991.   *Marvelous Possessions: The Wonder of the New World*. Chicago:
            University of Chicago Press.
Greenleaf, Jeremiah.
   1842.   *A New Universal Atlas*. Brattleboro, Vt.: French.
Gregory, Derek.
   1994.   *Geographical Imaginations*. Oxford: Blackwell.
Gregory, Derek, and D. Ley.
   1988.   "Editorial: Culture's Geographies." *Environment and Planning D:
            Society and Space* 6:115–16.
Gregson, Nicky.
   1993.   "'The Initiative': Delimiting or Deconstructing Social Geogra-
            phy?" *Progress in Human Geography* 17:525–30.

Grillet, Donat V.
  1991.   *Where on Earth: A Refreshing View of Geography.* New York: Prentice-Hall.
Gross, Jo-Ann, ed.
  1992.   *Muslims in Central Asia.* Durham, N.C.: Duke University Press.
Gross, Paul; Norman Levitt; and Martin W. Lewis, eds.
  1996.   *The Flight from Science and Reason. Annals of the New York Academy of Sciences* 775.
Grosvenor, Gilbert M.
  1995.   "In Sight of the Tunnel: The Renaissance of Geography Education." *Annals of the Association of American Geographers* 85:409–20.
Guénon, René.
  1930.   *Orient et Occident.* Paris: Éditions Didier et Richard.
Gulick, Sidney Lewis.
  1962.   *The East and West: A Study of Their Psychic and Cultural Characteristics.* Rutland, Vt.: Tuttle.
Gupta, A., and J. Ferguson.
  1992.   "Beyond 'Culture': Space, Identity, and the Politics of Difference." *Cultural Anthropology* 7:6–23.
Guyot, Arnold.
  1970 [1849].
           The Earth and Man. New York: Arno and New York Times.
Habib, Irfan.
  1990.   "Merchant Communities in Precolonial India." In James D. Tracy, ed. *The Rise of Merchant Empires: Long-Distance Trade in the Early Modern World, 1350–1750.* Cambridge: Cambridge University Press. Pp. 371–99.
Haddon, A. C.
  1925.   *The Races of Man and Their Distribution.* New York: Macmillan.
Hadjor, Kafi Buenor.
  1992.   *Dictionary of Third World Terms.* London: Tauris.
Halecki, Oscar.
  1950.   *The Limits and Divisions of European History.* New York: Sheed and Ward.
Hall, D. G. E.
  1955.   *A History of South-East Asia.* New York: St. Martin's Press.
Hall, John A.
  1985.   *Powers and Liberties: The Causes and Consequences of the Rise of the West.* Berkeley: University of California Press.
Hall, Kenneth.
  1985.   *Maritime Trade and State Development in Early Southeast Asia.* Honolulu: University of Hawaii Press.
Hall, Robert.
  1947.   *Area Studies with Special Reference to Their Application for Research in the Social Sciences.* New York: Social Science Research Council.
Hall, Sidney, and William Hughes.
  1856.   *General Atlas of the World.* Edinburgh: North Bridge.

Halperin, Charles J.
  1985.   *Russia and the Golden Horde: The Mongol Impact on Medieval
           Russian History*. Bloomington: Indiana University Press.
Hamashita Takeshi.
  1993.   "Chiiki kenkyu to Ajia" [Regional Research and Asia]. In Mi-
           zoguchi Yuzo et al., eds., *Ajia Kangaeru* [Thinking from Asia].
           Vol. 2, *Chiiki Shisutemu* [Regional Systems]. Tokyo: University
           of Tokyo Press. Pp. 1–12.
Hambly, Gavin.
  1969.   *Central Asia*. London: Weidenfeld and Nicolson.
Hammerton, J. A., ed.
  1927.   *Universal History of the World*. London: Educational Book
           Company.
Hammond's Ambassador World Atlas.
  1954.   Maplewood, N.J.: Hammond and Co.
Hammond's Modern Atlas of the World.
  1909.   New York: Hammond.
Hammond World Atlas, Ambassador Edition.
  1991.   Maplewood, N.J.: Hammond.
           Citation Edition.
  1992.   Maplewood, N.J.: Hammond.
Harding, Susan, and Fred Myers, eds.
  1994.   "Further Inflections: Toward Ethnographies of the Future"
           (theme issue). *Cultural Anthropology* 9, no. 3.
Harley, J. B., and David Woodward, eds.
  1992.   *The History of Cartography*. Vol. 2, Book 1, *Cartography in the
           Traditional Islamic and South Asian Societies*. Chicago: University of
           Chicago Press.
Harper, Robert A., and Theodore Schmudde.
  1978.   *Between Two Worlds: An Introduction to Geography*. Boston:
           Houghton Mifflin.
Harris, Chauncy.
  1993.   "New European Countries and Their Minorities." *Geographical
           Review* 83(3):301–20.
Harris, James F.
  1992.   *Against Relativism: A Philosophical Defense of Method*. LaSalle, Ill.:
           Open Court.
Harris, Nigel.
  1986.   *The End of the Third World: Newly Industrializing Countries and the
           Decline of an Ideology*. London: Penguin.
Harvey, David.
  1989.   *The Condition of Postmodernity: An Enquiry into the Origins of
           Cultural Change*. Oxford: Blackwell.
Hashikawa, Bunso.
  1980.   "Japanese Perspectives on Asia: From Dissociation to Coprosper-
           ity." In Akira Iriye, ed., *The Chinese and the Japanese*. Princeton:
           Princeton University Press. Pp. 328–55.

Hauner, Milan.
    1990.   *What Is Asia to Us? Russia's Asian Heartland Yesterday and Today.*
             London: Routledge.
    1991.   "Russia's Geopolitical and Ideological Dilemmas in Central Asia."
             In Robert Canfield, ed., *Turko-Persia in Historical Perspective.*
             Cambridge: Cambridge University Press. Pp. 189–216.
Hay, Denys.
    1957.   *Europe: The Emergence of an Idea.* Edinburgh: Edinburgh Univer-
             sity Press.
Hay, Stephen N.
    1970.   *Asian Ideas of East and West: Tagore and His Critics in Japan, China
             and India.* Cambridge: Harvard University Press.
Hegel, Georg Wilhelm Friedrich.
    1942 [1821].
             Hegel's Philosophy of Right. Translated by T. M. Knox. Oxford:
             Clarendon.
    1956 [1830–31].
             The Philosophy of History. Translated by J. Sibree. New York: Dover.
Heginbotham, Stanley J.
    1994.   "Rethinking International Scholarship: The Challenge of Transi-
             tion from the Cold War Era." *Items (Social Science Research Council)*
             48:33–40.
Helgren, David; Robert Sager; and Saul Israel.
    1985.   *World Geography Today.* New York: Holt, Rinehart, and Winston.
Henley, David E. F.
    1995.   "Ethnogeographic Integration and Exclusion in Anticolonial
             Nationalism: Indonesia and Indochina." *Comparative Studies in
             Society and History* 37 286–324.
Henry, William A., III.
    1994.   *In Defense of Elitism.* New York: Anchor Books.
Hepner, George F., and Jesse O. McKee.
    1992.   *World Regional Geography: A Global Approach.* St. Paul: West.
Herbertson, F. D., ed.
    1903.   *Asia.* London: Black.
Herder, J. G.
    1968.   *Reflections of the Philosophy of the History of Mankind.* Abridged and
             translated by Frank E. Manuel. Chicago: University of Chicago
             Press.
Hermassi, Elbaki.
    1980.   *The Third World Reassessed.* Berkeley and Los Angeles: University
             of California Press.
Herodotus.
    1954 [circa 446 B.C.E.].
             The Histories. London: Penguin.
Hettne, Björn.
    1995.   *Development Theory and the Three Worlds: Toward an International
             Political Economy of Development.* Burnt Mill, Harlow, Essex:
             Longman; New York: Wiley.

Hewes, Leslie.
   1983.   "Carl Sauer: A Personal View." *Journal of Geography* 82:140–45.
Hill, Hal.
   1994.   "ASEAN Economic Development: An Analytical Survey." *Journal of Asian Studies* 53:832–66.
Hill, Michael.
   1981.   "Positivism: A 'Hidden' Philosophy in Geography." In Milton E. Harvey and Brian P. Holly, eds., *Themes in Geographic Thought.* London: Croom Helm. Pp. 39–60.
Hobsbawm, Eric.
   1993.   "The New Threat to History." *New York Review of Books* 40 (16 December), pp. 62–64.
Hodgson, Marshall G. S.
   1974.   *The Venture of Islam: Conscience and History in a World Civilization,* 3 vols. Chicago: University of Chicago Press.
   1993.   *Rethinking World History: Essays on Europe, Islam, and World History.* Edited by Edmund Burke III. Cambridge: Cambridge University Press.
Hoekveld, G. A.
   1990.   "Regional Geography Must Adjust to New Realities." In R. J. Johnston, J. Hauer, and G. A. Hoekveld, eds., *Regional Geography: Current Developments and Future Prospects.* London: Routledge. Pp. 11–31.
Hoffman, George W., ed.
   1969.   *A Geography of Europe Including Asiatic USSR.* New York: Ronald Press.
Hogarth, D. G.
   1905.   *The Nearer East.* London: Frowde.
Holton, Gerald.
   1988 [1973].
            Thematic Origins of Scientific Thought: Kepler to Einstein. Cambridge: Harvard University Press.
Horkheimer, Max, and Theodor Adorno.
   1993 [1944].
            Dialectic of Enlightenment. Translated by John Cumming. New York: Continuum.
Hourani, Albert.
   1962.   *Arabic Thought in the Liberal Age, 1798–1939.* Oxford: Oxford University Press.
   1980.   *Europe and the Middle East.* Berkeley: University of California Press.
   1991.   *A History of the Arab Peoples.* New York: Warner.
Howell, David G.
   1989.   *Tectonics of Suspect Terranes: Mountain Building and Continental Growth.* New York: Chapman and Hall.
Howell, David L.
   1995.   *Capitalism from Within: Economy, Society, and the State in a Japanese Fishery.* Berkeley and Los Angeles: University of California Press.

Hudud al-'Alam (The Regions of the World).

    1937 [982].   Translated by V. Minorsky. London: Luzak.

Hunter, Michael.

    1990.   "Science and Heterodoxy: An Early Modern Problem Reconsidered." In David C. Lindberg and Robert S Westman, eds., *Reappraisals of the Scientific Revolution*. Cambridge: Cambridge University Press. Pp. 437–60.

Huntington, Ellsworth.

    1907.   *The Pulse of Asia*. New York: Houghton Mifflin.

    1945.   *Mainsprings of Civilization*. New York: Wiley and Sons.

Huntington, Samuel P.

    1993.   "The Clash of Civilizations?" *Foreign Affairs* 72:23–49.

Hupchick, Dennis P.

    1993.   *The Bulgarians in the Seventeenth Century: Slavic Orthodox Society and Culture under Ottoman Rule*. Jefferson, N.C.: McFarland.

Ibn Khaldun.

    1967.   *The Muqaddimah: An Introduction to History*. Edited and abridged by N. J. Dawood, translated Franz Rosenthal. Princeton: Princeton University Press.

Inden, Ronald.

    1990.   *Imagining India*. Oxford: Blackwell.

Independent Commission on International Development (Willy Brandt, chair).

    1980.   *North-South: A Program for Survival*. Cambridge: MIT Press.

Inkster, Ian.

    1988.   "The Other Side of Meiji: Conflict and Conflict Management." In Gavan McCormack and Yoshio Sugimoto, eds., *The Japanese Trajectory: Modernization and Beyond*. Cambridge: Cambridge University Press. Pp. 107–28.

International Centre for Minority Studies and Intercultural Relations Foundation, Sofia.

    1994.   *Relations of Compatibility and Incompatibility between Christians and Muslims in Bulgaria*. Sofia: International Centre for Minority Studies and Intercultural Relations Foundation.

Jackson, Richard H., and Lloyd E. Hudman.

    1982.   *World Regional Geography: Issues for Today*. New York: Wiley and Sons.

    1990.   *Cultural Geography: People, Places, and Environment*. St. Paul: West.

Jacobs, Margaret C.

    1988.   *The Cultural Meaning of the Scientific Revolution*. New York: Knopf.

James, Preston.

    1964.   *One World Divided*. New York: Blaisdell.

James, Preston, and Nelda Davis.

    1959.   *The Wide World: A Geography*. New York: Macmillan.

James, Preston, and Geoffrey Martin.

    1981.   *All Possible Worlds*. New York: Wiley and Sons.

Jameson, Frederic.
    1986.   "Third World Literature in the Era of Multinational Capital."
             Social Text 15:65–88.
Jean, Clinton M.
    1991.   *Behind the Eurocentric Veils: The Search for African Realities.*
             Amherst: University of Massachusetts Press.
Jenner, W. J. F.
    1992.   *The Tyranny of History: The Roots of China's Crisis.* London:
             Penguin.
Jerrold, Douglas.
    1954.   *The Lie about the West: A Response to Professor Toynbee's Challenge.*
             London: Dent and Sons.
Johnson, Paul.
    1991.   *The Birth of the Modern: World Society 1815–1830.* New York:
             HarperCollins.
Johnston, R. J.
    1991.   *Geography and Geographers: Anglo-American Human Geography*
             since 1945. London: Arnold.
Johnston, R. J.; J. Hauer; and G. A. Hoekveld, eds.
    1990.   *Regional Geography: Current Developments and Future Prospects.*
             London: Routledge.
Johnston, R. J.; Peter J. Taylor; and Michael J. Watts, eds.
    1995.   *Geographies of Global Change: Remapping the World in the Late*
             Twentieth Century. Oxford: Blackwell.
Johnston, W.
    1880.   *The World: An Atlas.* Edinburgh: Johnston.
Joll, James.
    1980.   "Europe: An Historian's View." *History of European Ideas*
             1:7–19.
Jones, Eric L.
    1988.   *Growth Recurring: Economic Change in World History:* Oxford:
             Clarendon.
Jones, Stephen B.
    1955.   "Views of the Political World." *Geographical Review* 45:309–26.
Jordan, Terry G.
    1973 and 1988.
             The European Culture Area: A Systematic Geography. Two editions.
             New York: Harper and Row.
Judt, Tony.
    1991.   "The Rediscovery of Central Europe." In Stephen R. Graubard,
             ed. *Eastern Europe . . . Central Europe . . . Europe.* Boulder, Col.:
             Westview Press. Pp. 23–58.
Kabbani, R.
    1986.   *Europe's Myths of Orient.* Bloomington: Indiana University Press.
Kaiser, Robert J.
    1994.   *The Geography of Nationalism in Russia and the USSR.* Princeton:
             Princeton University Press.

Kaiwar, Vasant.
    1991.   "On Provincialism and 'Popular Nationalism': Reflections on Samir Amin's Eurocentrism." *South Asia Bulletin* 11:69–78.

Kamiya Nobuyuki.
    1989.   "Toajia sekai to bakuhansei kokka—sekaishi no naka no nihon kinsei" [The East Asian World and the Bakuhan State: Early Modern Japan in World History]. In Murakami Tadashi, ed., *Nihon Kinseishi Kenkyu Jiten* [A Research Dictionary for Early Modern Japanese History]. Tokyo: Tokyodo Shuppan. Pp. 6–7.

    1991.   "Bakufu no taigai seisaku wa sakoku ka kaikoku ka" [Was the Bakufu's Foreign Policy "Close the Country or "Open the Country"?] In Aoki Michio and Hosaka Satoru, eds., *Soten Nihon no rekishi* [A Comprehensive History of Japan]. Vol. 5, *Kinsei hen* [Early Modern]. Tokyo: Shinjinbutsu Oraisha. Pp. 50–64.

Kandiyoti, Deniz.
    1992.   "Women, Islam, and the State: A Comparative Approach." In Juan R. I. Cole, ed., *Comparing Muslim Societies*. Ann Arbor: University of Michigan Press. Pp. 237–60.

Kang, Liu.
    1992.   "Subjectivity, Marxism, and Culture Theory in China." *Social Text* 31/32:114–40.

Kaplan, Robert D.
    1994.   "The Coming Anarchy." *Atlantic Monthly* 273, no. 2 (February): 44–76.

Karan, Pradyumna P., and Wilford A. Bladen.
    1985.   "The Geopolitical Base." In Ashok K. Dutt, ed., *Southeast Asia: Realm of Contrasts*. Boulder, Col.: Westview Press. Pp. 20–35.

Karl, Rebecca.
    1993.   "Global Connections: Liang Qichao and the 'Second World' at the Turn of the Twentieth Century." *Duke University Working Papers in Asian/Pacific Studies*. Asian/Pacific Institute, Duke University.

    1995.   "Secret Sharers: Chinese Nationalism and the Non-Western World at the Turn of the Twentieth Century." Ph.D. dissertation, Duke University.

Katzner, Kenneth.
    1995.   *The Languages of the World*. London: Routledge.

Kautilya.
    1915 [circa 300 B.C.E.].
        Arthasastra. Translated by R. Shamasastry. Mysore: Mysore Printing and Publishing.

Keane, A. H.
    1896.   *Asia*. London: Stanford.

Kennan, George F.
    1993.   "The Balkans Crisis: 1913 and 1993." *New York Review of Books* 40, no. 13 (15 July): 1–7.

Kennedy, Paul.
  1987.   *The Rise and Fall of the Great Powers: Economic Change and Military Conflict from 1500 to 2000.* New York: Vintage.
Keyes, Charles F.
  1977.   *The Golden Peninsula: Culture and Adaptation in Mainland Southeast Asia.* New York: Macmillan.
Khazanov, Anatoly M.
  1994 [1983].
           Nomads and the Outside World. Madison: University of Wisconsin Press.
Khodarkovsky, Michael.
  1992.   *Where Two Worlds Met: The Russian State and the Kalmyk Nomads, 1600–1771.* Ithaca, N.Y.: Cornell University Press.
Kidron, Michael, and Ronald Segal.
  1984.   *The New State of the World Atlas.* New York: Simon and Schuster.
Kiernan, V. G.
  1980.   "Europe in the Colonial Mirror." *History of European Ideas* 1:39–61.
Kimble, George H. T.
  1938.   *Geography in the Middle Ages.* New York: Russell and Russell.
Kinder, Hermann, and Werner Hilgemann.
  1974.   *The Anchor Atlas of World History,* vol. 1. Garden City, N.Y.: Anchor.
Kinross, J. P. D. B.
  1977.   *The Ottoman Centuries: The Rise and Fall of the Turkish Empire.* New York: Morrow Quill.
Kirkpatrick, F. A.
  1939.   *Latin America: A Brief History.* Cambridge: Cambridge University Press.
Kish, George, ed.
  1956.   *An Introduction to World Geography.* Englewood Cliffs, N.J.: Prentice-Hall.
Kitchen, Thomas.
  1771.   *Atlas of the World.* London.
  1773.   *General Atlas.* London: Sayer.
Kobe City Museum, ed.
  1989.   *Akioka Kochizu Korekushon Meihinten* [an exhibition of notable items from the Akioka small map collection]. Kobe: Shiritsu Hakubutsukan.
Kogawa, Tetsuo.
  1988.   "New Trends in Japanese Popular Culture." In Gavan McCormack and Yoshio Sugimoto, eds., *The Japanese Trajectory: Modernization and Beyond.* Cambridge: Cambridge University Press. Pp. 54–66.
Kohn, Hans.
  1934.   *Orient and Occident.* New York: Day.
  1960a.   *The Mind of Germany.* New York: Harper and Row.
  1960b.   *Pan-Slavism: Its History and Ideology.* New York: Vintage.

Kolakowski, Leszek.
   1990.   *Modernity on Endless Trial.* Chicago: University of Chicago
            Press.
Kopf, David.
   1969.   *British Orientalism and the Bengal Renaissance.* Berkeley: University
            of California Press.
Kopytoff, Igor.
   1987.   *The African Frontier: The Reproduction of Traditional African
            Societies.* Bloomington: Indiana University Press.
Krishna, Daya.
   1988.   "Comparative Philosophy: What Is and What Ought
            to Be." In Gerald James Larson and Elliot Deutsch, eds. 1988.
            Interpreting across Boundaries: New Essays in Comparative Philosophy.
            Princeton: Princeton University Press. Pp. 71–83.
Kulke, Hermann, and Dietmar Rothermund.
   1990.   *A History of India.* London: Routledge.
Kurian, George.
   1983.   *Atlas of the Third World.* New York: Facts on File.
Kurtén, Björn.
   1971.   *The Age of Mammals.* New York: Columbia University Press.
Lach, Donald F.
   1977.   *Asia in the Making of Europe.* Vol. 2, *A Century of Wonders* (Books
            2 and 3). Chicago: University of Chicago Press.
Lamb, Alastair.
   1968.   *Asian Frontiers: Studies on a Continuing Problem.* London: Pall
            Mall Press.
Lambert, Richard.
   1984.   *Beyond Growth: The Next Stage in Language and Area Studies.*
            Washington, D.C.: Association of American Universities.
   1990.   "Blurring the Disciplinary Boundaries: Area Studies in the United
            States." *American Behavioral Scientist* 33:712–32
Lane, Charles.
   1994.   "The Tainted Sources of 'The Bell Curve.'" *New York Review
            of Books* 41 (1 December), pp. 14–19.
Lapidus, Ira M.
   1988.   *A History of Islamic Societies.* Cambridge: Cambridge University
            Press.
Larson, Gerald James.
   1988.   "Introduction: The 'Age-Old Disjunction between the Self and
            the Other.' " In Gerald James Larson and Elliot Deutsch, eds.,
            Interpreting across Boundaries: New Essays in Comparative Philosophy.
            Princeton: Princeton University Press. Pp. 3–18.
Lasker, Bruno.
   1944.   *Peoples of Southeast Asia.* New York: Knopf.
Lattimore, Owen.
   1953.   "The New Political Geography of Inner Asia." *Geographical
            Journal* 119:17–32.

1988 [1940].
    Inner Asian Frontiers of China. Oxford: Oxford University Press.
Lawrence, Bruce B.
  1989.   *Defenders of God: The Fundamentalist Revolt against the Modern Age.* New York: Harper and Row.
  1993.   "Enough Said: Trying to Build Cultural Bridges Instead of Shoring Up Ideological Walls." Paper presented at Colloque sur la Croisement des Cultures: Monde Arabe-USA, Marrakesh, 15 April.
Ledyard, Gari.
  1994.   "Cartography in Korea." In J. B. Harley and David Woodward, eds., *The History of Cartography.* Vol. 2, Book 2, *Cartography in the Traditional East and Southeast Asian Societies.* Chicago: University of Chicago Press. Pp. 235–345.
Lee, Ki-baik.
  1984.   *A New History of Korea.* Cambridge: Harvard University Press.
Leedy, Loreen.
  1992.   *Blast Off to Earth: A Look at Geography.* New York: Holiday House.
Lefkowitz, Mary.
  1996.   *Not Out of Africa: How Afrocentrism Became an Excuse to Teach Myth as History.* New York: Basic Books.
Legge, J. D.
  1992.   "The Writing of Southeast Asian History." In Nicholas Tarling, ed., *The Cambridge History of Southeast Asia.* Vol. 1, *From Early Times to c. 1800.* Cambridge: Cambridge University Press. Pp. 1–54.
Levin, M. G. and L. P. Potapov, eds.
  1964.   *The Peoples of Siberia.* Chicago: University of Chicago Press.
Lewis, Archibald.
  1988.   *Nomads and Crusaders 1000–1368.* Bloomington: Indiana University Press.
Lewis, Bernard.
  1982.   *The Muslim Discovery of Europe.* New York: Norton.
  1990.   *Race and Slavery in the Middle East: An Historical Enquiry.* Oxford: Oxford University Press.
1993a. [1950].
    The Arabs in History. Oxford: Oxford University Press.
  1993b.   *Islam and the West.* Oxford: Oxford University Press.
Lewis, Bernard, and P. M. Holt, eds.
  1962.   *Historians of the Middle East.* London: Oxford University Press.
Lewis, Clinton, and J. D. Campbell, eds.
  1951.   *The Oxford Atlas.* Oxford: Oxford University Press.
Lewis, Martin W.
  1989.   "Comercialization and Community Life: The Geography of Exchange in a Small-Scale Society." *Annals of the Association of American Geographers* 79:390–410.
  1991.   "Elusive Societies: A Regional-Cartographical Approach to the

Study of Human Relatedness." *Annals of the Association of American Geographers* 81:605–26.

1992a.   *Green Delusions: An Environmentalist Critique of Radical Environmentalism.* Durham, N.C.: Duke University Press.

1992b.   *Wagering the Land: Ritual, Capital, and Environmental Degradation in the Cordillera of Northern Luzon 1900–1986.* Berkeley and Los Angeles: University of California Press.

1996.   "Radical Environmental Philosophy and the Assault on Reason." In Paul Gross, Norman Levitt, and Martin W. Lewis, eds., *The Flight from Science and Reason. Annals of the New York Academy of Sciences* 775:209–30.

Lewis, Pierce.

1992.   "Introducing a Cartographic Masterpiece." *Annals of the Association of American Geographers* 82:289–304.

Leyser, K. J.

1992.   "Concepts of Europe in the Early and High Middle Ages." *Past and Present* 137:25–47.

Lieberman, Victor.

1993a.   "Abu-Lughod's Egalitarian World Order: A Review Article." *Comparative Studies in Society and History* 35:544–50.

1993b.   "Local Integration and Eurasian Analogies: Structuring Southeast Asian History, c. 1350–c. 1830." *Modern Asian Studies* 27:475–572.

1995.   "An Age of Commerce in Southeast Asia? Problems of Regional Coherence." *Journal of Asian Studies* 54:796–807.

Lincoln, W. Bruce.

1994.   *The Conquest of a Continent: Siberia and the Russians.* New York: Random House.

Lindberg, David C.

1992.   *The Beginnings of Western Science: The European Scientific Tradition in Philosophical, Religious, and Institutional Context, 600 b.c. to a.d. 1450.* Chicago: University of Chicago Press.

Lindberg, David C., and Robert S. Westman, eds.

1990.   *Reappraisals of the Scientific Revolution.* Cambridge: Cambridge University Press.

Linnaeus, Carolus [von Linné, Karl].

1735.   *Systema Naturae.* Facsimile of First Edition. Nieukoop: De Graaf.

Livingston, Paisley.

1988.   *Literary Knowledge: Humanistic Inquiry and the Philosophy of Science.* Ithaca, N.Y.: Cornell University Press.

Livingstone, David N.

1992.   *The Geographical Tradition: Episodes in the History of a Contested Tradition.* Oxford: Blackwell.

L. L. Poates and Co. Complete Atlas of the World.

1912.   New York: Poates.

Longworth, Philip.

1994.   *The Making of Eastern Europe.* New York: St. Martin's Press.

Longxi, Zhang.
    1988.   "The Myth of the Other: China in the Eyes of the West." *Critical
             Inquiry* 15:108–31.
Lyde, Lionel.
    1926.   *The Continent of Europe.* London: Macmillan.
MacFadden, Clifford; Henry Kendall; and George Deasy.
    1946.   *Atlas of World Affairs.* New York: Crowell.
MacIntyre, Alasdair.
    1988.   *Whose Justice? Which Rationality?* Notre Dame, Ind.: University
             of Notre Dame Press.
Mackinder, Halford J.
    1904.   "The Geographical Pivot of History." *Geographical Journal*
             23:421–37.
Mahan, Alfred T.
    1968 [1902].
             Retrospect and Prospect. Port Washington, N.Y.: Kennikat Press.
             Originally published as an article in *National Review.*
Malcom, Howard.
    1845.   *Travels in South-Eastern Asia.* Boston: Gould, Kendall, and
             Lincoln.
Malleret, L.
    1961.   "The Position of Historical Studies in the Countries of Former
             French Indo-China in 1956." In D. G. E. Hall, ed., *Historians of
             South East Asia.* London: Oxford University Press. Pp. 301–12.
Maloney, Clarence.
    1974.   *Peoples of South Asia.* New York: Holt, Rinehart, and Winston.
Malte-Brun, M.
    1827.   *Universal Geography,* 6 vols. Translated anonymously. Philadelphia:
             Finley.
Manguin, Pierre-Yves.
    1993.   "The Vanishing *Jong:* Insular Southeast Asian Fleets in Trade
             and War (Fifteenth to Seventeenth Centuries)." In Anthony Reid,
             ed., *Southeast Asia in the Early Modern Era: Trade, Power and Belief.*
             Ithaca, N.Y.: Cornell University Press. Pp. 197–213.
Manicas, Peter T.
    1987.   *A History and Philosophy of the Social Sciences.* Oxford: Blackwell.
Mann, Michael.
    1986.   *The Sources of Social Power.* Vol. 1, *A History of Power from
             the Beginning to a.d. 1760.* Cambridge: Cambridge University
             Press.
Mann, Thomas.
    1983 [1918].
             Reflections of a Nonpolitical Man. Translated by Walter D. Morris.
             New York: Ungar.
Manning, Patrick.
    1990.   *Slavery and African Life: Occidental, Oriental, and African Slave
             Trades.* Cambridge: Cambridge University Press.

March, Andrew.
    1974.   *The Idea of China*. New York: Praeger.
Marcus, Harold G.
    1994.   *A History of Ethiopia*. Berkeley: University of California Press.
Marshall, P. J. and Glyndwr Williams.
    1982.   *The Great Map of Mankind: Perceptions of New Worlds in the Age of Enlightenment*. Cambridge: Harvard University Press.
Martin, Lawrence.
    1944.   "The Miscalled Middle East." *Geographical Review* 34:335–36.
Masica, Colin P.
    1976.   *Defining a Linguistic Area: South Asia*. Chicago: University of Chicago Press.
Mason, Jim.
    1993.   *An Unnatural Order: Uncovering the Roots of Our Domination of Nature and Each Other*. New York: Simon and Schuster.
Maunder, Samuel.
    1854.   *The History of the World*. New York: Hill.
Mauro, Frédéric.
    1990.   "Merchant Communities." In James D. Tracy, ed., *The Rise of Merchant Empires: Long-Distance Trade in the Early Modern World, 1350–1750*. Cambridge, Cambridge University Press. Pp. 255–86.
Mazzaoui, Michael.
    1991.   "Islamic Culture and Literature in Iran and Central Asia in the Early Modern Period." In Robert L. Canfield, ed., *Turko-Persia in Historical Perspective*. Cambridge: Cambridge University Press. Pp. 78–103.
McCormick, Barrett L., and David Kelly.
    1994.   "The Limits of Anti-Liberalism." *Journal of Asian Studies,* 53:804–31.
McCoy, Alfred W., ed.
    1980.   *Southeast Asia under Japanese Occupation*. New Haven: Yale University Southeast Asia Studies Monograph Series, no. 22.
McDowell, Linda.
    1993.   "Space, Place, and Gender Relations: Part II. Identity, Difference, Feminist Geometries and Geographies." *Progress in Human Geography* 17:305–18.
McElligott, Anthony.
    1994.   "Reforging Mitteleuropa in the Crucible of War: The Economic Impact of Integration under German Hegemony." In Peter Stirk, ed., *Mitteleuropa: History and Prospects*. Edinburgh: Edinburgh University Press. Pp. 129–59.
McEvedy, Colin.
    1972.   *The Penguin Atlas of Modern History*. New York: Penguin.
    1992.   *The New Penguin Atlas of Medieval History*. London: Penguin.
    1995.   *The Penguin Atlas of African History*. London: Penguin.
McMullin, Neil.
    1994.   "Communication to the Editor." *Journal of Asian Studies* 52:676–78.

McMurry, F. M., and A. E. Parkins.
 1921.   *Advanced Geography*. New York: Macmillan.
McNeill, William.
 1963.   *The Rise of the West: A History of the Human Community*. Chicago: University of Chicago Press.
 1974.   *The Shape of European History*. New York: Oxford University Press.
 1976.   *Plagues and Peoples*. New York: Anchor.
 1982.   *The Pursuit of Power: Technology, Armed Force, and Society Since a.d. 1000*. Chicago: University of Chicago Press.
 1986.   *Mythistory and Other Essays*. Chicago: University of Chicago Press.
 1990.   "*The Rise of the West* after Twenty-five Years." *Journal of World History* 1:1–21.
 1993.   "Foreword." In Andre Gunder Frank, and Barry K. Gills, eds., *The World System: Five Hundred or Five Thousand Years?* London: Routledge. Pp. vii–xiii.
 1995.   "The Changing Shape of World History." *History and Theory* 34:8–26.
Meinig, Donald W.
 1956.   "Cultural Blocks and Political Blocks: Emergent Patterns in World Affairs." *Western Humanities Review* 10:203–22.
 1969.   *Imperial Texas: An Interpretive Essay in Cultural Geography*. Austin: University of Texas Press.
Melko, Matthew, and Leighton R. Scott.
 1987.   *The Boundaries of Civilizations in Space and Time*. New York: University Press of America.
Mende, Tibor.
 1955.   *South-East Asia between Two Worlds*. London: Turnstile Press.
Merchant, Carolyn.
 1992.   *Radical Ecology: The Search for a Livable World*. London: Routledge.
Metcalf, Barbara D.
 1995.   "Too Little and Too Much: Reflections on Muslims in the History of India." *Journal of Asian Studies* 54:951–67.
Meyer, Henry Cord.
 1946.   "Mitteleuropa in German Political Geography." *Annals of the Association of American Geographers* 37:178–94.
Michalak, Wieslaw, and Richard Gibb.
 1992.   "Political Geography and Eastern Europe." *Area* 24:341–49.
Mignolo, Walter.
 1993.   "Misunderstanding and Colonization: The Reconfiguration of Memory and Space." *South Atlantic Quarterly* 92:209–60.
Mikesell, Marvin.
 1983.   "The Myth of the Nation-State." *Journal of Geography* 82:257–60.
Mill, Hugh Robert, ed.
 1922.   *The International Geography*. New York: Appleton.

Mill, John Stuart.
    1989 [1859].
        On Liberty (with the Subjection of Women and Chapters on Socialism).
        Cambridge: Cambridge University Press.
Millward, James A.
    1994.   "A Uyghur Muslim in Qianlong's Court: The Meaning
        of the Fragrant Concubine." *Journal of Asian Studies*
        53:427–58.
Minshull, R.
    1967.   *Regional Geography: Theory and Practice*. London: Hutchinson
        University Library.
Miquel, Andre.
    1967.   *La Géographie humaine du monde musulman jusqu'au milieu IIe*
        siècle. Paris: Mouton.
Mitchell, S. August.
    1849.   *A New Universal Atlas of the World*. Philadelphia: Mitchell.
Mitchell, Timothy.
    1988.   *Colonizing Egypt*. Berkeley and Los Angeles: University of
        California Press.
    1992.   "Orientalism and the Exhibitionary Order." In Nicholas B. Dirks,
        ed., *Colonialism and Culture*. Ann Arbor: University of Michigan
        Press. Pp. 289–317.
Mitsukuri Shogo.
    1845.   *Konyo Zushiki Ho* [Supplement to an Illustrated World Geogra-
        phy]. Woodblock original in Harvard-Yenching Library.
Miyoshi, Masao.
    1991.   *Off Center: Power and Culture Relations between Japan and the*
        United States. Cambridge: Harvard University Press.
Montagu, Ashley, ed.
    1964.   *The Concept of Race*. New York: Collier.
Montesquieu, Charles-Louis De Secondat, Baron de
    1949 [1748].
        The Spirit of the Laws, 2 vols. Translated by Thomas Nugent. New
        York: Hafner.
Moore, Barrington.
    1966.   *Social Origins of Dictatorship and Democracy: Lord and Peasant*
        in the Making of the Modern World. Boston: Beacon.
Moran, Warren.
    1993.   "The Wine Appellation as Territory in France and California."
        Annals of the Association of American Geographers 83:694–717.
Morgan, Theodore, and Nyle Spoelstra.
    1969.   *Economic Interdependence in Southeast Asia*. Madison: University
        of Wisconsin Press.
Morris, John. W., ed.
    1972.   *World Geography*. New York: McGraw-Hill.
Morton, H. V.
    1941.   *Middle East*. New York: Dodd, Mead.

Mudimbe, V. Y.
    1988.   *The Invention of Africa: Gnosis, Philosophy, and the Order of Knowledge.* Bloomington: Indiana University Press.

Murphey, Rhoads.
    1992.   *A History of Asia.* New York: HarperCollins.

Murphy, Alexander B.
    1991.   "Regions as Social Constructs: The Gap between Theory and Practice." *Progress in Human Geography* 15:22–35.
    1995.   "Economic Regionalization and Pacific Asia." *Geographical Review* 85:127–40.

Mutton, Alice F.
    1961.   *Central Europe: A Regional and Human Geography.* London: Longmans.

Myers, Ramon H., and Mark R. Peattie, eds.
    1984.   *The Japanese Colonial Empire 1895–1945.* Princeton: Princeton University Press.

Nakamura, Hajime.
    1960.   *The Ways of Thinking of Eastern Peoples.* Tokyo: Japanese Ministry of Education.
    1988.   "The Meaning of the Terms 'Philosophy' and 'Religion' in Various Traditions." In Gerald James Larson and Elliot Deutsch, eds., *Interpreting across Boundaries: New Essays in Comparative Philosophy.* Princeton: Princeton University Press. Pp. 137–51.

Nanda, Meera.
    1991.   "Is Modern Science a Western, Patriarchal Myth? A Critique of Populist Orthodoxy." *South Asia Bulletin* 11:32–61.
    Forthcoming.
        "The Epistemic Charity of Sociology of Scientific Knowledge and Why the Third World Must Decline the Offer." In Noretta Kortege, ed., *A House Built on Sand.* New York: Oxford University Press.

Narla, V. R.
    1981.   *East and West: Myth of Dichotomy.* Nagarjunanagar, India: Nagarjuna University Press.

Nash, Roderick F.
    1989.   *The Rights of Nature: A History of Environmental Ethics.* Madison: University of Wisconsin Press.

National Geographic Atlas of the World, Fifth Edition.
    1981.   Washington, D.C.: National Geographic.

Needham, Joseph.
    1954–84. *Science and Civilization in China,* 6 vols. Cambridge: Cambridge University Press.

Nehru, Jawaharlal.
    1942 [1934].
        Glimpses of World History. New York: Day.

Neill, W.
    1969.   *The Geography of Life.* New York: Columbia University Press.

The New Oxford Atlas.
   1978.   Oxford: Oxford University Press.
The New York Times Atlas of the World.
   1980.   New York: Times Books.
Nijim, Basheer K.
   1992.   "Southwest Asia-North Africa." In George F. Hepner and Jesse
            O. McKee, eds., *World Regional Geography: A Global Approach.* St.
            Paul: West. Pp. 426–79.
Nordholt, Jan Willem Schulte.
   1995.   *The Myth of the West: America as Last Empire.* Translated by
            Herbert Rowen. Grand Rapids, Mich.: Eerdmans.
Norris, Robert E.
   1990.   *World Regional Geography.* St. Paul: West.
Northrop, F. S. C.
   1960.   *The Meeting of East and West.* New York: Macmillan.
O'Connor, Richard A.
   1995.   "Agricultural Change and Ethnic Succession in Southeast Asian
            States: A Case for Regional Anthropology." *Journal of Asian
            Studies* 54:968–96.
O'Gorman, Edmundo.
   1961.   *The Invention of America: An Inquiry into the Historical Nature
            of the New World and the Meaning of Its History.* Bloomington:
            Indiana University Press.
O'Hanlon, Rosalind, and David Washbrook.
   1992.   "After Orientalism: Culture, Criticism, and Politics in the Third
            World." *Comparative Studies in Society and History* 34:141–67.
Okey, Robin.
   1992.   "Central Europe/Eastern Europe: Behind the Definitions."
            Past and Present 137:102–33.
O'Leary, Brendan.
   1989.   *The Asiatic Mode of Production: Oriental Despotism, Historical
            Materialism and Indian History.* Oxford: Blackwell.
Oliver, Roland, and Anthony Atmore.
   1981.   *The African Middle Ages 1400–1800.* Cambridge: Cambridge
            University Press.
O'Loughlin, John, and Herman van der Wusten.
   1990.   "Political Geography of Panregions." *Geographical Review*
            80:1–20.
Ortelius, Abrahamus.
   1570.   *Theatrum Orbis Terrarum.* Antwerp.
Ó Tuathail, Gearóid, and Timothy W. Luke.
   1994.   "Present at the (Dis)integration: Deterritorialization and Reterri-
            torialization in the New Wor(l)d Order." *Annals of the Association
            of American Scholars* 84:381–98.
Oxford English Dictionary, Compact Edition.
   1971 [1884–1928].
            Oxford: Oxford University Press.

Painter, Joe.
  1995.   *Politics, Geography, and "Political Geography."* London: Arnold.
Palairet, John.
  1775.   *Bowles' Universal Atlas.* London: Bowles.
Palmer, R. R. and Joel Colton.
  1971.   *A History of the Modern World.* Fourth Edition. New York: Knopf.
Pan, Lynn.
  1990.   *Sons of the Yellow Emperor: A History of the Chinese Diaspora.* New York: Kodansha International.
Panikkar, K. M.
  1969.   *Asia and Western Dominance.* New York: Collier.
Pankhurst, Richard.
  1992.   *A Social History of Ethiopia: The Northern and Central Highlands from Early Medieval Times to the Rise of Emperor Tewodros II.* Trenton, N.J.: Red Sea Press.
Parker, Geoffrey.
  1988.   *The Military Revolution: Military Innovation and the Rise of the West, 1500–1800.* Cambridge: Cambridge University Press.
Parker, W. H.
  1960.   "Europe: How Far?" *Geographical Journal* 126:278–97.
Parkinson, C. Northcote.
  1963.   *East and West.* Boston: Houghton Mifflin.
Parmelee, Maurice.
  1929.   *Oriental and Occidental Culture: An Interpretation.* London: Williams and Norgate.
Patai, Raphael.
  1952.   "The Middle East as a Culture Area." *Middle East Journal* 6:1–21.
  1962.   *Golden River to Golden Road: Society, Culture, and Change in the Middle East.* Philadelphia: University of Pennsylvania Press
Patten, William, and J. E. Homans.
  1910.   *New Encyclopedic Atlas and Gazetteer of the World.* New York: Collier and Sons.
Pearce, E. A., and C. G. Smith.
  1984.   *World Weather Guide.* New York: Times Books.
Peet, Richard.
  1991a.   "The End of History . . . Or Its Beginning?" *Professional Geographer* 43:512–19.
  1991b.   *Global Capitalism: Theories of Societal Development.* London and New York: Routledge.
Perlin, Frank.
  1983.   "Proto-industrialization and Pre-colonial South Asia." *Past and Present* 98:30–95
Perry, W. J.
  1918.   *The Megalithic Culture of Indonesia.* Manchester: Manchester University Press.
Peters Atlas of the World.
  1990.   New York: Harper and Row.

Phelan, John Leddy.
    1959.   *The Hispanization of the Philippines.* Madison: University of
                Wisconsin Press.
Phillips, Seymour.
    1994.   "Outer World of the European Middle Ages." In Stuart B. Schwartz,
                ed., *Implicit Understandings: Observing, Reporting, and Reflecting
                on the Encounters between Europeans and Other Peoples in the Early
                Modern Era.* Cambridge: Cambridge University Press. Pp. 23–63.
Pierce, Leslie P.
    1993.   *The Imperial Harem: Women and Sovereignty in the Ottoman
                Empire.* Oxford: Oxford University Press.
Pinker, Steven.
    1994.   *The Language Instinct.* New York: Morrow.
Pitt, Moses.
    1680.   *The English Atlas.* London: Theatre.
Platt, Robert S.
    1943.   "Regionalism in World Order." *Annals of the Association of
                American Geographers* 33:230–31.
Pletsch, Carl E.
    1981.   "The Three Worlds, or the Division of Social Scientific Labor,
                circa 1950–1975. *Comparative Studies in Society and History*
                23:565–90.
Plumwood, Val.
    1993.   *Feminism and the Mastery of Nature.* London: Routledge.
Ponting, Clive.
    1991.   *A Green History of the World: The Environment and the Collapse of
                Great Civilizations.* New York: Penguin.
Porter, Michael E.
    1990.   *The Competitive Advantage of Nations.* New York: Macmillan.
Poulton, Hugh.
    1991.   *The Balkans: Minorities and States in Conflict.* London: Minority
                Rights Publications.
Prakash, Gyan.
    1990.   "Writing Post-Orientalist Histories of the Third World: Perspec-
                tives from Indian Historiography." *Comparative Studies in Society
                and History* 32:383–408.
Pratt, Mary Louise.
    1992.   *Imperial Eyes: Travel Writing and Transculturation.* London:
                Routledge.
Price, Marie.
    1996.   "Competing Visions, Shifting Boundaries: The Construction of
                Latin America as a World Region." Paper presented at the Annual
                Meeting of the Association of American Geographers, Charlotte,
                North Carolina, 12 April.
Price, Marie, and Martin Lewis.
    1993.   "The Reinvention of Cultural Geography." *Annals of the Associa-
                tion of American Geographers* 83:1–17.

Pryor, John H.
1988.   *Geography, Technology, and War: Studies in the Maritime History of the Mediterranean, 649–1571.* Cambridge: Cambridge University Press.

Ptolemy, Claudius.
1932 [circa 135 C.E.].
Geography of Claudius Ptolemy. Translated by E. C. Stevenson. New York: New York Public Library.

Pudup, M. B.
1988.   "Arguments within Regional Geography." *Progress in Human Geography* 12:369–90.

Purcell, Victor.
1965.   *South and East Asia since 1800.* Cambridge: Cambridge University Press.

Pye, Lucian.
1967.   *Southeast Asia's Political Systems.* Englewood Cliffs, N.J.: Prentice-Hall.

Radhakrishnan, S.
1956.   *East and West.* New York: Harper and Brothers.

Rand McNally.
1881.   *Indexed Atlas of the World.* Chicago: Rand McNally.
1932.   *The Rand McNally World Atlas: International Edition.* Chicago: Rand McNally.
1984 and 1994.
New International Atlas. Two editions. Chicago: Rand McNally.
1991–92. *Educational Publishing Catalog.* Skokie, Ill.: Rand McNally.
1993a.   *Atlas of World History.* Chicago: Rand McNally.
1993b.   *Historical Atlas and Guide.* N.p.: Rand McNally.

Ranke, Leopold von.
1885.   *Universal History: The Oldest Historical Group of Nations and the Greeks.* Edited and translated by G. W. Prothero and D. C. Tovey. New York: Harper Brothers.

Ratzel, Friedrich.
1896.   *The History of Mankind.* Translated from the second German edition by A. J. Butler. New York: Macmillan.

Ravenhill, John.
1990.   "The North South Balance of Power." *International Affairs* 66:731–48.

Raychaudhuri, Tapan.
1992.   "Europe in India's Xenology: The Nineteenth Century Record." *Past and Present* 137:157–82.

Reader's Digest World Atlas.
1992.   Pleasantville, N.Y.: Reader's Digest Association.

Reclus, Élisée.
1891.   *The Earth and Its Inhabitants,* 19 vols. New York: Appleton.

Reid, Anthony.
1988.   *Southeast Asia in the Age of Commerce 1450–1680.* Vol. 1, *The Lands below the Winds.* New Haven: Yale University Press.

1994.   "Early Southeast Asian Categorization of Europeans." In Stuart
         B. Schwartz, ed., *Implicit Understandings: Observing, Reporting,
         and Reflecting on the Encounters between Europeans and Other Peoples
         in the Early Modern Era.* Cambridge: Cambridge University Press.
         Pp. 268–94.
ed. 1993. *Southeast Asia in the Early Modern Era: Trade, Power and Belief.*
         Ithaca, N.Y.: Cornell University Press.
Reither, Joseph.
1973.   *World History: A Brief Introduction.* New York: McGraw-Hill.
Reynolds, Craig. J.
1995.   "A New Look at Old Southeast Asia." *Journal of Asian Studies*
         54:419–46.
Richards, John F.
1993.   *The Mughal Empire.* Cambridge: Cambridge University Press.
1995.   "In Defense of Area Studies." Global Forum Series Occasional
         Papers, No. 95–01. Center for International Studies, Duke
         University.
Ridker, Ronald G., ed.
1976.   *Changing Resource Problems of the Fourth World.* Washington,
         D.C.: Resources for the Future.
Rippy, J. Fred.
1958.   *Latin America: A Modern History.* Ann Arbor: University of
         Michigan Press.
Ritter, Carl.
1863.   *Geographical Studies.* Translated and edited by W. L. Gage. Boston:
         Gould and Little.
1864.   *Comparative Geography.* Translated by William Gase. New York:
         American Book Company.
Rivlin, Benjamin, and Joseph Szyliowicz, eds.
1965.   *The Contemporary Middle East: Tradition and Innovation.* New
         York: Random House.
Robert de Vaugondy, Gilles, and Didier Robert de Vaugondy.
1798.   *Atlas Universal.* Paris: Delamarche.
Roberts, J. M.
1993.   *History of the World.* New York: Oxford University Press.
Robertson, William S.
1922.   *A History of the Latin-American Nations.* New York: Appleton.
Robinson, Francis.
1991.   "Perso-Islamic Culture in India from the Seventeenth to the Early
         Twentieth Centuries. In Robert L. Canfield, ed., *Turko-Persia in
         Historical Perspective.* Cambridge: Cambridge University Press.
         Pp. 104–31.
Rorlich, Azade-Ayse.
1986.   *The Volga Tatars: A Profile in National Resilience.* Stanford: Hoover
         Institution Press.
Ross, Andrew.
1992.   "New Age Technoculture." In L. Grossberg, C. Nelson, and P.
         Treichler, eds., *Cultural Studies.* London: Routledge. Pp. 531–55.

Rossabi, Ralph A.
  1990.   "The 'Decline' of the Central Asian Caravan Trade." In James D.
           Tracy, ed., *The Rise of Merchant Empires: Long-Distance Trade in the
           Early Modern World, 1350–1750*. Cambridge: Cambridge University
           Press. Pp. 351–70.
Rougier, Louis.
  1971.   *The Genius of the West*. Los Angeles: Nash.
Rowe, William T.
  1984.   *Hankow: Commerce and Society in a Chinese City, 1796–1889*.
           Stanford: Stanford University Press.
Roy, Oliver.
  1992.   "Ethnic Identity and Political Expression in Northern
           Afghanistan." In Jo-Ann Gross, ed., *Muslims in Central Asia*.
           Durham, N.C.: Duke University Press. Pp. 73–86.
Rozman, Gilbert, ed.
  1991.   *The East Asian Region: Confucian Heritage and Its Modern
           Adaptation*. Princeton: Princeton University Press.
Rupnik, Jacques.
  1991.   "Central Europe or Mitteleuropa." In Stephen R. Graubard, ed.,
           Eastern Europe . . . Central Europe . . . Europe. Boulder, Col.:
           Westview Press. Pp. 233–65.
Russell, Richard J; Fred B. Kniffen; and Evelyn L. Pruitt.
  1951.   *Culture Worlds*. New York: Macmillan.
Sack, Robert David.
  1986.   *Human Territoriality: Its Theory and History*. Cambridge: Cam-
           bridge University Press.
Sahlins, Marshall.
  1976.   *Culture and Practical Reason*. Chicago: University of Chicago
           Press.
Sahlins, Peter.
  1989.   *Boundaries: The Making of France and Spain in the Pyrenees*.
           Berkeley and Los Angeles: University of California Press.
Said, Edward W.
  1978.   *Orientalism*. New York: Pantheon.
  1985.   "Orientalism Reconsidered." In Francis Barker, Peter Hulme,
           Margaret Iversen, and Diana Loxley, eds., *Europe and Its Others*,
           vol. 1. Colchester, U.K.: University of Essex. Pp. 14–27
Sanchez, Luis A.
  1945.   *Existe América Latina?* Mexico City: Fondo de Cultura Economica.
Sanson, Nicolas.
  1674.   *Mappe-Monde: Geo-Hydrographique, ou Description generale du globe*.
           Paris.
Sara, Har Iqbal Singh.
  1983.   "The Super-Occidental Nature of Sikh Religion." *Journal of Sikh
           Studies* 10:3–36.
Sarda, Har Bilas.
  1906.   *Hindu Superiority: An Attempt to Determine the Position of the
           Hindu Race in the Scale of Nations*. Ajmer, India: Rajputana.

Sardesai, D. R.
    1981.   *Southeast Asia: Past and Present.* New Delhi: Vikas.
Savage, Victor; Lily Kong; and Brenda S. A. Yeoh.
    1993.   "The Human Geography of Southeast Asia: An Analysis of Post-War
             Developments." *Singapore Journal of Tropical Geography* 14:229–50.
Schevill, Ferdinand.
    1991 [1918].
             A History of the Balkans from the Earliest Times to the Present Day.
             New York: Dorset Press.
Schöpflin, George.
    1991.   "The Political Traditions of Eastern Europe." In Stephen R.
             Graubard, ed., *Eastern Europe . . . Central Europe . . . Europe.*
             Boulder, Col.: Westview Press. Pp. 59–94.
Schurz, William L.
    1963.   *Latin America: A Descriptive Survey.* New York: Dutton.
Schwab, Raymond.
    1984.   *The Oriental Renaissance: Europe's Rediscovery of India and the East
             1680–1880.* New York: Columbia University Press.
Schwartzberg, Joseph E.
    1992.   "Geographical Mapping." In J. B. Harley and David Woodward,
             eds., *The History of Cartography.* Vol. 2, Book 2, *Cartography in
             the Traditional Islamic and South Asian Societies.* Chicago: Univer-
             sity of Chicago Press. Pp. 388–493.
    1994.   "Cosmography in Southeast Asia." In J. B. Harley and David
             Woodward, eds., *The History of Cartography.* Vol. 2, Book 2,
             Cartography in the Traditional East and Southeast Asian Societies.
             Chicago: University of Chicago Press. Pp. 689–700.
The Scribner-Black Atlas of the World.
    1890.   New York: Scribner's Sons.
Semple, Ellen Churchill.
    1911.   *Influences of Geographic Environment.* New York: Holt.
Shaffer, Lynda.
    1994.   "Southernization." *Journal of World History* 5:1–22.
Shannon, Thomas Richard.
    1989.   *An Introduction to the World-System Perspective.* Boulder, Col.:
             Westview Press.
Shiozu Kan'ichiro.
    1875.   *Sekai Chizu Ryakusetsu* [A Short Explication of the Map of the
             World]. Tokyo: Takeoka Bun.
Simmons, I. G.
    1993.   *Interpreting Nature: Cultural Constructions of the Environment.*
             New York: Routledge.
Singer, Kurt.
    1973.   *Mirror, Sword and Jewel: The Geometry of Japanese Life.* Tokyo:
             Kodansha.
Sinnhuber, Karl A.
    1954.   "Central Europe—Mitteleuropa—Europe Centrale." *Institute of
             British Geographers, Transactions and Papers* 20:15–37.

Sinor, Denis, ed.
   1970 [1954].
      Orientalism and History. Bloomington: Indiana University Press.
   ed. 1990.   *The Cambridge History of Early Inner Asia.* Cambridge: Cambridge
      University Press.
Sion, Jules.
   1929.   *Asie Occidental, La Haute Asie, Asie des Moussons.* (Vols. 8 and 9 of
      Géographie Universelle, edited by P. Vidal de la Blache and L.
      Gallois.) Paris: Colin.
Slater, David.
   1995.   "Trajectories of Development Theory: Capitalism, Socialism, and
      Beyond." In R. J. Johnston, Peter J. Taylor, and Michael J. Watts,
      eds. 1995. *Geographies of Global Change: Remapping the World in the
      Late Twentieth Century.* Oxford: Blackwell. Pp. 63–81.
Smith, Anthony D.
   1986.   *The Ethnic Origins of Nations.* Oxford: Blackwell.
Smith, Benjamin.
   1899.   *The Century Atlas of the World.* New York: Century.
Smith, J. R., and C. B. Smith.
   1988.   *Essentials of World History.* Hauppauge, N.Y.: Barron's Educa-
      tional.
Smith, Jonathan M.
   1996.   "Geographical Rhetoric: Modes and Tropes of Appeal." *Annals of
      the Association of American Geographers* 86:1–20.
Smith, Neil.
   1987.   "'Academic War over the Field of Geography': The Elimination of
      Geography at Harvard, 1947–1951." *Annals of the Association of
      American Geographers* 77:155–72.
   1992.   "History and Philosophy of Geography: Real Wars, Theory
      Wars." *Progress in Human Geography* 16:257–71.
   1994.   "Geography, Empire and Social Theory." *Progress in Human
      Geography* 18:491–500.
Snowdon, Frank M.
   1983.   *Before Color Prejudice: The Ancient View of Blacks.* Cambridge:
      Harvard University Press.
Soja, Edward.
   1989.   *Postmodern Geographies: The Reassertion of Space in Critical Social
      Theory.* London: Verso.
Solheim, Wilhelm G., II.
   1985.   "'Southeast Asia': What's in a Name, Another Point of View."
      Journal of Southeast Asian Studies 16: 141–47.
Som, Tjan Tjoe.
   1953.   "The Meeting of East and West: The Oriental View." In S.
      Hofstra, ed., *Eastern and Western World.* The Hague: Van Hoeve.
      Pp. 13–23.
Sparke, Matthew.
   1994.   "White Mythologies and Anemic Geography." *Environment and
      Planning D: Society and Space* 12:105–23.

Spate, O. H. K., and A. T. A. Learmonth.
  1967.   *India and Pakistan: A General and Regional Geography.* London: Methuen.
Spence, Jonathon D.
  1990.   *The Search for Modern China.* New York. Norton.
Spencer, J. E., and William L. Thomas.
  1971.   *Asia, East by South: A Cultural Geography.* New York: Wiley and Sons.
Spengler, Oswald.
  1926.   *The Decline of the West,* 2 vols. New York: Knopf.
Spuler, Bertold.
  1994 [1969].
            The Mongol Period: A History of the Muslim World. Translated by F. Bagley. Princeton, N.J.: Wiener.
Staley, Eugene.
  1941.   "The Myth of the Continents." *Foreign Affairs,* April, pp. 1–16.
Stamp, L. Dudley, and George Kimble.
  1954.   *The World: A General Geography.* Toronto: Longmans, Green.
Stansfield, Charles, and Chester Zimolzak.
  1990.   *Global Perspectives: A World Regional Geography.* New York: Merrill.
Stavrianos, L. S.
  1975.   *The World since 1500: A Global History.* Englewood Cliffs, N.J.: Prentice-Hall.
Steadman, John M.
  1969.   *The Myth of Asia.* New York: Simon and Schuster.
Stein, Burton.
  1977.   "The Segmentary State in South Indian History." In Richard Fox, ed., *Realm and Region in Traditional India.* Durham, N.C.: Duke University Program in Comparative Studies on Southern Asia. Pp. 3–51.
  1989.   *Vijayanagara. The New Cambridge History of India* Part 1, vol. 2. Cambridge: Cambridge University Press.
Steinberg, David Joel, ed.
  1985.   *In Search of Southeast Asia: A Modern History.* Honolulu: University of Hawaii Press.
Stieler, A.
  1865.   *Schul-Atlas über alle Theile der Erde.* Gotha: Justus Perthes.
Stirk, Peter M. R.
  1994.   "The Idea of Mitteleuropa." In Peter Stirk, ed. *Mitteleuropa: History and Prospects.* Edinburgh: Edinburgh University Press. Pp. 1–35.
Stoianovich, Traian.
  1994.   *Balkan Worlds: The First and Last Europe.* Armonk, N.Y.: Sharp.
Strabo (of Amasya).
  1854 [circa 10 B.C.E.].
            The Geography. Translated by H. C. Hamilton. London: Bohn.

Strayer, Joseph E., and Hans Gatzke.
:   1984.   *The Mainstream of Civilization*. New York: Harcourt Brace Jovanovitch.
Stuart-Glennie, J. S.
:   1879.   *Europe and Asia*. London: Chapman and Hall.
Suleri, Sara.
:   1992.   *The Rhetoric of English India*. Chicago: University of Chicago Press.
Sweet, William W.
:   1919.   *A History of Latin America*. New York: Abingdon Press.
Szücs, Jenö.
:   1983.   "The Three Historical Regions of Europe." *Acta Historica Academiae Scientiarum* 19:160–80.
Tamamoto, Masaru.
:   1991.   "The Japanese Discovery of Islam: Japan's Pan-Asianism during the Great East Asian War." Paper presented at Washington/Southeast Regional Japan Seminar, Washington, D.C.
Tanaka, Stefan.
:   1993.   *Japan's Orient: Rendering Pasts into History*. Berkeley: University of California Press.
Tandon, Prakash.
:   1968.   *Punjabi Century, 1857–1947*. Berkeley: University of California Press.
Tarling, Nicholas.
:   1966.   *A Concise History of Southeast Asia*. New York: Praeger.
Tavakoli-Targhi, Mohamad.
:   1991.   "The Persian Gaze and Women of the Occident." *South Asia Bulletin* 11:21–31.
Taylor, Keith.
:   1993.   "Nguyen Hoang and the Beginning of Vietnam's Southward Expansion." In Anthony Reid, ed., *Southeast Asia in the Early Modern Era: Trade, Power and Belief*. Ithaca, N.Y.: Cornell University Press. Pp. 42–68.
Taylor, Peter J.
:   1992.   "Tribulations on Transition." *Professional Geographer* 44:10–12.
:   1994.   "The State as Container: Territoriality in the Modern World System." *Progress in Human Geography* 18:151–162.
TeBrake, William H.
:   1985.   *Medieval Frontier: Culture and Ecology in Rijnland*. College Station: Texas A&M University Press.
Terlouw, Cees P.
:   1990.   "Regions of the World System: Between the General and the Specific." In R. J. Johnston, J. Hauer, and G. A. Hoekveld, eds., *Regional Geography: Current Developments and Future Prospects*. London: Routledge. Pp. 50–66.
Thomas, Hugh.
:   1979.   *A History of the World*. New York: Harper and Row.

Thongchai, Winichakul.
>   1994.   *Siam Mapped: A History of the Geo-Body of a Nation*. Honolulu:
>   University of Hawaii Press.

Thornton, John.
>   1992.   *Africa and Africans in the Making of the Atlantic World*. Cam-
>   bridge: Cambridge University Press.

Threadgold, Warren.
>   1988.   *The Byzantine Revival 780–842*. Stanford: Stanford University
>   Press.

Thrift, Nigel.
>   1992.   "Muddling Through: World Orders and Globalization." *Profes-
>   sional Geographer* 44:3–7.
>   1993.   "For a New Regional Geography 3. " *Progress in Human Geogra-
>   phy* 17:92–100.

Thurow, Lester.
>   1992.   *Head to Head: The Coming Battle among Japan, Europe, and
>   America*. New York: Morrow.

Tilly, Charles.
>   1990.   *Coercion, Capital, and European States, ad 990–1990*. Oxford:
>   Blackwell.

Times Atlas of the World, Ninth Edition.
>   1992.   London: Times Books.

Toby, Ronald P.
>   1994.   "The 'Indianness' of Iberia and Changing Japanese Iconographies
>   of Other." In Stuart B. Schwartz, ed., *Implicit Understandings:
>   Observing, Reporting, and Reflecting on the Encounters between
>   Europeans and Other Peoples in the Early Modern Era*. Cambridge:
>   Cambridge University Press. Pp. 323–51.
>   Forthcoming.
>   *A Narcissism of Petty Differences: Identity and Alien in Early Modern
>   Japan*. Berkeley: University of California Press.

Toulmin, Stephen.
>   1990.   *Cosmopolis: The Hidden Agenda of Modernity*. New York: Free
>   Press.

Townsend, Meredith.
>   1921 [1901].
>   *Asia and Europe*. London: Constable.

Toynbee, Arnold J.
>   1934–1961.
>   *A Study of History*, 12 vols. New York: Oxford University Press.

Tozer, H. F.
>   1964 [1897].
>   *A History of Ancient Geography*. New York: Biblo and Tannen.

Trewarthwa, Glenn T.
>   1952.   "Chinese Cities: Origins and Functions." *Annals of the Association
>   of American Geographers* 42:69–93.
>   1965.   *Japan: A Geography*. Madison: University of Wisconsin Press.

Tripathi, Maya Prasad.
1969.   *Development of Geographic Knowledge in Ancient India*. Varanasi: Bharatiya Vidya Prakashan.
Tsao, Jiun-han, and Cheng-wen Tsai, eds.
1984.   *Northeast Asian and European Relations: New Dimensions and Strategies*. Taipei: Asia and the World Institute.
Tuan, Yi-Fu.
1996.   *Cosmos and Hearth: A Cosmopolite's View*. Minneapolis: University of Minnesota Press.
Turner, Bryan S.
1994.   *Orientalism, Postmodernism, and Globalism*. London: Routledge.
Unno, Kazutaka.
1994.   "Cartography in Japan." In J. B. Harley and David Woodward, eds., *The History of Cartography*. Vol. 2, Book 2, *Cartography in the Traditional East and Southeast Asian Societies*. Chicago: University of Chicago Press. Pp. 346–77.
Vadney, T. E.
1990.   "World History as an Advanced Academic Field." *Journal of World History* 1:209–23.
Vambéry, Arminius.
1906.   *Western Culture in Eastern Lands: A Comparison of the Methods Adopted by England and Russia in the Middle East*. London: Murry.
Vandenbosch, Amry.
1946.   "Regionalism in Southeast Asia." *Far Eastern Quarterly* 5:427–38.
van der Wee, Herman.
1990.   "Structural Changes in European Long-Distance Trade, and Particularly in the Re-export Trade from South to North, 1350–1750." In James D. Tracy, ed., *The Rise of Merchant Empires: Long-Distance Trade in the Early Modern World, 1350–1750*. Cambridge: Cambridge University Press. Pp. 14–33.
Van Doren, Charles.
1992.   *A History of Knowledge: Past, Present, and Future*. New York: Ballantine.
Van Loon, Hendrik Willem.
1937.   *Van Loon's Geography: The Story of the World*. Garden City, N.Y.: Garden City Publishing.
Van Paassen, C.
1957.   *The Classical Tradition of Geography*. Groningen: Wolters.
Vansina, Jan.
1990.   *Paths in the Rainforest: Toward a History of Political Tradition in Equatorial Africa*. Madison: University of Wisconsin Press.
Vidal-Naquet, Pierre, ed.
1987.   *The Harper Atlas of World History*. Translated from French edition. New York: Harper and Row.
Vlekke, B. M. H.
1953.   "The Meeting of East and West: The Western View." In S. Hofstra, ed., *Eastern and Western World*. The Hague: Van Hoeve. Pp. 27–35.

Voll, John O.
    1994.   "Islam as a Special World System." *Journal of World History*
             5:213–26.
Voltaire.
    1937 [circa 1760].
             Fragments on India. Translated by Freda Bedi. Lahore: Contempo-
             rary India Publishing.
    1963 [1756].
             Essai sur les moeurs et l'esprit des nations. Edited by René Pomeau.
             Paris: Classiques Garnier.
von Beckerath, Herbert.
    1942.   *In Defense of the West: A Political and Economic Study.* Durham,
             N.C.: Duke University Press.
Von Laue, Theodore H.
    1987.   *The World Revolution of Westernization: The Twentieth Century in
             Global Perspective.* Oxford: Oxford University Press.
Vryonis, Speros.
    1971.   *The Decline of Medieval Hellenism in Asia Minor and the Process of
             Islamization from the Eleventh through the Fifteenth Century.* Los
             Angeles: University of California Press.
Wagley, Charles.
    1948.   *Area Research and Training: A Conference Report on the Study of
             World Areas.* New York: Social Science Research Council.
Waldron, Arthur N.
    1994.   "Introduction." In Bertold Spuler, *The Mongol Period: A History of
             the Muslim World.* Princeton, N.J.: Wiener. Pp. vii–xxxiii.
Walicki, Andrzej.
    1989 [1964].
             *The Slavophile Controversy: History of a Conservative Utopia in
             Nineteenth Century Russian Thought.* Translated by Hilda Andrews-
             Rusiecka. Notre Dame, Ind.: University of Notre Dame Press.
Wallerstein, Immanuel.
    1974.   *The Modern World System I: Capitalist Agriculture and the Origins of
             the European World Economy in the Sixteenth Century.* New York:
             Academic Press.
    1993.   "The TimeSpace of World-Systems Analysis: A Philosophical
             Essay." *Historical Geography* 23:5–22.
Warntz, Christopher W.
    1968.   *The Continent Problem—Geography and Spatial Variance.* Harvard
             Papers in Theoretical Geography, no. 9. Cambridge: Laboratory
             for Computer Graphics, Harvard University.
Washbrook, David.
    1990.   "South Asia, the World System, and World Capitalism." *Journal of
             Asian Studies* 49:479–508.
Weber, Alfred.
    1948.   *Farewell to European History (Or the Conquest of Nihilism).* Trans-
             lated by R. F. C. Hull. New Haven: Yale University Press.

Weber, George.
   1853.   *Outline of Universal History from the Creation to the Present Time.*
            Translated by M. Behr. Boston: Little, Brown.
Weber, Max.
   1968.   *Economy and Society: An Outline of Interpretive Sociology,* 2 vols.
            Edited by Guenther Roth and Claus Wittich. New York: Bedmin-
            ster Press.
Webster's Geographical Dictionary.
   1964.   Springfield, Mass.: Merriam.
Weigert, Hans W.
   1942.   *Generals and Geographers: The Twilight of Geopolitics.* New York:
            Oxford University Press.
Weitzmann, Walter R.
   1994.   "Constantin Frantz, Germany and Central Europe." In Peter Stirk,
            ed., *Mitteleuropa: History and Prospects.* Edinburgh: Edinburgh
            University Press. Pp. 36–60.
Wheeler, Jesse H., and J. Trenton Kostbade.
   1993.   *Essentials of World Regional Geography.* Fort Worth: Harcourt Brace
            Jovanovich.
White, Hayden.
   1973.   *Metahistory: The Historical Imagination in Nineteenth Century
            Europe.* Baltimore: Johns Hopkins University Press.
White, Richard.
   1991.   *The Middle Ground: Indians, Empires, and Republics in the
            Great Lakes Region, 1650–1815.* Cambridge: Cambridge University
            Press.
Whittlesey, Derwent.
   1954.   "The Regional Concept and the Regional Method." In Preston
            James and Clarence Jones, eds., *American Geography: Inventory
            and Prospect.* Syracuse, N.Y.: Syracuse University Press. Pp. 19–69.
Wigen, Kären.
   1995a.   "Bringing the World Back In: Meditations on the Space-Time of
            Japanese Early Modernity." Research paper, Asian/Pacific Studies
            Institute, Duke University.
   1995b.   *The Making of a Japanese Periphery, 1750–1920.* Berkeley and Los
            Angeles: University of California Press.
Wilcox, Marrion, and George E. Rines.
   1917.   *Encyclopedia of Latin America.* New York: Encyclopedia Ameri-
            cana.
Wilgus, A. Curtis.
   1931.   *A History of Hispanic America.* Washington, D.C.: Mime-o-Form
            Service.
Wilkinson, David.
   1993.   "Civilizations, Cores, World Economies, and Oikumenes." In
            Andre Gunder Frank and Barry K. Gills, eds., *The World System:
            Five Hundred or Five Thousand Years?* London: Routledge. Pp.
            221–46.

Wilkinson, Robert.
    1794.   *A General Atlas*. London: Wilkinson.
Williams, Gwyn.
    1985.   *When Was Wales? A History of the Welsh*. London: Penguin.
Williams, Henry Smith, ed.
    1907.   *The Historian's History of the World,* 25 vols. New York and
             London: History Association.
Williams, Joseph E., ed.
    1960.   *Prentice-Hall World Atlas*. Englewood Cliffs, N.J.: Prentice-Hall.
Williams, Lea E.
    1976.   *Southeast Asia: A History*. New York: Oxford University Press.
Willis, F. Roy.
    1982.   *World Civilization*. Vol. 1, *From Ancient Times through the Sixteenth
             Century*. Lexington, Mass.: Heath.
Wilson, Edward O.
    1992.   *The Diversity of Life*. Cambridge: Harvard University Press.
Wink, André.
    1986.   *Land and Sovereignty in India: Agrarian Politics under the Eigh-
             teenth-Century Maratha Svarjya*. Cambridge: Cambridge Univer-
             sity Press.
    1991.   *Al-Hind: The Making of the Indo-Islamic World*. Leiden: Brill.
Wittfogel, Karl A.
    1956.   "The Hydraulic Civilizations." In William L. Thomas, Jr., ed.
             Man's Role in Changing the Face of the Earth, vol. 1. Chicago:
             University of Chicago Press. Pp. 152–64.
    1981 [1957].
             Oriental Despotism: A Comparative Study of Total Power. New York:
             Vintage.
Wolf, Eric R.
    1982.   *Europe and the People without History*. Berkeley: University of
             California Press.
Wolff, Larry.
    1994.   *Inventing Eastern Europe: The Map of Civilization on the Mind of the
             Enlightenment*. Stanford: Stanford University Press.
Wolpert, Stanley.
    1991.   *India*. Berkeley: University of California Press.
Woodbridge, William C.
    1824.   *Modern Atlas of Universal Geography*. Hartford: Cook.
Woodcock, George.
    1967.   *Kerala: A Portrait of the Malabar Coast*. London: Faber and Faber.
Woodside, Alexander.
    1993.   "The Asia-Pacific Idea as a Mobilization Myth." In Arif Dirlik, ed.,
             What Is in a Rim? Critical Perspectives on the Pacific Region Idea.
             Boulder, Col.: Westview Press. Pp. 13–28.
Woodward, David.
    1987.   "Medieval *Mappaemundi*" In J. B. Harley and David Woodward,
             eds., *The History of Cartography*. Vol. 1, *Cartography in Prehistoric,*

Ancient, and Medieval Europe and the Mediterranean. Chicago: University of Chicago Press. Pp. 286–370.

Woolf, Stuart.

1992.   "Construction of a European World-View in the Revolutionary-Napoleonic Years." *Past and Present* 137:72–101.

World Almanac and Book of Facts 1993.

1992.   New York: World Almanac.

World Bank.

1995.   *World Bank Atlas.* Washington, D.C.: World Bank.

Worster, Donald.

1985.   *Rivers of Empire: Water, Aridity, and the Growth of the American West.* New York: Pantheon.

Wriggins, W. Howard.

1992.   "The Dynamics of Regional Politics: An Orientation." In W. Howard Wriggins, ed., *Dynamics of Regional Politics: Four Systems on the Indian Ocean Rim.* New York: Columbia University Press. Pp. 1–21.

Wright, John Kirtland.

1925.   *The Geographical Lore of the Time of the Crusades.* New York: American Geographical Society.

1928.   *The Geographical Basis of European History.* New York: Holt.

Yapp, M. E.

1992.   "Europe in the Turkish Mirror." *Past and Present* 137:134–55.

Yee, Cordell D. K.

1994.   "Cartography in China" [organized into six discrete articles]. In J. B. Harley and David Woodward, eds., *The History of Cartography.* Vol. 2, Book 2, *Cartography in the Traditional East and Southeast Asian Societies.* Chicago: University of Chicago Press. Pp. 35–202.

Yiengpruksawan, Mimi Hall.

1993.   "Japanese War Paint: Kawabata Ryushi and the Emptying of the Modern." *Archives of Asian Art* 46:76–90.

Yoshimoto, Mitsuhiro.

1991.   "The Difficulty of Being Radical: The Discipline of Film Studies and the Postcolonial World Order." *boundary 2* 18:242–57.

Young, Robert.

1990.   *White Mythologies: Writing History and the West.* London: Routledge.

Yúdice, George.

1992.   "We Are *Not* the World." *Social Text* 10: 202–16.

"Z" [Martin Malia].

1991.   "To the Stalin Mausoleum." In Stephen R. Graubard, ed., *Eastern Europe . . . Central Europe . . . Europe.* Boulder, Col.: Westview Press. Pp. 283–338.

Zerubavel, Eviatar.

1992.   *Terra Cognita: The Mental Discovery of America.* New Brunswick, N.J.: Rutgers University Press.

Index

rationality in, 87; and Sinocentrism, 132–134; and Southeast Asia, 171, 173, 174; traditional geographical concepts in, 69, 71, 133, 236n102; and withdrawal from oceanic trade, 138; in world history, 107, 109. *See also* East Asia

Chirol, Valentine, 65

Christianity and Christendom, 24, 25, 49, 55, 68, 88, 92, 170; according to Toynbee, 129; in Arabic world, 153; and the East-West split, 135; and the Enlightenment, 80; Nestorian, 179; in the Philippines, 174; in relation to Islam, 146; and the West, 74

Cipolla, Carlo, 93, 230n50, 247nn. 88, 95

civilizations, x, 16, 36, 54, 71, 102, 110, 115, 124, 139, 146, 153, 157, 190, 229n33; according to Braudel, 263n31; according to S. Huntington, 134–135; according to Toynbee, 125–130, 154, 260n14; boundaries among, 141–142; in contemporary world history, 130–131; and pedagogy, 155–156; and world regions, 188; in world systems theory, 137, 140

climatic variation, 42, 45, 184–185, 257n68

Cohn, Bernard, 6

Cold War, the, ix, 4, 51, 60, 96, 134, 191, 192, 207n3; and the emergence of area studies, 166, 167; and Southeast Asia, 173

colonialism. *See* Imperialism

Columbia University, 110

communism, 55, 60, 95, 175, 209n9

Comte, August, 108

Condorcet, 80, 241n37

Confucianism and Confucian world, 130–131, 143; according to S. Huntington, 135; and democracy, 246n86; in Southeast Asia, 174

Confucius, 79, 80

continents, x, 1, 8, 9, 16, 157, 196, 198, 217n59; abandonment of, 33; in Afrocentrism, 105, 116; in ancient Greek thought, 21–23; in Chinese thought, 71; contemporary views of, 33; contrasted with world regions, 186; defined, 21, 29; differentiated from islands, 34; and early Christianity, 23; and environmental determinism, 42–45; and Eurocentrism, 104; in geology, 34, 221–222n89; in Japanese thought, 221nn. 80, 81; in Korean thought,

221n84; in medieval thought, 23–24; myth of, xiii, 2–3, 6, 11, 15, 33–35, 220n76; and race, 120, 218n64; in Renaissance thought, 24–25; secularization of in early modern period, 27; in South Asian thought, 220n79, 221n84; teleological view of, 30; transition from threefold to fourfold model, 25–26; twentieth-century views, 32

Coon, Carleton, 121–122, 258n74

Cornell University, 110

Croatia, 68

Cronon, William, 15

cultural regions. *See* regions and regionalization, cultural

Curtin, Philip, 152, 266n59

cyberspace, 154

Dar al-Islam, 69, 148–149

deconstruction, 17, 81, 103

deep ecology, 88. *See also* radical environmentalism

democracy, 51, 75, 101; in ancient Greece, 91; and the East, 92, 246n86; in Western civilization, 91–92

dependency theory, 114

Derrida, Jacques, 82

despotism, 61–62, 75, 77, 93–98, 106

diaspora, 142, 151–153, 192

Diderot, Denis, 80, 240n34

diffusionism, 7, 119, 254n26

Diop, Cheikh Anta, 118

Dirlik, Arif, 283n19

disease, 139

Djaït, Hichem, 147, 240n26

Dravidians, 118

Duke University, 41

Durant, Will and Ariel, 108

East, the, xiii, 1, 6, 7, 107; according to Toynbee, 127; and area studies, 163, 166, 171; changing definitions of, 55–60; and communism, 60–62; cultural constructions of, 73–82; and democracy, 92; and despotism, 61, 82, 93–98; distinction from West, 48, 54–55, 73–82, 195, 239nn. 1, 12; divided into "hither" and "farther," 54; duality of definition of, 62; and economic stagnation, 6, 75, 82, 92, 97, 127; and Germany, 59–60; and Japan, 70; and the Middle East, 63; and Orientalism, 102; and rationality, 86–88;